Edited by
Alexei A. Lapkin

Handbook of Green Chemistry

**Volume 12
Green Chemical Engineering**

Related Titles

Islam, M.R., Islam, J.S., Zatzman, G.M., Rahman, M., Mughal, M.

The Greening of Pharmaceutical Engineering
Practice, Analysis, and Methodology

2016

Print ISBN: 978-0-470-62603-0 (Also available in a variety of electronic formats)

Brinck, T. (ed.)

Green Energetic Materials

2014

Print ISBN: 978-1-119-94129-3 (Also available in a variety of electronic formats)

Marteel-Parrish, A., Abraham, M.A.

Green Chemistry and Engineering
A Pathway to Sustainability

2014

Print ISBN: 978-0-470-41326-5 (Also available in a variety of electronic formats)

Boodhoo, K.V., Harvey, A.P. (eds.)

Process Intensification for Green Chemistry - Engineering Solutions for Sustainable Chemical Processing

2013

Print ISBN: 978-0-470-97267-0 (Also available in a variety of electronic formats)

Edited by Alexei A. Lapkin

Handbook of Green Chemistry

Volume 12: Green Chemical Engineering

Verlag GmbH & Co. KGaA

The Editor

Prof. Paul T. Anastas
Yale University
Center for Green Chemistry & Green Engineering
225 Prospect Street
New Haven, CT 06520
USA

Volume Editor

Prof. Alexei A. Lapkin
Cambridge University
Department of Chemical Engineering and Biotechnology
Philippa Fawcett Drive
Cambridge, CB3 0AS
United Kingdom

Cover
The cover picture contains images from Corbis Digital Stock (Dictionary) and PhotoDisc, Inc./Getty Images
(Flask containing a blue liquid).

Handbook of Green Chemistry
Set (12 volumes):
ISBN: 978-3-527-31404-1
oBook ISBN: 978-3-527-62869-8

All books published by **Wiley-VCH** are carefully produced. Nevertheless, authors, editors, and publisher do not warrant the information contained in these books, including this book, to be free of errors. Readers are advised to keep in mind that statements, data, illustrations, procedural details or other items may inadvertently be inaccurate.

Library of Congress Card No.: applied for

British Library Cataloguing-in-Publication Data
A catalogue record for this book is available from the British Library.

Bibliographic information published by the Deutsche Nationalbibliothek
The Deutsche Nationalbibliothek lists this publication in the Deutsche Nationalbibliografie; detailed bibliographic data are available on the Internet at <http://dnb.d-nb.de>.

© 2018 Wiley-VCH Verlag GmbH & Co. KGaA, Boschstr. 12, 69469 Weinheim, Germany

All rights reserved (including those of translation into other languages). No part of this book may be reproduced in any form – by photoprinting, microfilm, or any other means – nor transmitted or translated into a machine language without written permission from the publishers. Registered names, trademarks, etc. used in this book, even when not specifically marked as such, are not to be considered unprotected by law.

Print ISBN: 978-3-527-32643-3
ePDF ISBN: 978-3-527-68841-8
ePub ISBN: 978-3-527-68840-1

Cover Design Adam-Design, Weinheim, Germany
Typesetting Thomson Digital, Noida, India
Printing and Binding betz-druck GmbH, Darmstadt, Germany

Printed on acid-free paper
10 9 8 7 6 5 4 3 2 1

Contents

About the Editors *XIII*
List of Contributors *XV*
Preface *XIX*

1   **Chemical Engineering Science and Green Chemistry – The Challenge of Sustainability** *1*
Alexei A. Lapkin
1.1   Sustainability Challenge for the Chemical Industry *1*
1.2   From Green to Sustainable Chemistry *5*
1.3   Chemical Engineering Science for Sustainability *7*
1.4   Trends in Chemical Engineering Science *9*
1.5   Topics Covered in This Book *11*
      Acknowledgment *13*
      References *13*

Part One: **Molecular Engineering of Materials, Reactions, and Processes** *17*

2   **Recent Advances in the Molecular Engineering of Solvents for Reactions** *19*
Eirini Siougkrou, Amparo Galindo, and Claire S. Adjiman
2.1   Introduction *19*
2.2   Solvent Effects on Reactions *22*
2.3   Design or Selection of Solvents for Chemical Reactions *26*
2.3.1   Model-Based Screening Methods *27*
2.3.2   Generate-and-Test Methods *28*
2.3.3   Optimization-Based Methods *30*
2.3.4   Discussion *34*
2.4   A Case Study *35*
2.5   Conclusions *38*
      Acknowledgments *38*
      References *39*

3	**Hierarchically Structured Pt and Non-Pt-Based Electrocatalysts for PEM Fuel Cells** *47*
	Panagiotis Trogadas and Marc-Olivier Coppens
3.1	Introduction *47*
3.2	Pure Hollow Pt Nanoparticles *49*
3.3	Hollow Pt Metal Alloys *51*
3.3.1	PtAu *52*
3.3.2	PtAg *53*
3.3.3	PtCo *56*
3.3.4	PtNi *58*
3.3.5	PtRu *59*
3.3.6	PtPd *61*
3.3.7	PtCu *62*
3.4	Non-Pt Alloy Nanostructures *63*
3.5	Conclusions and Outlook *64*
	Acknowledgment *65*
	References *65*

4	**New Frontiers in Biocatalysis** *73*
	John M. Woodley and Nicholas J. Turner
4.1	Introduction *73*
4.2	Recent Advances in Biocatalysis *74*
4.3	Biocatalytic Retrosynthesis *75*
4.4	Process-Driven Protein Engineering *80*
4.5	Process Developments *83*
4.5.1	Continuous Processes *83*
4.5.2	Kinetic Analysis *84*
4.6	Future Perspectives *84*
	References *85*

Part Two: Innovations in Design, Unit Operations, and Manufacturing *87*

5	**Conceptual Process Design and Process Optimization** *89*
	Alexander Mitsos, Ung Lee, Sebastian Recker, and Mirko Skiborowski
5.1	Introduction *89*
5.2	Mathematical Background *89*
5.2.1	System of Nonlinear Equations *90*
5.2.2	Nonlinear Programming (NLP) *90*
5.2.3	Mixed Integer Programming *92*
5.3	Synthesis *93*
5.3.1	Reactor Networks *93*
5.3.2	Separation Systems *95*
5.3.3	Overall Flowsheets *97*
5.4	Superstructure-Based Techniques *101*

5.4.1	Heat Exchange Networks	*101*
5.4.2	Process Flowsheet Optimization	*103*
5.5	Integrated Process Design, Operation, and Control	*105*
5.6	Water and Energy Processes	*105*
5.7	Conclusions and Outlook	*107*
	References	*107*

6 Development of Novel Multiphase Microreactors: Recent Developments and Future Challenges *115*
Evgeny Rebrov

6.1	Principles and Features	*115*
6.1.1	Continuous Phase Multiphase Microreactors	*115*
6.1.1.1	Falling Film Microreactor	*115*
6.1.1.2	Mesh Contactor	*116*
6.1.2	Dispersed Phase Multiphase Microreactors	*116*
6.1.2.1	Segmented Flow Microreactors	*116*
6.1.2.2	Microstructured Packed Beds	*117*
6.1.2.3	Prestructured Microreactors	*118*
6.1.2.4	Foam Microreactors	*120*
6.1.2.5	Microreactors with Fibrous Internal Structures	*120*
6.2	Experimental Practice	*121*
6.2.1	Flow Regimes	*121*
6.2.1.1	Capillary Microreactors	*121*
6.2.1.2	Structured Packed Beds	*122*
6.2.2	Dispersion and Holdup in Microstructured Packed Bed Reactors	*123*
6.2.2.1	Liquid Holdup	*123*
6.2.2.2	Hydrodynamic Dispersion	*124*
6.3	Modeling Features	*125*
6.3.1	Hydrodynamics	*125*
6.3.1.1	Falling Films Microreactors	*125*
6.3.2	Pressure Drop in Capillary Microreactors	*127*
6.3.2.1	Gas–Liquid Microreactors	*127*
6.3.2.2	Liquid–Liquid Microreactors	*130*
6.3.3	Mass Transfer	*131*
6.3.3.1	Capillary Microreactors	*131*
6.3.3.2	Falling Film Microreactors	*133*
6.3.4	Two-Phase Flow Distribution	*133*
6.4	Applications	*136*
6.4.1	Falling Film Microreactors	*136*
6.4.2	Capillary Microreactors	*137*
6.4.2.1	Wall Coated Catalytic Microreactors	*137*
6.4.2.2	Phase Transfer Catalysis in Microreactors	*139*
6.4.2.3	Microstructured Packed Bed Reactors	*142*
6.5	Conclusions and Outlook	*144*
	References	*144*

7	Process Intensification through Continuous Manufacturing: Implications for Unit Operation and Process Design *153*
	Sebastian Falß, Nicolai Kloye, Manuel Holtkamp, Angelina Prokofyeva, Thomas Bieringer, and Norbert Kockmann
7.1	Continuous Processes as a Means of Process Intensification *153*
7.2	Equipment for Continuous Processes *158*
7.2.1	Upstream Equipment *159*
7.2.1.1	Reactors without Active Mixing *159*
7.2.1.2	Reactors with Dynamic Mixing *161*
7.2.2	Downstream Equipment *163*
7.2.3	Process Integration *165*
7.2.4	Continuous Equipment as Enabling Technology *166*
7.3	Process Development and Implementation for Continuous Processes *168*
7.3.1	Process Development and Scale-Up *168*
7.3.2	Flexible Implementation of Continuous Processes *172*
7.4	Selected Case Studies *174*
7.5	Conclusion and Outlook *180*
	References *182*
8	How Technical Innovation in Manufacturing Is Fostered through Business Innovation *191*
	Nicolas Eghbali, Marianne Hoppenbrouwers, Steven Lemain, Gert De Bruyn, and Bart Vander Velpen
8.1	General Introduction *191*
8.2	Concept of Chemical Leasing and Take Back Chemicals *192*
8.2.1	The Concept of Take Back Chemicals *194*
8.2.2	Advantages and Challenges of the Take Back Chemicals Model *195*
8.2.2.1	What Are the Advantages of Implementing TaBaChem *196*
8.2.2.2	What Are the Impediments in Implementing the New Business Models? *197*
8.3	General Economic, Technical, and Management Aspects *198*
8.3.1	Economic Aspects *198*
8.3.1.1	Direct Gains, Indirect Gains, and Investments *198*
8.3.1.2	Pricing *199*
8.3.1.3	Conclusion on the Economic Aspects *200*
8.3.2	Technical Aspects *201*
8.3.2.1	Reuse of Chemicals *201*
8.3.2.2	Process Optimization *201*
8.3.2.3	Conclusion on the Technical Aspects *201*
8.3.3	Organizational/Managerial Aspects *202*
8.3.3.1	Sales *202*
8.3.3.2	Quality Assurance *202*
8.3.3.3	Tendering and Rewarding *202*

8.3.3.4	Knowledge Sharing	*202*
8.3.3.5	Logistics	*203*
8.3.3.6	Conclusion on the Organizational/Managerial Aspects	*203*
8.4	Compatibility of the Service Model with the Actual Legislation: Some Important Aspects	*203*
8.4.1	Transition from Sales to Providing a Service to the Customer	*204*
8.4.1.1	The Supplier Retains Ownership of the Chemical	*204*
8.4.1.2	Result-Oriented Services Lead to Different Pricing of a Chemical	*204*
8.4.1.3	A Transparent and Elaborated Contract Is Necessary	*205*
8.4.2	Closing the Life Cycle and Preventing Waste	*205*
8.4.3	Business Confidentiality and the Protection of Competition	*208*
8.4.3.1	Intellectual Property Rights	*208*
8.4.3.2	Competition	*208*
8.5	General Conclusion	*211*
	References	*211*

9 Applications of 3D Printing in Synthetic Process and Analytical Chemistry *215*

Victor Sans, Vincenza Dragone, and Leroy Cronin

9.1	Introduction	*215*
9.1.1	Polymerization-Based Additive Manufacturing (AM)	*216*
9.1.1.1	Stereolithography (SLA)	*217*
9.1.1.2	Photopolymer Jetting (PJ)	*217*
9.1.1.3	Physical Binding	*217*
9.1.2	Melting-Based Techniques	*218*
9.1.2.1	Selective Laser Melting (SLM)	*218*
9.1.2.2	Electron Beam Melting (EBM)	*218*
9.1.2.3	Fused Deposition Modeling (FDM)	*219*
9.1.2.4	Laser Sintering (LS)	*219*
9.1.2.5	Material Jetting (MJ)	*219*
9.2	Chemical Reactors Manufacturing by Additive Manufacturing Techniques	*220*
9.2.1	3D Printing Technologies in Chemistry	*220*
9.3	3D Printing Applied to Flow Chemistry	*226*
9.3.1	Mesoscale Reactors	*226*
9.3.2	3D Printed Membranes	*235*
9.4	Applications of 3D Printed Flow Devices in Analytical Chemistry	*239*
9.4.1	3D Printing of Valves, Pumps and Actuators	*239*
9.4.2	Modular Devices Based on SL	*242*
9.5	Future Trends	*248*
9.5.1	Ultrafast Printing	*249*
9.5.2	Smart Materials through 4D Printing	*250*

9.6 Conclusions *251*
References *252*

Part Three: Enabling Technologies *257*

10 Process Analytical Chemistry and Nondestructive Analytical Methods: The Green Chemistry Approach for Reaction Monitoring, Control, and Analysis *259*
Miriam Fontalvo Gómez, Boris Johnson Restrepo, Torsten Stelzer, and Rodolfo J. Romañach
10.1 Green Chemistry and Chemical Analysis in Manufacturing *259*
10.2 Process Analytical Chemistry: Concept and Objectives *260*
10.3 Vibrational Spectroscopy *264*
10.4 Challenges to Overcome *268*
10.5 Applications of Process Analytical Chemistry and Nondestructive Analyses *270*
10.5.1 Dairy Industry *270*
10.5.2 Synthesis of Active Pharmaceutical Ingredients *271*
10.5.3 Preparation of Polymeric Strip Film Unit Dosage Forms *273*
10.5.4 Polymer Industry *274*
10.5.5 Process Analytical Chemistry for Biodiesel Production *276*
10.6 Future Trends in PAC *279*
Acknowledgments *281*
References *281*

11 NMR Spectroscopy and Microscopy in Reaction Engineering and Catalysis *289*
Carmine D'Agostino, Mick D. Mantle, and Andrew J. Sederman
11.1 Introduction *289*
11.2 Basic Principles of NMR *290*
11.2.1 Nuclear Spins and Bulk Magnetization *290*
11.2.2 NMR Spectroscopy of Liquids *293*
11.2.3 NMR Relaxation *295*
11.2.3.1 Spin–Lattice Relaxation *295*
11.2.3.2 Spin–Spin Relaxation *296*
11.2.4 Pulsed Field Gradient NMR *297*
11.3 The NMR Toolkit in Reaction Engineering and Catalysis *299*
11.3.1 NMR Spectroscopy in Catalysis and Reaction Engineering *300*
11.3.2 Diffusion of Fluids Confined in Porous Catalysts *306*
11.3.2.1 Catalyst Deactivation Studies Using PFG NMR *311*
11.3.3 NMR Relaxation Time Analysis in Porous Catalytic Materials *314*
11.3.4 Combining NMR Spectroscopy with Magnetic Resonance Imaging *319*
11.4 Summary *324*
References *324*

12	**An Introduction to Closed-Loop Process Optimization and Online Analysis** *329*	

Christopher S. Horbaczewskyj, Charlotte E. Willans, Alexei A. Lapkin, and Richard A. Bourne

12.1	Introduction *329*	
12.2	Principles of Self-Optimization and Requirements for Experimental Systems *330*	
12.3	Analytical Techniques for Closed-Loop Optimization *332*	
12.4	Decision Algorithms in Closed-Loop Optimization *334*	
12.4.1	Algorithms for Discovery *335*	
12.4.2	Algorithms for Developing Process Understanding *337*	
12.4.3	Algorithms for Automated Process Optimization *338*	
12.5	Application Examples of Closed-Loop Discovery and Optimization *341*	
12.5.1	Discovery in Closed-Loop Self-Optimization *341*	
12.5.2	High-Throughput Screening *342*	
12.5.3	Examples of One-Variable-at-a-Time Reaction Optimization *344*	
12.5.4	Examples of Application of Design of Experiments *346*	
12.5.5	Rate-Based/Physical Organic Approaches *350*	
12.5.6	Examples of Algorithm-Based Self-Optimization *364*	
12.6	Conclusions and Future Directions *368*	
	Acknowledgments *369*	
	References *369*	

Index *375*

About the Editors

Series Editor

Paul T. Anastas joined Yale University as Professor and serves as the Director of the Center for Green Chemistry and Green Engineering there. From 2004–2006, Paul was the Director of the Green Chemistry Institute in Washington, D.C. Until June 2004 he served as Assistant Director for Environment at the White House Office of Science and Technology Policy where his responsibilities included a wide range of environmental science issues including furthering international public-private cooperation in areas of Science for Sustainability such as Green Chemistry. In 1991, he established the industry-government-university partnership Green Chemistry Program, which was expanded to include basic research, and the Presidential Green Chemistry Challenge Awards. He has published and edited several books in the field of Green Chemistry and developed the 12 Principles of Green Chemistry.

Volume Editor

Alexei A. Lapkin, originally trained in biochemistry at Novosibirsk State University, specialized his master thesis in catalysis and membrane separation. He then worked at Boreskov Institute of Catalysis (Novosibirsk, Russian Federation) in the area of membrane catalysis, before joining the University of Bath in 1997, first as a research assistant. He obtained his Ph.D. from the University of Bath (2000) under the supervision of Professor John W. Thomas and then began his independent academic career. He joined the University of Cambridge as Professor of Sustainable Reaction Engineering in 2013. His research focuses on methods of developing cleaner chemical manufacturing processes, in particular in the areas of specialty and pharmaceutical chemistry, but also works on the process intensification concepts and chemical reactor technologies suitable for many application areas, from scaled manufacture of controlled-functionality nanomaterials to inherently safe methods of catalytic oxidation of hydrocarbons. The focus on methods allows his group to branch into diverse application areas, while retaining the core specialism in process development. His group is also contributing to the work on the methods of evaluation of sustainability of chemical processes and products.

List of Contributors

Claire S. Adjiman
Department of Chemical Engineering
Centre for Process Systems Engineering
Imperial College London
London SW7 2AZ
United Kingdom

Thomas Bieringer
Bayer AG
Engineering & Technology
Kaiser-Wilhelm-Allee 1
51368 Leverkusen
Germany

Richard A. Bourne
University of Leeds
School of Chemistry
Leeds LS6 9JT
UK

Gert De Bruyn
Royal Haskoning DHV
Laan 1914 nr 35
3818 EX Amersfoort
The Netherlands

Marc-Olivier Coppens
University College London
EPSRC "Frontier Engineering"
Centre for Nature Inspired Engineering
Torrington Place
London WC1E 7JE
UK

Leroy Cronin
University of Glasgow
School of Chemistry
University Avenue
Glasgow G12 8QQ
UK

Carmine D'Agostino
The University of Manchester
School of Chemical Engineering and Analytical Science
The Mill, Sackville Street
Manchester M13 9PL
UK

Vincenza Dragone
University of Glasgow
School of Chemistry
University Avenue
Glasgow G12 8QQ
UK

Nicolas Eghbali
Royal Haskoning DHV
Laan 1914 nr 35
3818 EX Amersfoort
The Netherlands

Sebastian Falß
Invite GmbH
Kaiser-Wilhelm-Allee 50
51373 Leverkusen
Germany

Amparo Galindo
Department of Chemical Engineering
Centre for Process Systems Engineering
Imperial College London
London SW7 2AZ
United Kingdom

Miriam Fontalvo Gómez
Universidad del Atlántico
Department of Chemistry
km 7 Antigua via Puerto Colombia
Barranquilla
Colombia

Manuel Holtkamp
Invite GmbH
Kaiser-Wilhelm-Allee 50
51373 Leverkusen
Germany

Marianne Hoppenbrouwers
University of Hasselt
Faculty of Law
Martelarenlaan 42
3500 Hasselt
Belgium

Christopher S. Horbaczewskyj
University of Leeds
School of Chemistry
Leeds LS6 9JT
UK

Nicolai Kloye
Bayer AG
Engineering & Technology
Kaiser-Wilhelm-Allee 1
51368 Leverkusen
Germany

Norbert Kockmann
TU Dortmund University
Biochemical and Chemical Engineering
Emil-Figge-Strasse 68
44227 Dortmund
Germany

Alexei A. Lapkin
University of Cambridge
Department of Chemical Engineering and Biotechnology
Philippa Fawcett Drive
Cambridge CB3 0AS
UK

and

Cambridge Centre for Advanced Research and Education in Singapore Ltd.
1 Create Way, CREATE Tower #05-05
138602 Singapore
Singapore

Ung Lee
RWTH Aachen University
Aachener Verfahrenstechnik
Process Systems Engineering (SVT)
Turmstr. 46
52056 Aachen
Germany

List of Contributors | XVII

Steven Lemain
Royal Haskoning DHV
Laan 1914 nr 35
3818 EX Amersfoort
The Netherlands

Mick D. Mantle
University of Cambridge
Department of Chemical
Engineering and Biotechnology
Philippa Fawcett Drive
West Cambridge Site
Cambridge CB3 0AS
UK

Alexander Mitsos
RWTH Aachen University
Aachener Verfahrenstechnik
Process Systems Engineering (SVT)
Turmstr. 46
52056 Aachen
Germany

Angelina Prokofyeva
Bayer AG
Engineering & Technology
Kaiser-Wilhelm-Allee 1
51368 Leverkusen
Germany

Evgeny Rebrov
University of Warwick
School of Engineering
UK

and

Tver State Technical University
Laboratory of Biotechnology and
Chemistry
Komsomolsky Pr. 5, Tver, 170026
Russia

Sebastian Recker
RWTH Aachen University
Aachener Verfahrenstechnik
Process Systems Engineering (SVT)
Turmstr. 46
52056 Aachen
Germany

Boris Johnson Restrepo
Universidad de Cartagena
Environmental Chemistry Research
Group
School of Exact and Natural
Sciences
Campus of San Pablo
130015 Cartagena
Colombia

Rodolfo J. Romañach
University of Puerto Rico
Center for Structured Organic
Particulate Systems (C-SOPS)
Mayagüez Campus
Mayagüez 00681-9000
Puerto Rico

Victor Sans
The University of Nottingham
Department of Chemical and
Environmental Engineering
Faculty of Engineering
University Park
Nottingham
NG7 2RD

Andrew J. Sederman
University of Cambridge
Department of Chemical
Engineering and Biotechnology
Philippa Fawcett Drive
West Cambridge Site
Cambridge CB3 0AS
UK

Eirini Siougkrou
Department of Chemical Engineering
Centre for Process Systems Engineering
Imperial College London
London SW7 2AZ
United Kingdom

Mirko Skiborowski
Technical University of Dortmund
Department of Biochemical and Chemical Engineering
Emil-Figge-Str. 70
44227 Dortmund
Germany

Torsten Stelzer
University of Puerto Rico
Department of Pharmaceutical Sciences
Medical Sciences Campus
San Juan 00936
Puerto Rico

and

University of Puerto Rico
Crystallization Design Institute
Molecular Sciences Research Center
San Juan 00926
Puerto Rico

Panagiotis Trogadas
University College London
EPSRC "Frontier Engineering"
Centre for Nature Inspired Engineering
Torrington Place
London WC1E 7JE
UK

Nicholas J. Turner
University of Manchester
Manchester Institute of Biotechnology
School of Chemistry
131 Princess Street
Manchester M1 7DN
UK

Bart Vander Velpen
Royal HaskoningDHV
Laan 1914 nr 35
3818 EX Amersfoort
The Netherlands

Charlotte E. Willans
University of Leeds
School of Chemistry
Leeds LS6 9JT
UK

John M. Woodley
Technical University of Denmark
Department of Chemical and Biochemical Engineering
Søltofts Plads
2800 Lyngby
Denmark

Preface

The volume on *Green Chemical Engineering* was envisaged as an outlook at the possible future of chemical engineering – where the discipline with long traditions and enormous societal impact and importance is likely to develop. The discipline of chemical engineering is extremely diverse due to the universal applicability of the chemical engineering toolbox. As a result, this volume does not cover all the topics that might be associated with the practice of chemical engineering. Many important traditional areas, such as safety, control, separations, industrial reactor design, are not included in this book. Instead, its focus is on the emerging capabilities and the expanding links to the neighboring basic and applied sciences – biology, chemistry, physics, and applied mathematics. In most areas of chemical engineering, we observe the same trends – increased use of sensors, extensive use of computer-aided tools based on advances in fundamental understanding of physical, biological, and chemical phenomena, much better predictive power of models, and increased capability in integration of knowledge across multiple timescales and length scales. At the same time, some problems of green engineering and sustainability are better solved by other means, for example, by devising new business models or adopting new manufacturing capabilities within chemical engineering, such as additive manufacturing and robotics. The distinctive character of chemical engineering is its *systems* perspective on the problems. This is most fitting to the current societal challenges of climate change, access to water, sustainable production of food and energy, and closed material cycles. In this volume, our shared attempt is to perform "system expansion" for our discipline and demonstrate the potential advances that are offered at new scientific interfaces, especially in solution of the sustainability challenge.

University of Cambridge, Cambridge, UK　　　　　　　　　　　Alexei A. Lapkin
2018

1
Chemical Engineering Science and Green Chemistry – The Challenge of Sustainability

Alexei A. Lapkin

1.1
Sustainability Challenge for the Chemical Industry

The challenge of sustainability is well illustrated by the global ecological footprint "wedge" (Figure 1.1). It shows that the global footprint is exceeding the ecological reserve since 1960s [1]. Footprint is only one aspect of sustainability, reflecting the use and availability of resources and damage to natural environment. The sustainability challenge could be paraphrased as achieving economic prosperity, while ensuring social equality and well-being and not destroying the environment [2]. The ecological footprint is one of the more critical aspects for today: it reflects several important relationships that must be understood in order to identify the path to feasible solutions of the sustainability challenge. The key relationship is between human aspirations and our ways of attaining them. Here, we will not go into the topic of aspirations. This is discussed in detail in sustainability literature and includes satisfaction of basic needs, aspirations of a certain standard of living, aspirations toward self-fulfillment and self-realization, which are highly context specific. What is important for us is that there appears to be a clear positive link between the standard of living and the energy intensity per capita. This ultimately stems from two factors: (i) the way we attain the increases in our standard of living and (ii) basic thermodynamics. Our standard of living today depends on access to manufactured goods used in production of building materials, clothing, food, modern education and entertainment technology, modern healthcare, and so on. In turn, manufacturing necessarily involves expenditure of energy, as we transform matter into more complex and ordered forms [3]. As population grows and a larger proportion of the population increases its standard of living, the demand for energy and resources will grow, due to the link between the standard of living and the energy required to achieve it. At present, the developed countries with a high human development index (HDI) [4] and the ecological footprint far exceeding their own resources, export their waste to, and import resources from, the

Handbook of Green Chemistry Volume 12: Green Chemical Engineering, First Edition. Edited by Alexei A. Lapkin.
© 2018 Wiley-VCH Verlag GmbH & Co. KGaA. Published 2018 by Wiley-VCH Verlag GmbH & Co. KGaA.

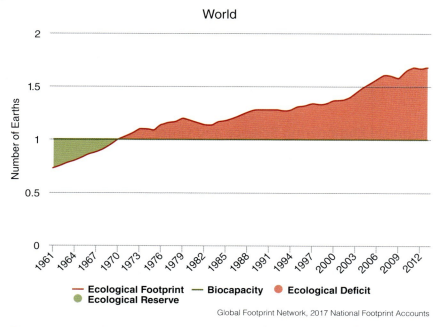

Figure 1.1 A comparison of ecological footprint, represented as a number of Earths, versus biocapacity. (Global Footprint Network [5].)

countries with low HDI and low ecological footprint. This leads to inequality and exploitation, and this situation of course cannot continue indefinitely.

The second reason why it is critical to decouple HDI from energy demand is the apparent anthropogenic effect on climate. The vary rapid increase in global population of humans and their aspirations toward better life result in a rapidly increasing demand for energy [6–8]. At present, this results in the increase in the rate of emissions of carbon dioxide that today far exceeds the rate of biological and geological sequestration of atmospheric carbon [9], resulting in the observed climate change.

The fundamental challenge of sustainability, therefore, is to decouple the quality of life and human aspirations from energy and material intensities of achieving them.

In this respect, chemical industry will play a significant role. Chemistry is ubiquitous in everyday life and is responsible for the majority of innovations in all aspects of life, from agriculture and food to healthcare, sport, entertainment, and fashion. It is also the major source of innovation in the energy-efficient technologies. Calculation of the environmental impact of the manufacture of materials for energy generation and energy-efficient technologies, and comparison with the saved emissions by these technologies (Figure 1.2) shows the significance of chemistry to finding ways of decoupling HDI and ecological footprint. These data focus on carbon emissions, which today are regarded as the most important

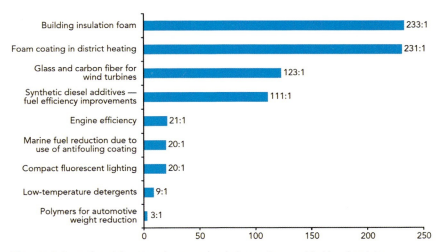

Figure 1.2 Examples of the green house gas emission savings enabled by chemistry, represented as a ratio of emission savings to emissions incurred. (Adapted from Refs [10, 11].)

target as the driver of climate change [7]. Among the larger contributions to the reduction of carbon emissions by innovation in chemistry are development of insulation materials and materials for construction of wind turbines; fuel technology, including fuel additives for better combustion; shift to lower carbon fuels; novel materials for energy-efficient lighting, and detergents for low-temperature cleaning. Thus, the manufacture of novel materials is an important technological solution for reduction of carbon emissions, as it directly affects two of the major anthropogenic sources of carbon emissions, namely, transport and housing. Of course, the manufacture of materials is, basically, chemistry.

These solutions are not affecting the way how energy is being produced, but are directed at efficiency of the use of energy: A reduction in the waste of energy should result in lowering the demand for energy production. The last decade saw a rapid uptake of renewable energy technologies, in particular wind and solar power. As a result, a new energy technology paradigm has emerged – chemistry as a major industrial energy storage system (Figure 1.3). This paradigm is based on the proposition that excess power will be available from renewable energy installations at low-demand off-peak times during a day. This low-cost electricity could be used to convert unreactive molecules, CO_2 and

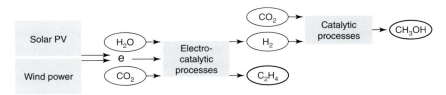

Figure 1.3 Chemical industry as an energy storage solution.

H_2O, into basic chemical feedstocks, such as methanol and olefins, in what has been termed "CO_2 recycling technology" [12]. In a way this approach parallels the biological conversion of carbon dioxide into molecules of higher *exergy* [13], driven by constant supply of solar energy, but at a higher intensity (defined as production rate per unit area).

An estimate of the potential scale of the CO_2 recycling technology has been given for the example of using the potentially available electricity to synthesize methanol from CO_2: This technology can produce 1.2×10^9 metric tones of methanol from CO_2 per year [12]. This is an order of magnitude larger amount than the global methanol production via current conventional technology. Of course, in this case the CO_2-derived methanol is not a final product, but a convenient to transport source of an activated C1 group for further conversion to bulk chemical products. This technology may have a very significant impact on the overall chemical supply chain and, consequently, reduce emissions from the manufacture of chemicals.

To understand the impact of CO_2 recycling technology on global emissions, it is necessary to compare life cycle impacts from conventional routes to methanol, with the proposed CO_2-based technologies. The global warming potential or green house gas (GHGs) emissions, evaluated in the units of kilogram CO_2 equivalents per kilogram of product, is one of the most important indicators. The total contribution of the CO_2 recycling technology to GHG is comprised of three main contributions: direct use of CO_2 as a feedstock, avoiding CO_2 emissions from conventional routes to methanol, and CO_2 emissions due to the new processes. These values would vary depending on the CO_2 recycling technology used. As an example, the estimate of GHG reduction from introduction of the solar–thermal methanol synthesis is -1.71 kg CO_2 equiv. kg^{-1} compared to GHG emissions from conventional methanol synthesis of 0.67 [14]. Here, the negative sign means the reduction of carbon emissions in the wider system of the chemical supply chain.

The second route to utilization of CO_2, shown in Figure 1.3, is via electrocatalytic reduction of CO_2 to ethylene [15]. A much larger reduction in GHG emissions compared to that of methanol synthesis would be expected for the case of the production of ethylene from CO_2, since GHG emissions from conventional ethylene production range between 2.5 and 8.9 [16].

The merger of chemical and energy industries is a major opportunity to rapidly and significantly reduce global carbon emissions from the three main contributions to anthropogenic emissions of carbon dioxide: energy generation, transport (through fuel substitution), and chemical manufacturing itself.

As manufacturing of molecules and materials might in the future be integrated into new energy technologies and become one of critical solutions to the reduction of carbon emissions, *what* is being manufactured and the impact of chemical products during and after their use comprise the second aspect of the sustainable chemical technologies. Chemical industry is producing tens of thousands of chemical compounds. Very few of these are bulk platform chemicals produced in millions of tones per year, with the majority being produced at an annual rate of approximately 1000 tonnes. Today, chemical manufacturing is experiencing its fastest growth in the developing countries. It is already projected that developing countries will be responsible for manufacturing of 37% of high-volume industrial chemicals by 2030 [10]. Thus, emissions to atmosphere and pollution of natural environment associated with the manufacture and use of chemicals are set to increase, unless significant changes are introduced into the industry through use of green and sustainable chemistry. World Health Organization (WHO) estimates that 25% of the burden of disease is linked to environmental factors, including chemical pollution [17]. As controls of the use, storage, and disposal of toxic chemicals are lacking mainly in the developing countries, the rapid increase in the size of chemical industries in these countries may lead to the increase in the damage to human health and the environment.

The impact of chemicals on the environment is increasingly better documented, however, the lack of knowledge of the interactions of chemicals with the environment is a significant problem, especially for the large number of existing chemicals, introduced into manufacturing and products before current toxicity, and environmental impact testing regimes were introduced into legislations [18]. Thus, consideration of *what* is being produced and used in the final products, especially the products that are highly distributed and end up in the environment, is an urgent priority for chemical and chemistry-using industries.

1.2
From Green to Sustainable Chemistry

The 12 principles of green chemistry formulated at the end of 1990s (Table 1.1) have marked an important step in developing our understanding of the relationship of chemistry research with the environmental impact of chemical industry and chemical products [19]. The focus of the principles is on delivery of the *target useful functions* without the specific to the chemicals' negative effects, such as toxicity or material intensity (through use of auxiliary substances in the synthesis). This is the principle of *ideality*, well known in the innovation literature [20–22]. The very simple idea behind green chemistry principles is that if materials being released into environment do not possess inherent hazards, are benign toward environment, and have low material and energy intensities in manufacturing and use, than the problems we are forced to solve now would not be further exacerbated. Green chemistry has seen remarkable successes, especially with regard to developing solvent guides and processes using alternative solvent media [23–29], application of catalysis and biocatalysis in organic synthesis [30], and the significant progress of the chemistry of biofeedstocks [31–34], among many others. It is safe to state that today the principles of green chemistry are embedded in everyday work of most chemists. Furthermore, there is an increasing appreciation of the intimate link of green chemistry with engineering. This stems from the basic idea that green chemistry is a

Table 1.1 Green chemistry principles [19].

It is better to prevent waste than to treat or cleanup waste after it is formed

Synthetic methods should be designed to maximize the incorporation of all materials used in the process into the final product

Wherever practicable, synthetic methodologies should be designed to use and generate substances that possess little or no toxicity to human health and the environment

Chemical products should be designed to preserve efficacy of function while reducing toxicity

The use of auxiliary substances (e.g., solvents, separation agents, etc.) should be made unnecessary wherever possible and innocuous when used

Energy requirements should be recognized for their environmental and economic impacts and should be minimized. Synthetic methods should be conducted at ambient temperature and pressure

A raw material of feedstock should be renewable rather than depleting wherever technically and economically practicable

Unnecessary derivatization (blocking group, protection/deprotection, temporary modification of physical/chemical processes) should be avoided whenever possible

Catalytic reagents (as selective as possible) are superior to stoichiometric reagents

Chemical products should be designed so that at the end of their function they do not persist in the environment and break down into innocuous degradation products

Analytical methodologies need to be further developed to allow for real-time, in-process monitoring and control prior to the formation of hazardous substances

Substances and the form of a substance used in a chemical process should be chosen so as to minimize the potential for chemical accidents, including releases, explosions, and fires

system's approach [35], and decisions about molecules, modes of activation, solvents, feedstocks, and operating conditions will necessarily affect decisions about processes, mass and energy integration, supply chain, and business models. Green chemistry has evolved into sustainable chemistry.

In a broader context of sustainability, OECD defined *sustainable chemistry* as "the design, manufacture and use of efficient, effective, safe and more environmentally benign chemical products and processes" [36]. The technical areas of sustainable chemistry closely match the principles of green chemistry, but the impact of these is evaluated with respect to human health and the environment, safety of workers and users, energy and resources consumption, and economic viability. The emphasis on evaluation of the impacts over the life cycle of products and processes and at different levels – local, country, regional, and global – warrants the use of the term *sustainable*, rather than the narrower definition of the scientific and technical challenges defined by the green chemistry principles. The problem of developing better chemistry has become truly a system-level problem and no discipline is better suited to contribute the development of sustainable chemistry as chemical/biochemical engineering, the discipline deeply routed in systems analysis and tools.

1.3 Chemical Engineering Science for Sustainability

Understanding behavior of hierarchical interacting systems is the core of chemical engineering discipline. The manufacture of chemicals is a structured hierarchical complex system with multiple dynamic interactions at different levels of the hierarchy and between the levels [37]. As a result, any changes proposed to such a system, for example, replacing a stoichiometric synthesis with a catalytic process, replacing elements of the supply chain based on the new criteria of inherent safety or the origin of the feedstocks, and so on, require system-level changes. Any solution proposed in isolation from the rest of the system would not be successful until its effects on the overall system's behavior are understood, and are *positive*. We can trace the history of discourse on systems in chemical manufacturing to the development of ideas of industrial ecology [38, 39], system-level process design [40, 41], hierarchical indicators for environmental impact of green chemistry and technology [42], and the adoption of life cycle assessment (LCA) in chemistry and chemical technology [43–49]. In particular, indicators and LCA are the methods that allow quantification of the outcome of new developments in terms of their *positive* contribution to the reduction of the environmental impact of the economically feasible processes. This requires a brief explanation.

In many cases there is a compromise between economically optimal or environmentally optimal solutions. We can use as an illustration a recent study of conversion of a biowaste-derived terpene feedstock into a useful but not easily accessed platform molecule (Scheme 1.1). The conceptual process was optimized toward two simultaneous objectives, the minimum values of a cost function and a CO_2 emissions indicator, with process operating conditions being the optimization variables [46]. The results are reproduced in Figure 1.4. The optimization rapidly converges to a minimum of both objectives, as shown in Figure 1.4a. Upon expanding the minimum solution, we see that it is, in fact, a series of Pareto optimal solutions with a trade-off between cost and GHG. The set of Pareto solutions is of interest as these solutions correspond to the lowest environmental

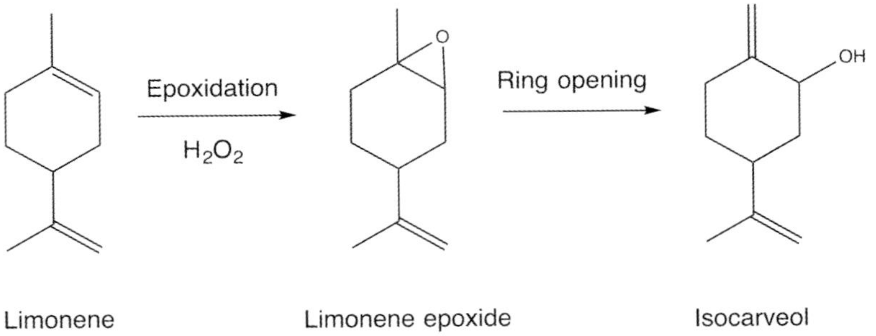

Scheme 1.1 Conversion of biowaste-derived limonene to isocarveol.

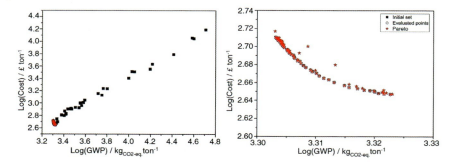

Figure 1.4 (a) A course of optimization over 100 iterations in attaining a simultaneous minimum of a cost function and global warming potential in the synthesis of isocarveol from terpenes and (b) the expanded optimal region showing the set of simultaneously optimal Pareto solutions. (Reproduced with permission from Ref. [46]. Copyright 2017, John Wiley & Sons, Inc.)

impact (in terms of CO_2 emissions) and the best economic potential. Any of the solutions on the Pareto front are optimal, and further criteria, considering in detail the operating conditions of each of the solutions, should be considered to select the process for implementation.

The consideration of the process itself, within the system boundary of the conversion of limonene to isocarveol is not enough to justify that this process gives a positive contribution to sustainable chemical supply chain. The results given in Figure 1.4 are obtained using life cycle assessment cradle-to-gate system boundary, starting from how the specific biowaste is separated to give pure limonene as a feedstocks. Within this system boundary it was clear that using the waste of paper manufacturing as a source of limonene is significantly more attractive than orange peel, due to the much lower CO_2 emissions and lower cost [46]. Here, comparative LCA allows the quantification of the potential impact of the introduction of new technology.

The transition from green to sustainable chemistry is the inclusion of the wider system in the domain of problems of chemistry and chemical engineering. Using the same example with which we began this chapter, the CO_2 recycling technology, the only reason why manufacturing of bulk feedstocks from CO_2 may become a serious proposition is the rapid drop in the cost of renewable electricity, which occurred mainly due to political and economic decisions. The development at the system level of energy infrastructure translates into changes in the supply chains, business models, process technology, and ultimately, requires novel chemistry and novel engineering solutions.

The desired attributes of green engineering solutions have been well articulated by the principles of green *chemical* engineering, reproduced in Table 1.2 [50]. These attributes closely follow the principles of green chemistry [19], with the addition of the system-level attributes, such as conservation of exergy [13], expressed in Table 1.2 as embedded entropy and complexity, and

Table 1.2 Principles of green engineering [50].

Designers need to strive to ensure that all material and energy inputs and outputs are as inherently nonhazardous as possible
It is better to prevent waste than to treat or cleanup waste after it is formed
Separation and purification operations should be designed to minimize energy consumption and materials use
Products, processes, and systems should be designed to maximize mass, energy, space, and time efficiency
Products, processes, and systems should be "output pulled" rather than "input pushed" through the use of energy and materials
Embedded entropy and complexity must be viewed as an investment when making design choices on recycle, reuse, or beneficial disposition
Targeted durability, not immortality, should be a design goal
Design for unnecessary capacity or capability (e.g., "one size fits all") solutions should be considered a design flaw
Material diversity in multicomponent products should be minimized to promote disassembly and value retention
Design of products, processes, and systems must include integration and interconnectivity with available energy and materials flows
Products, processes, and systems should be designed for performance in a commercial "afterlife"
Material and energy inputs should be renewable rather than depleting

many features that have recently been formulated in the concept of "circular economy" [51].

The challenges of sustainability require system-level solutions and, in particular, ability to model interactions within complex dynamical systems, such as the developing integrated chemistry-energy systems or biorefining systems, among many others. Within chemical engineering science, the ability to identify the critically important mechanisms within a chemical system, for example, what factors affect stability of catalysts under industrial operating conditions, and the ability to explore scenarios of technology development within broad system boundaries are some of the current important areas of development of the field. We can now look at some current trends in chemical engineering science.

1.4 Trends in Chemical Engineering Science

The new challenges of sustainability require new tools and solutions from the chemical engineering science. If we take as an example, how current grand challenges are viewed by an authoritative and representative engineering community, US National Academy of Engineering (Table 1.3), access to bulk chemical products or fossil energy are not on the list, but clean water, renewable energy,

Table 1.3 NAE 2017 grand challenges for engineering [52].

Make solar energy economical
Provide energy from fusion
Develop carbon sequestration methods
Manage the nitrogen cycle
Provide access to clean water
Restore and improve urban infrastructure
Advance health informatics
Engineer better medicines
Reverse engineer the brain
Prevent nuclear terror
Secure cyberspace
Enhance virtual reality
Advance personalized learning
Engineer the tools of scientific discovery

carbon sequestration, and better medicines are. I would argue that solutions to these challenges require much closer integration of chemical engineering with the neighboring sciences, and developing the capability for transcending many length scales that connect molecular systems with manufacturing systems and the final applications. These trends – (i) of merger of chemical engineering with physical, biological, and medical sciences and (ii) setting much broader system boundaries for problems – are evident in current chemical engineering literature and the subject matter of research in leading chemical engineering departments in universities around the world.

Recently, several publishers launched journals that are explicitly aimed at the interdisciplinary space between sciences and chemical engineering. These include *ACS Sustainable Chemistry and Engineering*, RSC's *Reaction Chemistry and Engineering*, RSC's *Molecular Systems Design and Engineering*, Elsevier's *Sustainable Chemistry and Pharmacy*, and so on, as well as the already well-established interdisciplinary chemistry–chemical engineering–material science journals, such as *ChemSusChem* and *Green Chemistry*. If we look at research in the top chemical engineering university departments using, for example, a QS ranking to define "top," we find biomedical research, nanomaterial's engineering, artificial intelligence, data science, robotics, sensors, and so on. It appears that principles of chemical engineering are becoming an integral part of discovery sciences while the neighboring sciences are becoming an integral part of the chemical engineering design toolbox. For this reason, it seems rather timely to present in a single volume the new topics and capabilities within the field of chemical engineering that have emerged recently.

1.5 Topics Covered in This Book

This volume of the Green Chemistry series provides an outlook on recent and new trends in chemical engineering, emerging in response to the challenge of *sustainable chemistry*, as a next chapter in the evolution of this field. Chemical engineering is becoming increasingly linked with molecular sciences and tools of molecular and materials design. Part One of the book deals with two large topics in molecular design and engineering – engineering of solvents and design of functional nanomaterials. Solvents is a critical topic for green chemistry, as in many areas of chemical industry that suffer from large waste-to-product output ratios, the E-factor [53]; it is large because of the significant use of solvents in both synthesis and reactor cleaning. With the increasing attention to bio-based chemical supply chains, the role of solvents is likely to further increase. However, replacing solvents is a challenging problem since solvents frequently are not inert in the reactions and new solvents need to be designed to not only provide the favorable solvation properties at reasonable cost and with little environmental impact, but also to favorably affect the reaction outcome. Nano-structured functional materials are becoming increasingly used in the most wide ranging applications. Ensuring the control over structure, composition and morphology at nanoscale, and especially in bulk manufacture of nanomaterials has become a significant barrier for commercialization of functional nanomaterials. Design of scalable manufacturing of nanomaterials should account for the need to control nanoscale processes via manipulated variables many orders of magnitude larger in the length scale. Hence, a significant attention is being paid to various novel synthetic methods that allow such control. This also emphasizes the need for new approaches to modeling that link multiple length and time scales for processes.

Applied mathematics has always been a critical component of chemical engineering curriculum and practice. However, today new mathematical methods are being adopted as chemical engineering faces new challenges and merges with new disciplines. Part Two deals with the state of the art in conceptual process design and optimization. Many of the current challenges in process design are system-level problems: reactor networks, optimization of complete flow sheets, optimization of heat integration networks, and super-structure optimization. In addition to traditional optimization tools, some areas of chemical process design have turned to data-driven methods, such as machine learning and Bayesian statistics-based design of experiments. These are some of new enabling technologies that are described in Part Three.

Part Two also deals with the innovations in unit operations and manufacturing in chemical industry. Process intensification [54] (PI) has become a standard tool within the chemical engineering design toolbox. However, we also observe new trends, such as wide adoption of PI within pharmaceutical and specialty chemical industries, where adoption of continuous flow manufacturing [55] is opening new business opportunities [56], for example, the potential to manufacture drugs on demand at a point of sale. To achieve this a radical increase in productivity, of the traditionally highly inefficient complex multistep syntheses, is required. This could be achieved in microreactors. Chapter 6 is dedicated to the state of the art in microreactor design, including their modeling. The problems and opportunities of adoption of PI in industry are discussed in detail in Chapter 7. Further advances on new business models that are emerging with the increased adoption of green chemistry solutions is the subject of Chapter 8. Chemical leasing is a concept that has already found applications in leasing of noble metals and some solvents. However, a series of pilot projects undertaken recently in Europe show how this approach may be adopted to other chemical products in new types of business-to-business relationships. Another aspect of PI is the rapid adoption of additive manufacturing technology in chemical industry. Chapter 9 describes the new opportunities that are being opened up for chemical synthesis and process design by additive manufacturing.

Biotechnology is deliberately given a very narrow focus in this book. Traditional areas of biotechnology, such as fermentation, tissue engineering, bioseparations, and biorefining, have been the subject of significant attention. However, one area of biotechnology has seen little coverage in the chemical engineering literature, but is likely to be the most disruptive – synthetic biology. Being able to design new reaction pathways using non-native biocatalytic reaction pathways is a major step forward for synthetic chemistry. Already, several successes in green chemistry are due to developing much shorter reaction sequences via adoption of enzymatic or whole-cell transformations. Chapter 4 deals with the systematic approach to the development of non-native biocatalytic transformations.

The final part of the book, Part Three, is dedicated to several key enabling technologies. Out of many, only three are included in this volume. Spectroscopy as a tool for process monitoring is well known and, in principle, well-studied area, but it is only now that it is becoming an essential tool for industrial chemical processes. There are still very few suppliers of industrial-grade spectroscopic equipment for real-time analysis under operating conditions, but the range of potential applications is vast, and the potential for energy and materials savings must not be underestimated. Chapter 10 describes the state of the art and current challenges of spectroscopic process monitoring. Chapter 11 deals with the more advanced technique of magnetic resonance imaging, which has seen remarkable developments over the last decade. The possibility to measure flow velocity maps under operating conditions, measure diffusion coefficients inside porous catalysts, and distinguish between bulk and adsorbed phases within catalysts and other materials, are some of recent highlights from MRI. As we progressively tackle more complex chemical problems, the opportunities to increase the number of state variables that could be directly observed are critical to our ability to design new processes using rational design principles, rather than trial and error. This opens up a debate about the role of physical and surrogate models in process development, addressed to some extent in the final chapter of the book, Chapter 12. This chapter deals with another recent addition to chemical engineering-enabling technologies: robotics and machine learning and artificial intelligence (AI).

Acknowledgment

This research was, in part, supported by the National Research Foundation, Prime Minister's Office, Singapore, under its CREATE program.

References

1. Wackernagel, M., Schulz, N.B., Deumling, D., Linares, A.C., Jenkins, M., Kapos, V., Monfreda, C., Loh, J., Myers, N., Norgaard, R., and Randers, J. (2002) Tracking the ecological overshoot of the human economy. *Proceedings of the National Academy of Sciences of the United States of America*, **99**, 9166–9271.
2. Brundtland, G.H. (1987) Our Common Future. Report of the World Commission on Environment and Development, United Nations.
3. Chenery, H.B. (1953) Process and production functions from engineering data, in *Studies in Structure in the American Economy* (ed. W.W. Leontieff), Oxford University Press, p. 299.
4. Jahan, S. (2016) Human Development Report 2016. Available at http://hdr.undp.org/sites/default/files/2016_human_development_report.pdf (accessed October 23, 2017).
5. Global Footprint Network http://data.footprintnetwork.org/#/countryTrends?type=earth&cn=5001 (accessed October 23, 2017).
6. Slesser, M., King, J., and Crane, D.C. (1997) *The Management of Greed*, Resource Use Institute, Ltd., Dunblane, Scotland.
7. IPCC (2015) Climate Change 2014: Synthesis Report. Contribution of Working Groups I, II and III to the Fifth Assessment Report of the Intergovernmental Panel on Climate Change. Available at http://www.ipcc.ch/pdf/assessment-report/ar5/syr/SYR_AR5_FINAL_full_wcover.pdf (accessed October 27, 2017).
8. HM Treasury (2006) Stern review: the economics of climate change, United Kingdom. Available at http://www.hm-treasury.gov.uk/independent_reviews/stern_review_economics_climate_change/stern_review_Report.cfm (accessed November 8, 2017).
9. Gorshkov, V.G., Gorshkov, V.V., and Makarieva, A.M. (2000) *Biotic Regulation of the Environment: The Issue of Global Change*, Springer, Chichester, UK.
10. UN Department of Economic and Social Affairs (2010) Trends in sustainable development. chemicals, mining, transport and Waste Management. Available at https://sustainabledevelopment.un.org/content/documents/28Trends_chem_mining_transp_waste.pdf (accessed September 15, 2017).
11. ICCA (2009) Innovations for Greenhouse Gas Reductions: A Life Cycle Quantification of Carbon Abatement Solutions Enabled by the Chemical Industry, International Council of Chemical Associations. Available at https://www.americanchemistry.com/Policy/Energy/Climate-Study/Innovations-for-Greenhouse-Gas-Reductions.pdf. (accessed October 23, 2017).
12. Perathoner, S. and Centi, G. (2014) CO_2 recycling: a key strategy to introduce green energy in the chemical production chain. *ChemSusChem*, **7**, 1274–1282.
13. Dewulf, J., Van Langenhove, H., Muys, B., Bruers, S., Bakshi, B.R., Grubb, G.F., Paulus, D.M., and Sciubba, E. (2008) Exergy: its potential and limitations in environmental science and technology. *Environmental Science and Technology*, **42**, 2221–2232.
14. Kim, J., Henao, C.A., Johnson, T.A., Dedrick, D.E., Miller, J.E., Stechel, E.B., and Maravelias, C.T. (2011) Methanol production from CO_2 using solar–thermal energy: process development and techno-economic analysis. *Energy & Environmental Science*, **4**, 3122–3132.
15. Gurudayal, G., Bullock, J., Srankó, D.F., Towle, C.M., Lum, Yanwei, Hettick, M.,

Scott, M.C., Javey, A., and Ager, J. (2017) Efficient solar-driven electrochemical CO_2 reduction to hydrocarbons and oxygenates. *Energy & Environmental Science*, **2017**, 2222–2230.

16 Chen, Q., Lv, M., Wang, D., Tang, Z., Wei, W., and Sun, Y. (2017) Eco-efficiency assessment for global warming potential of ethylene production processes: a case study of China. *Journal of Cleaner Production*, **142**, 3109–3116.

17 WHO (2009) Strategic Approach to International Chemicals Management. Available at http://apps.who.int/gb/ebwha/pdf_files/A62/A62_19-en.pdf (accessed October 23, 2017).

18 OECD (2001) Environmental Outlook for the Chemicals Industry, Organisation for Economic Cooperation and Development. Available at https://www.oecd.org/env/ehs/2375538.pdf (accessed October 28, 2017).

19 Anastas, P.T. and Warner, J.C. (1998) *Green Chemistry: Theory and Practice*, Oxford University Press.

20 Altshuller, G. (1984) *Creativity as an Exact Science*, Gordon & Breach Scientific Pub.

21 Salamatov, Y. (1999) *TRIZ: the Right Solution at the Right Time*, INSYTEC B.V., The Netherlands.

22 Mann, D.L. (2003) Better technology forecasting using systematic innovation methods. *Technological Forecasting and Social Change*, **70**, 779–795.

23 Sheldon, R.A. (2005) Green solvents for sustainable organic synthesis: state of the art. *Green Chemistry*, 7, 267–278.

24 Constable, D.J.C., Dunn, P.J., Hayler, J.D., Humphrey, G.R., Leazer, J.J.L., Linderman, R.J., Lorenz, K., Manley, J., Pearlman, B.A., Wells, A., Zaks, A., and Zhang, T.Y. (2007) Key green chemistry research areas – a perspective from pharmaceutical manufacturers. *Green Chemistry*, **9**, 411–420.

25 Horvath, I.T. and Anastas, P.T. (2007) Innovations and green chemistry. *Chemical Reviews*, **107**, 2169–2173.

26 Reinhardt, D., Ilgen, F., Kralisch, D., König, B., and Kreisel, G. (2008) Evaluating the greenness of alternative reaction media. *Green Chemistry*, **10**, 1170–1181.

27 Esteves, C. (2009) Sustainable solutions – green solvents for chemistry, in *Sustainable Solutions for Modern Economies* (ed. R. Höfer), Royal Society of Chemistry, Cambridge, pp. 403–420.

28 Anastas, P. and Eghbali, N. (2010) Green chemistry: principles and practice. *Chemical Society Reviews*, **39**, 301–312.

29 Henderson, R.K., Jimenez-Gonzalez, C., Constable, D.J.C., Alston, S.R., Inglis, G.G.A., Fisher, G., Sherwood, J., Binks, S.P., and Curzons, A.D. (2011) Expanding GSK's solvent selection guide – embedding sustainability into solvent selection starting at medicinal chemistry. *Green Chemistry*, **13**, 854–862.

30 Sheldon, R., Arends, I.W.C.E., and Hanefeld, U. (2007) *Green Chemistry and Catalysis*, Wiley-VCH Verlag GmbH, Weinheim, Germany.

31 Curran, M.A. (2003) Do bio-based products move us towards sustainability? A look at three USEPA case studies. *Environmental Progress*, **22**, 277–292.

32 Perlack, R.D., Wright, L.L., Turhollow, A.F., Graham, R.L., Stokes, B.J., and Erbach, D.C. (2005) Biomass as Feedstocks for a Bioenergy and Bioproducts Industry: the Technical Feasibility of a Billion-Ton Annual Supply (US DOE, USDA). Available at www.osti.gov/bridge.

33 Corma, A., Iborra, S., and Velty, A. (2007) Chemical routes for the transformation of biomass into chemicals. *Chemical Reviews*, **107**, 2411–2502.

34 Tuck, C.O., Perez, E., Horvath, I.T., Sheldon, R.A., and Poliakoff, M. (2012) Valorization of biomass: deriving more value from waste. *Science*, **337**, 695–699.

35 Graedel, T.E. (2001) Green chemistry as systems science. *Pure and Applied Chemistry*, **73**, 1243–1246.

36 OECD (2002) Sustainable Chemistry. Available at http://www.oecd.org/chemicalsafety/risk-management/29361016.pdf. (accessed October 29, 2017).

37 Lapkin, A., Voutchkova, A., and Anastas, P. (2011) A conceptual framework for description of complexity in intensive chemical processes. *Chemical Engineering*

and Processing: Process Intensification, **50**, 1027–1034.

38 Graedel, T.E. and Allenby, B.R. (1995) *Industrial Ecology*, Prentis Hall International, London.

39 Jelinski, L.W., Graedel, T.E., Laudise, R.A., McCall, D.W., and Patel, C.K.N. (1992) Industrial ecology: concepts and approaches. *Proceedings of the National Academy of Sciences of the United States of America*, **89**, 793–797.

40 Douglas, J.M. (1985) A hierarchical decision procedure for process synthesis. *AIChE*, **31**, 353–362.

41 Douglas, J. (1992) Process synthesis for waste minimisation. *Industrial & Engineering Chemistry Research*, **31**, 238–243.

42 Lapkin, A., Joyce, L., and Crittenden, B. (2004) A framework for evaluating the "greenness" of chemical processes: case studies for a novel VOC recovery technology. *Environmental Science and Technology*, **38**, 5815–5823.

43 Azapagic, A. (1999) Life cycle assessment and its application to process selection, design and optimisation. *Chemical Engineering Journal*, **73**, 1–21.

44 Anastas, P.T. and Lankey, R.L. (2000) Life cycle assessment and green chemistry: the yin and yang of industrial ecology. *Green Chemistry*, **2**, 289–295.

45 Ott, D., Kralisch, D., Dencic, I., Hessel, V., Laribi, Y., Perrichon, P.D., Berguerand, C., Kiwi-Minsker, L., and Loeb, P. (2014) Life cycle analysis within pharmaceutical process optimization and intensification: case study of active pharmaceutical ingredient production. *ChemSusChem*, **7**, 3521–3533.

46 Helmdach, D., Yaseneva, P., Heer, P.K., Schweidtmann, A.M., and Lapkin, A.A. (2017) A multiobjective optimization including results of life cycle assessment in developing biorenewables-based processes. *ChemSusChem*, **10**, 3632–3643.

47 Yaseneva, P., Plaza, D., Fan, X., Loponov, K., and Lapkin, A. (2015) Synthesis of the antimalarial API artemether in a flow reactor. *Catalysis Today*, **239**, 90–96.

48 Kralisch, D., Ott, D., and Gericke, D. (2015) Rules and benefits of life cycle assessment in green chemical process and synthesis design: a tutorial review. *Green Chemistry*, **17**, 123–145.

49 Gerber, L., Gassner, M., and Maréchal, F. (2011) Systematic integration of LCA in process systems design: application to combined fuel and electricity production from lignocellulosic biomass. *Computers and Chemical Engineering*, **35**, 1265–1280.

50 Anastas, P. and Zimmermann, J. (2003) Design through the 12 principles of green engineering. *Environmental Science and Technology*, **37**, 94A–101A.

51 Clark, J.H., Farmer, T.J., Herrero-Davila, L., and Sherwood, J. (2016) Circular economy design considerations for research and process development in the chemical sciences. *Green Chemistry*, **18**, 3914–3934.

52 N.A.o. Engineering (2017) NAE Grand Challenges for Engineering, National Academy of Engineering. Available at www.engineeringchallenges.org/File.aspx?id=11574&v=34765dff (accessed September 15, 2017).

53 Sheldon, R.A. (2000) Atom utilisation, E factors and the catalytic solution. *Comptes Rendus de l'Academie des Sciences Paris C, Serie IIc, Chimie*, **3**, 541–551.

54 Stankiewicz, A.I. and Moulijn, J.A. (2000) Process intensification: transforming chemical engineering. *Chemical Engineering and Processing*, **96**, 22–34.

55 Gutmann, D., Cantillo, D., and Kappe, C.O. (2015) Continuous-flow technology – a tool for the safe manufacturing of active pharmaceutical ingredients. *Angewandte Chemie International Edition*, **54**, 6688–6728.

56 Srai, J.S., Badman, C., Krumme, M., Futran, M., and Johnston, C. (2015) Future supply chains enabled by continuous processing – opportunities and challenges. *Journal of Pharmaceutical Sciences*, **104**, 840–849.

Part One
Molecular Engineering of Materials, Reactions, and Processes

2
Recent Advances in the Molecular Engineering of Solvents for Reactions

Eirini Siougkrou, Amparo Galindo, and Claire S. Adjiman

2.1
Introduction

The importance of solvents in the chemical industry is widely acknowledged; millions of tons of solvents are used in industrial processes annually and their impact on the environment and on energy consumption cannot be neglected. For example, in the pharmaceutical industry, 20 million tons of volatile organic compounds (VOCs) are released per year and solvent use typically accounts for 50% of greenhouse gas emissions [1]. Furthermore, solvents have been found to be linked directly to 60% of the energy consumption in the production of an active pharmaceutical ingredient [1]. Thus, the need not only to minimize the amount of solvent used in industrial processes, but also to search for alternative, more environment-friendly solvents, is pressing. According to Ref. [2], reactions and solvent selection/optimization are two of the major green engineering research areas for sustainable manufacturing, as there is much room for improvement. Given that reactive steps constitute the most frequently used process operations in the pharmaceutical industry [2], tools that can help to understand and optimize this class of transformations can have a high impact on efficiency and environmental performance.

Many fine chemical companies, such as GlaxoSmithKline (GSK), Eli Lilly, or Syngenta, and academic research groups have developed methods to measure the relative greenness of solvents and created databases with health and safety information to guide solvent selection [3–6]. Several criteria are taken into account in these tools, for example, in the case of GSK's guide, solvents are rated according to waste production, environmental impact, effects on health, flammability and explosion potential, reactivity and stability, life cycle assessment, regulatory constraints, and melting and boiling points. Chemometrics approaches have also been deployed to identify good solvents for reactions [7, 8]. While very useful, database-driven approaches are dependent upon the availability of large sets of experimental data and therefore have limited applicability in the development of novel chemistries and processes.

Handbook of Green Chemistry Volume 12: Green Chemical Engineering, First Edition. Edited by Alexei A. Lapkin.
© 2018 Wiley-VCH Verlag GmbH & Co. KGaA. Published 2018 by Wiley-VCH Verlag GmbH & Co. KGaA.

Over the last few decades, computer-aided molecular design (CAMD) approaches have emerged as a useful alternative to address the issue of solvent selection and design in a predictive and systematic way and to broaden the range of solvents and solvent mixtures that can be considered. The majority of CAMD approaches and applications have been focused on designing optimal solvents for separation processes [9–20]. However, in recent years, CAMD approaches have been proposed or extended to enable the design solvents for reactive processes for a variety of reactions, including Menschutkin reactions [21–28], solvolysis [22, 25, 29, 30], Kolbe–Schmitt reactions [31], glycerolysis [32], *cis*-glycol production [33], fermentation for bioethanol production [34–36], Diels–Alder reactions [25, 37–39], hydroformylation [40], an extractive biphasic esterification reaction [41], and selectivity control in lithiation and nucleophilic substitution [28].

This extensive body of work has shown CAMD to be a valuable tool for the design of solvents compared to traditional empirical methods for solvent selection, despite the sometimes limited accuracy of predictive models. Indeed, CAMD methods should be seen as a step in the search for the most suitable solvent rather than as a tool that provides a single final answer: they lead to the generation of a list of candidates that can then be analyzed further through targeted experiments [42, 43]. The use of systematic CAMD methods opens up a wider set of possible designs compared to an exclusively experimental investigation, while reducing the time and cost of process development. It also makes it possible to take multiple criteria into account at an early stage, including environmental impact metrics [43]. CAMD techniques thereby contribute to greener chemistry through a combination of enhanced productivity (e.g., greater process intensity and/or atom economy) and improved system-level metrics such as global warming potential [44].

Given the contributions that CAMD can make to green chemistry, we aim in this chapter to provide an overview of CAMD methodologies for the design of solvents that enhance chemical reactions and reactive processes. Some of the latest advances in this area are discussed and the methodology developed in our group at Imperial College London is presented in more detail as an example of what can be achieved with CAMD. We conclude by discussing possible future directions. Before turning our attention to reactive processes, however, we give some brief background information on the development of the field of CAMD.

In the context of process design and development, computer-aided molecular design first appeared as a concept in the early 1980s [45] and has since been developed extensively both in theory and practice [46–50]. CAMD can be defined as set of tools and methods to address the following problem [46]:

> Given a set of building blocks and a specified set of target performance measures, determine the molecule or molecular structure that best attains these targets.

We note that there may be other variables in the design problem, including composition and operating conditions. The performance measures can either be focused on the performance of the molecule to be designed, for example, maximize/minimize some physical properties or achieve a value as close as possible to the desired target values [19, 47, 51–55], or on a host of product performance metrics [56], or on process performance, for example, minimize process cost [9, 10, 14, 16, 18, 37, 57–60], maximize production [23, 35]. As can be seen from the definition, CAMD techniques can be applied when the desired properties/performance characteristics of the product or process are given, but the structure of at least one of the molecules in the product or process is unknown. It is thus the reverse of property prediction, where given the identity of the molecule/molecular structure, a set of properties are calculated or predicted.

CAMD can be applied to a whole range of different systems, with various sizes and complexity, such as solvents, refrigerants, and polymers. Much of the work in the field of CAMD has been dedicated to formulating design problems (identifying objectives and constraints and appropriate modeling approaches) for different applications and to devising solution approaches for the resulting large-scale and typically highly nonlinear problems. Some recent efforts have been directed at the design of blends or mixtures as a means to achieve even greater performance and functionality [27, 28, 37, 60–69].

The methods that have been proposed for solving CAMD problems can be categorized in two general groups [46]: (a) enumeration or generate-and-test approaches and (b) mathematical programming or optimization-based approaches. In the generate-and-test approaches, feasible molecular structures are first identified *in silico*, ensuring that only chemically feasible compounds are generated [45], and the properties of the resulting structures are then evaluated, leading to a ranked list of candidates. These approaches can be extended so that a hierarchy of increasingly accurate models is used to screen candidate molecules and reduce the size of the list [70]. With the mathematical programming techniques, the CAMD problem is expressed as an optimization problem, which usually falls in the family of mixed integer nonlinear programming (MINLP) problems. Some measure of performance is maximized or minimized, subject to a number of constraints (e.g., structural, physical property, and process model constraints). Similar to the case of generate-and-test approaches, the structural constraints ensure that the combinations of functional groups (i.e., the molecular structures) generated are meaningful [71]. This class of approaches has also led to the formulation of mixed integer linear programming (MILP) problems [29, 51], mixed integer dynamic optimization problems (MIDO) [57], or multiobjective optimization problems (MOO) [18, 59, 60]. As the plethora of approaches demonstrates, CAMD brings significant flexibility to the design of molecules for a given purpose and are well-suited to take into account a range of decision criteria; there is, therefore, strong potential to help achieve greener chemistry.

The remainder of this chapter is organized as follows. In Section 2.2, the effects of solvents on chemical reactions are discussed, and some of the ways in which these effects can be modeled and predicted are reviewed. In Section 2.3, selected existing CAMD methodologies for the design of solvents for reactions are presented. In Section 2.4, an example of solvent design for a Menschutkin reaction is shown. Finally, in Section 2.5, perspectives and current challenges are outlined.

2.2
Solvent Effects on Reactions

Solvents can play many roles in liquid-phase reactions: they can affect the reaction rate and selectivity, help to control the reactor temperature, enable transport of the reactants/products in and out of the reactor, or separate the products in extractive reaction processes [72]. The effects of solvents on the rates of chemical reactions were reported for the first time by Berthelot and Pean de Saint-Gilles [73], while they were studying the esterification of acetic acid with ethanol. A few years later, Menschutkin [74, 75] studied the reactions between trialkylamines and haloalkanes and noted that solvents *can greatly influence the course of reactions* [76]. Solvent effects on reactions and other physicochemical processes have been widely studied since, mainly but not exclusively through experimental investigations, and a good understanding of the underlying chemistry and physics has been attained [77]. In this section, we introduce only those concepts that are relevant to the remainder of the chapter.

In Figure 2.1a, the scheme for the Menschutkin reaction between phenacyl bromide and pyridine is shown. The concentration profile of the product varies considerably depending on the solvent, as can be seen in Figure 2.1b, even when all other conditions (starting concentrations and temperature) are held constant. The reaction rate constants, given in Table 2.1, vary by several orders of magnitude depending on the solvent used. Selectivity is another property that is considerably influenced by the choice of solvent and which in turn can impact on the sustainability of a process as it affects atom economy and the amount of waste material or side products generated. In Table 2.2, the selectivity of the cyclization reaction of ortho-trisubstituted 2-hydroxybenzophenones with cesium carbonate in various solvents is shown based on the ratio of the alkoxyxanthone product to the chlorozanthone product [78]. The relative amount of alkoxyxanthone is seen to decrease when the reaction medium changes from a polar aprotic solvent to a polar protic solvent. Finally, a representative example of multiple solvent effects is the ring-closing metathesis of diethyl diallylmalonate catalyzed by the 1,3-dimesityl-4,5-dihydroimidazol-2-ylidene ruthenium complex (Grubbs II catalyst [79]). The reaction rate constant, the solubility of the catalyst, and the deactivation of the catalyst are all altered by the nature of the solvent [80, 81].

In order to understand the effects of solvents on the reaction rate, one can start from the definition of the rate. For a bimolecular reaction, with reactants A and B, and second-order kinetics, the concentration-based reaction rate r at a fixed temperature is given by the following expression:

$$r = k \cdot C_A \cdot C_B, \tag{2.1}$$

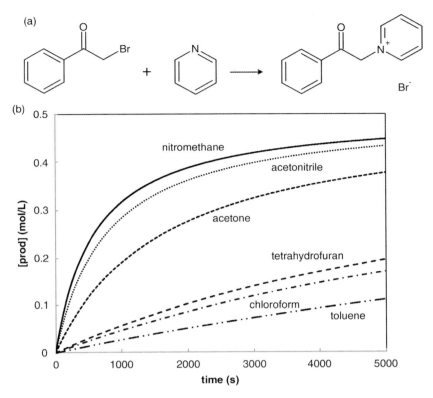

Figure 2.1 (a) Menschutkin reaction of phenacyl bromide and pyridine to form phenacylpyridinium bromide (prod). (b) Simulated concentration of the product of the Menschutkin reaction in different solvents, based on ^{1}H NMR spectroscopy data at 298 K [24], with initial concentrations of 0.5 M phenacyl bromide and 0.5 M pyridine. (Adapted from Ref. [24]. Copyright 2013, Nature Publishing Group.)

where k is the reaction rate constant at the reaction temperature, and C_A and C_B are the concentrations of reactants A and B, respectively. The maximum concentrations of the reactants that can be achieved depend on the solubility of the reactants in the solvent. Thus, the higher the solubility of the reactants in the solvent, the higher the maximum reaction rate for a given value of k. The rate constant k also depends on the solvent, but this relation is somewhat more complex to analyze. A theoretical interpretation of the effects of the solvent on the rate constant can be derived based on the well-known *transition state theory* (TST) [83, 84]. According to TST, the reactants must overcome an energy barrier, the activation free energy, $\Delta^{\ddagger}\Delta G$, in order to transform into products, with a maximum value at the *transition state*, TS. The conformation adopted by the molecules at that maximum is referred to as the *activated complex* (Figure 2.2a). The activation free energy is defined as the difference between the free energy of the activated complex and the free energies of the reactants, weighted by the

Table 2.1 The rate constant k of the Menschutkin reaction of phenacyl bromide and pyridine in various solvents at 298 K [24, 82] and the dielectric constant or relative permittivity, ε, for these solvents. ε is shown as an indicator of solvent polarity.

Solvent	k (L · mol^{-1} · s^{-1})	ε
Nitromethane	3.43×10^{-3}	36.56
Acetonitrile	2.61×10^{-3}	35.69
Acetone	1.23×10^{-3}	20.49
Tetrahydrofuran	2.59×10^{-4}	7.43
Chloroform	2.08×10^{-4}	4.81
Toluene	1.15×10^{-4}	2.38

stoichiometric coefficients. In general, the solvent affects the reaction rate constant through stabilization/destabilization of the reactants and/or transition state, thereby altering the activation free energy (Figure 2.2b). Therefore, a solvent that results in a decrease in the activation free energy, which implies that the activated complex is stabilized relative to the reactants, is found to accelerate the reaction. On the other hand, a solvent that results in an increase in the activation barrier, and therefore stabilizes the reactants relative to the activated complex, lowers the reaction rate constant. For example, in Figure 2.2b, the reaction is faster in solvent S1 as the activation free energy is lower. When there are two competing reaction pathways at play, the two activated complexes can be stabilized to a greater or lesser extent by different solvents, resulting in variations in selectivity.

The free energies needed to calculate reaction rate constants and solubilities can be computed in many ways. The prediction of the solubility of the reactants can often be conveniently performed by group contribution models such as UNIFAC [85] or SAFT-γ [86, 87] or by continuum solvation models such as PCM [88], the Minnesota solvation model SMD [89], COSMO-RS [90], or COSMO-SAC [91]. The most significant challenge is posed by the calculation of the free energy of the activated complex as it requires the modification of the electronic structure and conformations relative to the reactants. In principle, it

Table 2.2 Cyclization reactions of chlorinated substrates with cesium carbonate (Cs$_2$CO$_3$) to give mixtures of alkoxyxanthones and chloroxanthone [78].

Solvent	T (°C)	t (h)	A:C
1,3-Dimethyl-3,4,5,6-tetrahydro-2(1H)-pyrimidinone	50	5	95:5
Dimethylformamide	50	3	91:9
Formamide	80–90	70	23:77
Tetrahydrofuran	80	30	15:85
Methanol	100	20	11:89

T indicates the reaction temperature in °C and t the reaction time in hours. The last column (A:C) indicates the percentage of reaction product that consists of alkoxyxanthones (A) and that that consists of chloroxanthone (C).

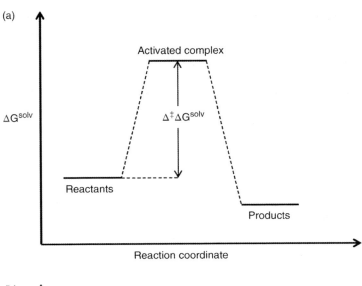

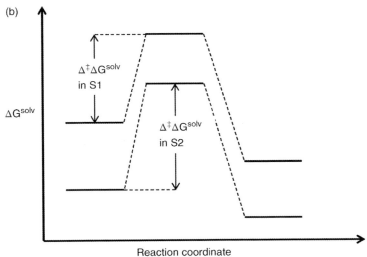

Figure 2.2 (a) Schematic representation of reaction energetics showing the relative free energies of reactants, activated complex, and products for a typical reaction. ΔG^{solv} denotes the free energy of solvation and $\Delta^{\ddagger}$ denotes an activation energy barrier. (b) Schematic representation of the activation free energy in two different solvents, S1 and S2.

is possible to use electronic structure methods or molecular modeling tools, such as reactive force fields, to evaluate the free energies of the reactants and activated complex in different solvents and evaluate the rate constant based on TST. In particular, continuum solvation models have been combined with quantum mechanical methods to obtain reaction rate constants as part of emerging

CAMD approaches to reaction solvent design. The polarizable continuum model (PCM) [88] has been used by Stanescu and Achenie [31], the SMD by our group [24, 92], and the COSMO-based models by Zhou et al. [25] and Austin et al. [28]. PCM and SMD models require as input the values of specific solvent properties such as the dielectric constant. The required properties can usually be obtained via predictive models such as group contribution methods [29, 93–95]. More rigorous approaches such as transition path sampling [96] can also be deployed to calculate rate constants but have not yet been used within the CAMD context.

In an attempt to provide a simpler quantitative analysis of solvent effects on reaction rates, empirical models, such as linear free-energy relationships (LFER), have been developed. In the context of solvent effects on reactions, LFERs are linear correlations between the logarithm of the rate constant or equilibrium constant and solvent properties. A commonly-used LFER is the solvatochromic equation initially proposed by Kamlet and Taft [97, 98] and further developed by Abraham et al. [99] and Abraham [100], where multiple solvent properties are taken into account, such as the solvatochromic parameters for polarity, acidity, and basicity. For instance, the following expression can be used to model the dependence of a reaction rate constant k on a solvent:

$$\log k = c_0 + c_A A + c_B B + c_S S + c_\delta \delta + c_H \delta_H^2, \tag{2.2}$$

where S, A, and B are the polarizability/polarity, the acidity, and the basicity of the solvent, respectively, as measured by solvatochromic shifts (also referred to as Abraham descriptors), δ is a polarizability correction term that depends on the chemical class of the solvent (e.g., aromatic or halogen), and δ_H is the Hildebrand solubility parameter. The c_i, $i = \{0, A, B, S, \delta, H\}$, constants are coefficients that depend on the specific reaction being studied. The solvatochromic equation has been widely used in the literature [23, 24, 30, 37, 92, 99, 101]. In the context of solvent design, it can be seen that once the coefficients in the solvatochromic equation have been obtained for a given reaction, the rate constant in other solvents can be predicted provided that their properties are known experimentally or can be calculated.

2.3
Design or Selection of Solvents for Chemical Reactions

In model-based methods for the design or selection of solvents, an understanding of the underlying physical chemistry, as captured in predictive models that offer a balance between computational cost and accuracy, is combined with molecular design concepts to arrive at a solvent or set of solvents that best meet the desired performance criteria. Many of the methods put forward to deal with reactive systems thus build on the theories and models mentioned in Section 2.2. They also need to account for the fact that many factors beyond the reaction rate contribute to reaction performance, including the cost of the solvent, environmental impact, and energy consumption (often governed by the cost of solvent recovery). Thus, the approaches proposed must be flexible enough to allow these different considerations to be taken into account. In this section, we consider three categories of methods: a set of solvent selection techniques that consists of model-based screening methods in which a list of solvents is evaluated using predictive models, and two sets of solvent design methods – optimization-based CAMD approaches and generate-and-test CAMD approaches.

2.3.1
Model-Based Screening Methods

In screening methods, the solvents are chosen from a list or database, according to specific requirements. Thus, there is no scope to generate new solvent molecules. In these approaches, correlative models have been used in order to model the impact of solvent on reaction rate or selectivity. A key choice is, therefore, what method to use in order to develop an empirical model that provides the most reliable predictions and that allows a greater set of solvent molecules to be tested.

The solvatochromic equation was used by Wicaksono et al. [30] in a database-screening approach. In the first step of the proposed methodology, an initial set of solvents is selected from a database containing rate information to build the solvatochromic equation for the reaction rate constant, using concepts from experiment design (the condition number) to select the data points to be included in fitting. The resulting solvatochromic equation is then used to identify the solvent in a database with the greatest predicted rate constant. Using the solvolysis of *tert*-butyl chloride as a case study (also used in Ref. [29]), the authors found high-performance solvents and determined that choosing the initial set of solvents based on the condition number is more effective than using rules of thumb to choose the initial set.

In order to enable a more predictive approach, Zhou et al. [25] proposed a method for the screening of solvents for reactions combining COSMO-RS [102] and principal component analysis (PCA) [103]. In this approach, a preselection of 136 common solvents is made and COSMO-RS is applied to calculate the σ-potentials of all solvents, obtaining twelve area parameters for each solvent. The area parameters are next projected onto a small number of independent variables using PCA and an expression is developed for the reaction rate constant or selectivity, as appropriate, based on experimental reaction data for a few solvents. A screening of the database of 136 solvents is then performed, based on the PCA model, to identify the best solvent. The solvent selection methodology was applied to three chemical reactions: the solvolysis of t-butyl chloride, the S_N2 reaction of phenacyl bromide and n-butylamine, and the Diels–Alder reaction of 2-methyl-1,3-butadiene and methyl acrylate. The authors showed that four principal components were sufficient in all cases to obtain a statistically meaningful model.

The use of the solvatochromic equation and COSMO-RS in these approaches [25, 30] extends the applicability of traditional database approaches by allowing missing data to be replaced by predicted rate constants, thereby expanding the set of solvents that can be considered. In the context of reactions, it is often necessary to take into account properties other than the reaction rate constant, such as phase behavior. To this end, COSMO-RS has been used as a screening for properties of relevance to chemical reactions other than reaction rates [40]. A computational solvent database can be used to identify alternative thermomorphic solvents for a catalytic hydroformylation reaction. While kinetic information was not included in the study of McBride *et al.* [40], several other criteria were taken into consideration, including solubility and miscibility. These thermodynamic properties were derived directly from the COSMO-RS model without modification.

The idea of a systematic screening approach to solvent selection has also been extended to a reactive–adsorptive process for the production of glycerol ethyl acetal by condensation of glycerol with acetaldehyde in a simulated moving bed reactor by Faria *et al.* [104]. Here, the selection method entails two steps: first, a hybrid screening based on predictive models and database information is performed; this is then followed by an experimental-based screening. In the initial screening, a number of criteria are taken into account, including physical properties related to process performance and environmental impact. The key physical properties of interest in the reactive–adsorptive process studied include the melting and boiling point of the solvent and its miscibility solvent with the initial reactant mixture. Through qualitative arguments, it was postulated that best process performance can be achieved using solvents with a polarity between that of water and glycerol ethyl acetal, where polarity was quantified by a combination of dielectric constant and dipole moment. The environmental, health, and safety (EHS) impact of the process was assessed by using the lethal dose (LD_{50}), the NFPA health hazard classification [105], the octanol–water partition coefficient, and the persistence time as indicators. Polarity and EHS impacts were obtained from databases, while miscibility was predicted using UNIFAC. This enabled the initial list of 100 solvents to be reduced to 3. In the last stage of the method, experimental tests for miscibility, adsorption, and reaction rates were performed on these 3 solvents.

2.3.2
Generate-and-Test Methods

The generate-and-test approach, which was first developed for solvent selection and design [45, 106], consists of two basic parts: "generation", in which the *in silico* synthesis of feasible molecular structures takes place based on specific rules, and, "testing", in which the properties of the resulting structures are evaluated and a ranked set of candidate structures is obtained.

The first instance of the application of the generate-and-test concept to the design of reaction solvents was put forward by Gani *et al.* [33], who focused on organic reactions and who combined industrial practice and computational tools

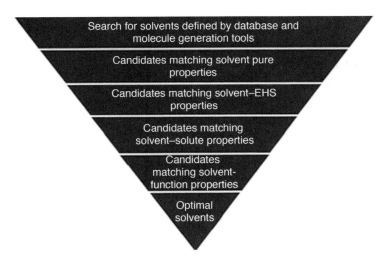

Figure 2.3 The hybrid solvent selection methodology proposed by Gani et al. [33] and also used in Refs [32, 108]. (Adapted from Ref. [108]. Copyright 2008, Elsevier.)

for property estimation in order to develop their methodology. In this approach, they use databases to screen for solvent properties and for reaction kinetics and apply the hybrid computer-aided molecular design technique of Harper et al. [70] to generate solvent candidates, as illustrated in Figure 2.3. In this approach, the database solvents and the solvents generated from CAMD are ranked in a first stage according to scores derived from a user-specified list of properties and targets. In a second stage, these candidate solvents are further evaluated using more detailed models to produce a final set of candidate solvents that not only satisfy chemical properties, but also EHS requirements. All calculations can be performed through the Integrated Computer-Aided System (ICAS) software [107]. The authors tested their methodology on several case studies: the biotransformation of toluene to toluene cis-glycol in an aqueous biocatalyst, the esterification of alcohols, the validation of an industrial reaction step after extraction, and the replacement of dichloromethane as a solvent in oxidation reactions. The solvent selection methodology in Ref. [33] was extended to handle multistep reactive systems and also solvent substitution for specific reactive steps in existing processes by Gani et al. [108]. The methodology was applied to solvent selection for an enzymatic glycerolysis reaction and to a multistage reactive system (also studied in Ref. [32]). This provided an evaluation of the methodology in industrial practice; it was found to be successful both in predicting suitable solvent substitutes and in predicting potential candidates for the different reaction steps. Promising solvent candidates that could be used in all reaction steps were also proposed.

As a further indication of the broad applicability and usefulness of CAMD approaches when seeking trade-offs between process and EHS performance, we note their application to several biocatalytic reactions [109]. Here, the method consists of first generating a large number of structures, calculating their properties with group contribution methods, and finally checking the property values against design constraints. Several solvent performance criteria were considered in the application of the approach, such as reaction equilibrium and solubility, as calculated using UNIFAC [85], as well as toxicity in terms of the lethal concentration [110]. The method was successfully applied to design solvents for the transesterification of octanol/vinyl laurate, the esterification of octanol/acrylic acid, and the transesterification of inulin/vinyl laurate.

In all the generate-and-test approaches described so far, reaction kinetics were either ignored or assessed via experiments. In the first work to offer a predictive assessment of reaction kinetics within a solvent design framework, Stanescu and Achenie [31] proposed an alternative generate-and-test approach that includes quantum mechanical calculations of the reaction rate constant in different solvents, using the Kolbe–Schmitt reaction as an example. In this approach, density functional theory (DFT) calculations using the PCM model [88] are used to study the reaction mechanism and the impact of the solvent. The Pro-CAMD [111] generate-and-test methodology within the ICAS software [107] are then applied to generate solvents in addition to those proposed in the literature for this reaction. Only the most promising solvents generated in this way are finally tested using DFT + PCM calculations to compute the reaction rate constants in these different media. The use of a two-stage approach is beneficial in this approach due to the comparatively high cost of the DFT + PCM calculations. The idea of predicting reaction rate constants after an initial design or screening could of course be adopted in other generate-and-test approaches.

2.3.3
Optimization-Based Methods

In optimization-based methods, the CAMD problem is formulated as a mathematical programming problem. The mathematical problem that usually arises is an MINLP, which can take the following form:

$$\min_{\mathbf{x},\mathbf{y}} f_{obj}(\mathbf{x}, \mathbf{y}) \qquad (2.3)$$

subject to

$$\mathbf{g}_1(\mathbf{y}) \leq \mathbf{0} \qquad (2.4)$$

$$\mathbf{g}_2(\mathbf{y}) \leq \mathbf{0} \qquad (2.5)$$

$$\mathbf{g}_3(\mathbf{x}, \mathbf{y}) \leq \mathbf{0} \qquad (2.6)$$

$$\mathbf{h}(\mathbf{x}, \mathbf{y}) = \mathbf{0} \qquad (2.7)$$

where f_{obj} is the performance objective function which needs to be minimized, $\mathbf{g}_1$, $\mathbf{g}_2$, $\mathbf{g}_3$, and $\mathbf{h}$ correspond to structural constraints, pure component property constraints, mixture property constraints, and process/EHS model constraints, respectively, $\mathbf{y}$ is a vector of binary variables related to the identities of the atom groups in the solvent, and **x** is a vector of continuous variables related to the mixture and/or process variables.

To the best of our knowledge, the work of Folić et al. [21], a hybrid experimental/computer-aided approach, constitutes the first attempt to apply a CAMD methodology to the design of solvents for reactions while taking solvent effects on the reaction rate constant into account. Starting with a small number of solvents and the relevant experimental data for these solvents (i.e., rate constants and several pure solvent properties), the method is based on the development of a reaction model using the solvatochromic equation (Eq. (2.2)). A computer-aided solvent design problem is then formulated as an MILP or MINLP as appropriate [21, 29], based on a combination of reaction model and group contribution methods [93, 94], in order to generate a number of candidate solvents that maximize/increase the rate constant. The suitability and performance of the candidate solvents can then be checked by comparison to experimental data; if the CAMD results are not compatible with the data, the measured rate constants for the candidate solvents can be added to the initial set of solvents and a new reaction model built. The procedure is repeated until the prediction accuracy is satisfactory. This methodology was applied to the solvolysis of *tert*-butyl chloride. An important feature of this framework is that while kinetic data of a small number of solvents is required, a very large solvent design space can be investigated. However, the predictive capabilities of the reaction rate model can be affected by the small size of the data set and Folić et al. [29] investigated this issue using sensitivity analysis. In subsequent work [23], the methodology was extended to more complex reaction schemes by including process model constraints in the problem formulation in order to take into consideration competing or consecutive reactions. The authors also focused on the subject of sensitivity of parameters and a systematic framework to handle the uncertainty was proposed and applied to the Menschutkin reaction of tripropylamine and methyl iodide.

Models other than the solvatochromic equation can be used to derive an empirical relationship between rate constants and solvent properties; the COSMO σ-profiles have received significant attention as a set of solvent properties [38, 39]. In their work, Zhou et al. [38, 39] used a linear equation based on six COSMO solvent descriptors (from the σ-profiles) as an empirical model of the reaction kinetics. They developed a group contribution method to relate the solvent descriptors to the constituent atom groups and embedded this approach within an MINLP CAMD formulation, which enables molecular, thermodynamic, and process considerations to be taken into account. They applied this approach to the design of a solvent for a Diels–Alder reaction and the corresponding process [38] and showed that a robust design approach [39] can be adopted to take parametric uncertainty into account. The combination of COSMO descriptors and a group contribution method to compute these descriptors thus constitutes a more predictive method than the earlier screening technique of Zhou et al. [25].

The optimization-based CAMD framework also enables issues of chemical and phase equilibrium to be taken into account in solvent design, as demonstrated by Zhou et al. [41] who have studied reactive extraction using UNIFAC-based models and taking a holistic view of mixture thermodynamics and their impact on process performance.

In recent optimization-based CAMD studies on reaction kinetics [24, 27, 28, 92], there has been increasing emphasis on embedding QM calculations of the rate constant based on transition state theory [83] within the design framework. Different strategies have been deployed to deal with the added computational cost of such a model using either surrogate models [24, 92] or decomposition approaches [27, 28].

The solvatochromic equation has thus been used as a surrogate model in the *ab initio* methodology to the design of solvents for reactions proposed by Struebing *et al.* [24, 92], thus extending significantly the applicability of the work of Folić et al. [23, 29]. The key innovations in this approach, referred to as QM-CAMD, are that no experimental data are needed to generate a set of solvent candidates for verification and that quantum mechanical calculations are used as input to the low-cost "surrogate" solvatochromic equation. The methodology proposed is as follows (Figure 2.4):

Step 1. The design problem is defined, a set of atom groups is selected and an initial set of solvents is chosen (six to seven solvents).
Step 2. The reaction rate constant in specific solvent(s) is calculated using transition state theory and quantum mechanical (QM) calculations with a continuum solvation model.
Step 3. A reaction model is built by regressing the solvatochromic equation to the QM "data" generated in step 2.

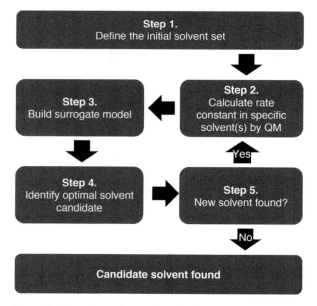

Figure 2.4 The solvent design algorithm for reactions proposed by Struebing *et al.* [24]. (Reproduced with permission from Ref. [24]. Copyright 2013, Nature Publishing Group.)

Step 4. The optimal solvent is found by using an optimization-based CAMD formulation, which embeds the reaction model from step 3.

Step 5. Convergence test: If a new solvent has been designed, go to step 2, otherwise terminate.

The algorithm iterates until no new solvent is found. The predictive power of the approach stems from the combination of QM calculations and group contribution methods. In step 2, instead of using experimental data to build the reaction model, the rate constants are predicted from TST using QM calculations with the SMD model [89], with input from group contribution methods for the solvent properties. Furthermore, in steps 3 and 4, group contribution methods are used to predict the solvent properties needed for the solvatochromic equation and for the CAMD optimization problem. The application of the methodology to the design of an optimal solvent that maximizes the rate constant for the Menschutkin reaction of phenacyl bromide and pyridine led to the identification of nitromethane as a solvent that provided a notable improvement in the rate constant in acetonitrile, the best solvent in the initial solvent set of step 1. Importantly, good agreement between the QM-predicted rate constants and experimental data was found and the best solvent identified computationally was tested experimentally by the authors, verifying that the improvement achieved relative to the best initial solvent was on the order of 40%. The methodology was recently extended [92] to design solvents with optimal reaction rates by taking into account the solubility of the reactants in the solvent. Thus, instead of choosing the best solvent by maximizing the reaction rate constant in step 4 (Figure 2.4), the reaction rate is maximized. The UNIFAC [85] group contribution method was used to calculate the solubility of the reactants in the solvent, while the method of Constantinou and Gani [93] was used to compute the solvent molar volume needed to convert solubility into concentration. The use of the reaction rate, rather than the rate constant, as the objective function led to a re-ranking of the best candidate solvents.

In a decomposition approach that incorporates QM calculations, Austin *et al.* [27, 28] made use of a combination of the COSMO-RS and COSMO-SAC methods, with TST, to predict reaction rate constants. A key difference with the surrogate-based approach is that the CAMD problem was solved as a projection in the space of σ-profiles (i.e., the method by which individual molecules are represented in the COSMO-RS/SAC formalism) and that molecules were sought that match the σ-profiles of interest, using derivative free optimization. Group contribution methods that relate σ-profiles as well as other relevant properties (e.g., σ-moments, molecular volumes) to atom groups [112] were used in the course of the optimization. The approach can readily be embedded in an overall methodology to design solvent mixtures and has been deployed on a Menschutkin reaction and on a lithiation reaction. Both pure and mixed solvents were designed. A tuning of the computational methods was necessary in order to study these different reactions – the identification of QM models (including levels of theory and basis sets) that provide a reliable estimation of reaction rates remains a topic of debate even for gas phase reactions [113], but especially when continuum solvation models, such as COSMO-RS and SMD, are used.

In the context of using CAMD methodologies to develop greener processes, the use of solvent mixtures and novel solvents, such as gas-expanded liquids, offers a promising direction for research. Siougkrou *et al.* [37] investigated this concept by developing a methodology for the design of gas-expanded liquids (GXLs) for reactions, based on process performance. They formulated an optimization-based CAMD problem for mixtures, based on the simultaneous consideration of solvent and process design. The solvatochromic equation was used to predict the rate constant and a preferential solvation model [114, 115] was applied to predict the composition dependence of the solvent properties. The number of mixtures that could be considered in the design space was limited by the empirical nature of this model. The phase equilibrium in the reactor was modeled with the group contribution VTPR (volume-translated Peng-Robinson) equation of state [116]. The application of the approach to a Diels–Alder reaction showed that when designing gas-expanded liquids, multiple process performance indicators need to be taken into account, such as the process cost and the amount of organic cosolvent. These results highlight the need for more predictive models of the effect of solvent mixtures on kinetics. Austin *et al.* [27, 28] also developed an approach for the design of mixtures (which they refer to as CAMxD) for reaction solvents. The use of COSMO-RS/COSMO-SAC thermodynamics makes it possible to predict the impact of mixed solvents on the reaction rate, although no validation of the accuracy of these predictions was presented for the reaction of interest. Nevertheless, the method makes it possible to identify binary solvent mixtures in principle.

2.3.4
Discussion

As is evident from the variety of methods now available for the design of reaction solvents, there has been a growing interest in this area of research in recent years. Many approaches have been proposed to address the challenge of identifying solvents that give the best performance despite the challenges arising when predicting key physical properties and the frequent need to find a trade-off between model accuracy and solution efficiency.

The model-based screening methods are relatively simple to implement and to use, but they are limited by their reliance on databases of known solvents, especially for reaction kinetics. On the other hand, the optimization-based and generate-and-test CAMD methods allow the exploration of a wider design space, but they entail larger computational costs, particularly where an attempt is made to predict the rate constants in the absence of experimental data [24, 27, 28, 31, 92]. Looking for a compromise between computational cost and accuracy, some researchers combine demanding theoretical models that require quantum mechanical calculations with simpler empirical models. The use of *ab initio* methods in solvent selection methods or CAMD formulations [24, 25, 27, 28, 31] is a step forward in computer-based solvent design, because it enables the incorporation of the prediction of reaction kinetics and thus eliminates the need for experimental data prior to the verification stage.

Environmental, health, and safety impacts of the solvent and processes have been taken into account and even minimized as part of the design process in several publications [33, 104, 109]. Other approaches where these considerations have not been included can easily be extended by making use of existing predictive models [110]. There is a need to consider alternative, "green" solvents in design approaches, such as GXLs (studied by Siougkrou et al. [37]), and other tunable solvents [117], such as ionic liquids.

Environmental benefits can also be derived by using solvent mixtures other than GXLs, making it possible to access a much broader range of properties. Mixtures have been the subject of growing attention in the CAMD literature (as computer-aided blend design, CAM^{b}D, or computer-aided mixture design (CAMxD) [27, 28, 60, 63, 66, 68, 69], but relatively little has been done in the context of the design of solvent mixtures for reactions. Mixtures are generally difficult to model and the need to predict compositional effects on kinetics introduces more challenges. There are limited experimental kinetic data in mixtures, while *ab initio* methods become more complex (and less reliable) when mixtures are considered. Nevertheless, this is an area that needs to be developed given the potential benefits of mixtures.

2.4
A Case Study

An example of the design of the optimal solvent for a chemical reaction is presented in this section in order to illustrate the key features of a systematic computer-based solvent design method.

The methodology of Struebing *et al.* [92] has been chosen (Figure 2.4) and its application to the Menschutkin reaction of pyridine (denoted here as reactant A) with phenacyl bromide (denoted here as reactant B) (Figure 2.1a) is presented. The kinetics of this particular reaction have been studied experimentally by many authors [82, 118–126] and have been investigated computationally [24, 92]. A detailed description of this application can be found in Ref. [92].

In the first step, the design problem is specified. Here, the objective is to maximize the reaction rate, subject to

- structure–property constraints,
- chemical feasibility and molecular complexity constraints, and
- design constraints.

The reactants are assumed to be dilute and the maximum allowable solubility of the solid phenacyl bromide x_B and the mole fraction of pyridine x_A are set to be equal to 0.25:

$$x_{B,max} = x_{A,fix} = 0.25. \tag{2.8}$$

Table 2.3 The structural groups used in the CAMD formulation for solvent design for the case study presented.

Structural groups used in CAMD		
CH_3	CH_3O	CH_3CN
CH_2	CH_2O	$CHCl_3$
CH	$CH-O$	CH_3NO_2
C	CH_2CN	
$C=C$	$COOH$	
aCH	CH_2Cl	
aC	$CHCl$	
$aCCH_3$	$CHCl_2$	
$aCCH_2$	$aCCl$	
$aCCH$	CH_2NO_2	
OH	$CHNO_2$	
$aCOH$	I	
CH_3CO	aCF	
CH_2CO		
CHO		
CH_3COO		
CH_2COO		

The groups in the second column are single-group molecules.

A list of the functional groups that are used in the design is given in Table 2.3. The initial set of solvents is also chosen in this stage. It consists of six solvents with diverse functional groups and a wide range of dielectric constants as shown in Table 2.4.

Table 2.4 Solvents and their properties for the initial solvent set used for the regression of the solvatochromic equation at the first iteration.

Solvent	ε [−]	A [−]	B [−]	S [−]	δ [−]	$\frac{\delta_H^2}{100}$ (cal cm^{-3})	k^{QM} (l mol^{-1} s^{-1})
Toluene	2.38	0.00	0.15	0.50	1.0	0.76	1.63×10^{-5}
Chlorobenzene	5.70	0.00	0.00	0.63	1.0	0.91	2.82×10^{-4}
Ethyl acetate	5.99	0.00	0.48	0.57	0.0	0.81	3.47×10^{-4}
Tetrahydrofuran	7.43	0.00	0.48	0.52	0.0	0.86	5.81×10^{-4}
Acetone	20.49	0.07	0.32	0.90	0.0	1.38	2.34×10^{-3}
Acetonitrile	35.69	0.00	0.49	0.69	0.0	0.79	1.54×10^{-3}

ε: dielectric constant, A: Abraham acidity, B: Abraham basicity, S: Abraham polarity, δ: polarizability correction parameter, δ_H: Hildebrand's solubility parameter, k^{QM} is the reaction rate constant calculated with the QM model.

In the second step of the algorithm, the reaction rate constants for all solvents in the initial set are calculated using quantum mechanical calculations, specifically the SMD solvation model [89] and the M05-2X/6-31G(d) level of theory. The predicted rate constants k^{QM} and the solvent properties used in the solvatochromic equation are shown in Table 2.4. The solvent properties required throughout the course of the algorithm are calculated with group contribution methods [29, 95], as described in Refs [24, 92].

As a third step, the surrogate model, that is, the solvatochromic equation, is regressed based on the quantum-mechanically derived rate constants for the initial solvent set. The resulting solvatochromic equation is as follows:

$$\log k^{CAMD} = -18.82 - 87.31A + 6.98B + 6.46S + 1.80\delta + 10.33 \frac{\delta_H^2}{100}. \quad (2.9)$$

Next, in the fourth step, Eq. (2.9) is used within a CAMD formulation, together with other property models such as UNIFAC [85] for the calculation of the solubility of phenacyl bromide, to design a first candidate solvent that maximizes the rate of the reaction. The designed solvent contains two structural groups, $1 \times CH_2NO_2$ and $1 \times I$, that is, iodo(nitro)methane, and has a predicted rate constant of $k^{CAMD} = 2.013 \cdot 10^{10}$ L · mol^{-1} s^{-1}. The surrogate model significantly overestimates the rate constant of the designed solvent in this first iteration (Table 2.5), indicating that the model needs improvement.

In the fifth step, the convergence criterion is applied and since the designed solvent was not used in the regression of the solvatochromic equation, the corresponding QM rate constant is calculated and added to the initial set of solvents to update the surrogate model. The algorithm iterates until no new solvent is found in the CAMD step (step 4). The solvents designed at each iteration of the algorithm are shown in Table 2.5.

The algorithm converges after five iterations to nitromethane, $1 \times CH_3NO_2$, with a predicted rate constant $k^{CAMD} = 3.192 \cdot 10^{-3}$ L · mol^{-1} · s^{-1}. The rate constant in nitromethane, according to the QM model, is $k^{QM} = 3.021 \cdot 10^{-3}$ L · mol^{-1} · s^{-1}, thus the deviation between the two models is 5.5%. It is interesting that iodo(nitro)methane, which was designed at the first iteration, has a higher

Table 2.5 Summarized results of the solvent design for the Menschutkin reaction.

Iter.	Designed solvent	x_A	x_B	c_A (mol l^{-1})	c_B (mol l^{-1})	k^{QM} (l mol^{-1} s^{-1})	r^{QM} (mol s l^{-1})
1	Iodo(nitro)methane	0.25	0.25	2.778	2.778	3.700×10^{-3}	2.855×10^{-2}
2	Nitroethane	0.25	0.25	2.806	2.806	2.172×10^{-3}	1.710×10^{-2}
3	Propanenitrile	0.25	0.25	2.815	2.815	1.731×10^{-3}	1.372×10^{-2}
4	Nitromethane	0.25	0.25	3.127	3.127	3.021×10^{-3}	2.954×10^{-2}
5	Nitromethane	0.25	0.25	3.127	3.127	3.021×10^{-3}	2.954×10^{-2}

x_A, x_B are the mole fractions of pyridine and phenacyl bromide, c_A and c_B are the concentrations of pyridine and phenacyl bromide, k^{QM} is the reaction rate constant calculated with the QM model, and r^{QM} is the reaction rate calculated with the QM model.

rate constant than nitromethane, while nitromethane has a higher reaction rate. This is due to the higher density of iodo(nitro)methane.

The algorithm, starting from only six diverse solvents, leads to the identification of several solvents that provide a high reaction rate, with the best solvent giving a 29% increase in the reaction rate constant and 65% increase in the reaction rate, compared to acetone, the best solvent in the initial set ($r_{\text{acetone}}^{\text{QM}} = 1.785 \cdot 10^{-2}$ mol $\cdot$ L^{-1} $\cdot$ s^{-1}).

Naturally, the accuracy of the method depends on the accuracy and effectiveness of the quantum mechanical and group contribution methods that are used, but since the algorithm is generic, more accurate models can easily be implemented as they become available.

2.5
Conclusions

The choice of solvent for a chemical reaction is significant since it can affect the performance of the process in numerous ways, including by changing productivity, atom economy, or EHS impact; however, making the right choice is a challenging task because of the complexity of the physical and chemical phenomena that determine the link between solvent structure and performance. A range of systematic methods for the selection or design of solvents for chemical reactions were presented and discussed critically in this chapter. An application of one of these methods was demonstrated to highlight the potential benefits of deploying systematic methods for solvent design. We found that only a few attempts have been made at embedding EHS metrics within the design problem beyond the inherent positive effect of increased productivity and selectivity that derive from a judicious choice of solvent.

Despite growing efforts in the area of solvent design for reactions, there is room for improvement in terms of the effectiveness, computational cost, and accuracy of existing methods but also in the way the sustainability of the proposed designs is taken into account. This can be improved via the inclusion of additional constraints and metrics that capture sustainability aspects, but also through the explicit consideration of alternative solvents that may be more benign by some metrics, such as gas-expanded liquids and ionic liquids. Computer-aided molecular approaches also offer the possibility of widening the design envelope to explore overall process performance and the best trade-offs between optimizing individual reaction stages and solvent recovery. In this context, the use of process-wide design methods with the aim to reduce the number of steps or to achieve stage telescoping could bring great benefits.

Acknowledgments

Financial support for this work from the Engineering and Physical Sciences Research Council of the United Kingdom via a Leadership Fellowship (EP/J003840/1) and a platform grant (EP/J014958/1) is gratefully acknowledged.

References

1 Jiménez-González, C., Curzons, A.D., Constable, D.J.C., and Cunningham, V.L. (2005) Expanding GSK's solvent selection guide: application of the life cycle assessment to enhance solvent selections. *Clean Technologies and Environmental Policy*, **7**, 42–50.

2 Jiménez-González, C., Poechlauer, P., Broxterman, Q.B., Yang, B., Ende, D., Baird, J., Bertsch, C., Hannah, R.E., Dell'Orco, P., Noorman, H., Yee, S., Reintjens, R., Wells, A., Massonneau, V., and Manley, J. (2011.) Key green engineering research areas for sustainable manufacturing: a perspective from pharmaceutical and fine chemicals manufacturers. *Organic Process Research and Development*, **15**, 900–911.

3 Curzons, A.D., Constable, D.C., and Cunningham, V.L. (1999) Solvent selection guide: a guide to the integration of environmental, health and safety criteria into the selection of solvents. *Clean Products and Processes*, **1**, 82–90.

4 Henderson, R.K., Jimenez-Gonzalez, C., and Constable, D.J.C. (2011) Expanding GSK's solvent selection guide: embedding sustainability into solvent selection starting at medicinal chemistry. *Green Chemistry*, **13**, 854–862.

5 Capello, C., Fischer, U., and Hungerbuehler, K. (2007) What is a green solvent? A comprehensive framework for the environmental assessment of solvents. *Green Chemistry*, **9**, 927–934.

6 Alfonsi, K., Colberg, J., Dunn, P.J., Fevig, T., Jennings, S., Johnson, T.A., Kleine, H.P., Knight, C., Nagy, M.A., Perry, D.A., and Stefaniak, M. (2008) Green chemistry tools to influence a medicinal chemistry and research chemistry based organisation. *Green Chemistry*, **10**, 31–36.

7 Carlson, R. (1992) *Design and Optimization in Organic Synthesis*, Elsevier, Amsterdam, The Netherlands.

8 Buncel, E., Stairs, R., and Wilson, H. (2003) *The Role of the Solvent in Chemical Reactions*, Oxford University Press.

9 Schilling, J., Tillmanns, D., Lampe, M., Hopp, M., Gross, J., and Bardow, A. (2017) From molecules to dollars: integrating molecular design into thermo-economic process design using consistent thermodynamic modeling. *Molecular Systems Design and Engineering*, **2**, 301–320.

10 Gopinath, S., Jackson, G., Galindo, A., and Adjiman, C.S. (2016) Outer approximation algorithm with physical domain reduction for computer-aided molecular and separation process design. *AIChE Journal*, **620** (9), 3484–3504.

11 El-Halwagi, A., Kazantzi, V., El-Halwagi, M., and Kazantzis, N. (2013) Optimizing safety-constrained solvent selection for process systems with economic uncertainties. *Journal of Loss Prevention in the Process Industries*, **26**, 495–498.

12 Bokhove, J., Schuur, B., and deHaan, A. B. (2012) Solvent design for trace removal of pyridines from aqueous streams using solvent impregnated resins. *Separation and Purification Technology*, **98**, 410–418.

13 Zhao, Y., Xiao, S., Liu, W., and Wu, Z. (2012) Computer aided solvent scanning for the separation of 2-methoxynaphthene and 2-acetyl-6-methoxynaphthalene. *Journal of Chemical Engineering Data*, **57**, 200–203.

14 Pereira, F.E., Keskes, E., Galindo, A., Jackson, G., and Adjiman, C.S. (2011) Integrated solvent and process design using a SAFT-VR thermodynamic description: high-pressure separation of carbon dioxide and methane. *Computers and Chemical Engineering*, **35**, 474–491.

15 Lapkin, A.A., Peters, M., Greiner, L., Chemat, S., Leonhard, K., Liauwb, M.A., and Leitner, W. (2010) Screening of new solvents for artemisinin extraction process using *ab initio* methodology. *Green Chemistry*, **12**, 241–251.

16 Bardow, A., Steur, K., and Gross, J. (2010) Continuous-molecular targeting for integrated solvent and process design. *Industrial and Engineering Chemistry Research*, **49**, 2834–2840.

17 Lek-Utaiwan, P., Suphanit, B., Mongkolsir, N., and Gani, R. (2008) Integrated design of solvent-based extractive separation processes. *Computer-Aided Chemical Engineering*, **25**, 121–126.

18 Papadopoulos, A.I. and Linke, P. (2005) Multiobjective molecular design for integrated process-solvent systems synthesis. *AIChE Journal*, **52**, 1057–1070.

19 Samudra, A.P. and Sahinidis, N.V. (2013) Optimization-based framework for computer-aided molecular design. *AIChE Journal*, **59**, 3686–3701.

20 Karunanithi, A.T., Achenie, L.E.K., and Gani, R. (2006) A computer-aided molecular design framework for crystallization solvent design. *Chemical Engineering Science*, **61**, 1247–1260.

21 Folić, M., Adjiman, C.S., and Pistikopoulos, E.N. (2004) The design of solvents for optimal reaction rates. *Computer-Aided Chemical Engineering*, **18**, 175–180.

22 Folić, M., Adjiman, C.S., and Pistikopoulos, E.N. (2005) A computer-aided methodology for optimal solvent for reactions with experimental verification. *Computer-Aided Chemical Engineering*, **20a–20b**, 1651–1656.

23 Folić, M., Adjiman, C.S., and Pistikopoulos, E.N. (2008) Computer-aided solvent design for reactions: maximizing product formation. *AIChE Journal*, **47**, 5190–5202.

24 Struebing, H., Ganase, Z., Karamertzanis, P.G., Siougkrou, E., Haycock, P., Piccione, P.M., Armstrong, A., Galindo, A., and Adjiman, C.S. (2013) Computer-aided molecular design of solvents for accelerated reaction kinetics. *Nature Chemistry*, **5**, 952–957.

25 Zhou, T., Qi, Z., and Sundmacher, K. (2014) Model-based method for the screening of solvents for chemical reactions. *Chemical Engineering Science*, **115**, 177–185.

26 Struebing, H., Obermeier, S., Siougkrou, E., Adjiman, C.S., and Galindo, A. (2017) A QM-CAMD approach to solvent design for optimal reaction rates. *Chemical Engineering Science*, **159**, 69–83.

27 Austin, N.D., Sahinidis, N.V., and Trahan, D.W. (2017) A COSMO-based approach to computer-aided mixture design. *Chemical Engineering Science*, **159**, 93–105.

28 Austin, N.D., Sahinidis, N.V., Konstantinov, I.A., and Trahan, D.W. (2017) COSMO-based computer-aided molecular/mixture design: A focus on reaction solvents. *AIChE Journal*, doi: 10.1002/aic.15871.

29 Folić, M., Adjiman, C.S., and Pistikopoulos, E.N. (2007) Design of solvents for optimal reaction rate constants. *AIChE Journal*, **53**, 1240–1256.

30 Wicaksono, D.S., Mhamdi, A., and Marquardt, W. (2014) Computer-aided screening of solvents for optimal reaction rates. *Chemical Engineering Science*, **115**, 167–176.

31 Stanescu, I. and Achenie, L.E.K. (2006) A theoretical study of solvent effects on Kolbe–Schmitt reaction kinetics. *Chemical Engineering Science*, **61**, 6199–6212.

32 Folić, M., Gani, R., Jiménez-González, C., and Constable, D.J.C. (2008) Systematic selection of green solvents for organic reacting systems. *Chinese Journal of Chemical Engineering*, **16**, 376–383.

33 Gani, R., Jiménez-González, C., and Constable, D.J.C. (2005) Method for selection of solvents for promotion of organic reactions. *Computers and Chemical Engineering*, **29**, 1661–1676.

34 Cheng, H.C. and Wang, F.S. (2007) Trade-off optimal design of a biocompatible solvent for an extractive fermentation process. *Chemical Engineering Science*, **62**, 4316–4324.

35 Cheng, H.C. and Wang, F.S. (2010) Computer-aided biocompatible solvent design for an integrated extractive fermentation-separation process. *Chemical Engineering Journal*, **162**, 809–820.

36 Chávez-Islas, L.M., Vasquez-Medrano, R., and Flores-Tlacuahuac, A. (2011) Optimal molecular design of ionic liquids for high-purity bioethanol production.

Industrial and Engineering Chemistry Research, **50**, 5153–5168.

37 Siougkrou, E., Galindo, A., and Adjiman, C.S. (2014) On the optimal design of gas-expanded liquids based on process performance. *Chemical Engineering Science*, **115**, 19–30.

38 Zhou, T., McBride, K., Zhang, X., Qi, Z., and Sundmacher, K. (2015) Integrated solvent and process design exemplified for a Diels–Alder reaction. *AIChE Journal*, **61**, 147–158.

39 Zhou, T., Lyu, Z., Qi, Z., and Sundmacher, K. (2015) Robust design of optimal solvents for chemical reactions: a combined experimental and computational strategy. *Chemical Engineering Science*, **137**, 613–625.

40 McBride, K., Gaide, T., Vorholt, A., Behr, A., and Sundmacher, K. (2016) Thermomorphic solvent selection for homogeneous catalyst recovery based on COSMO-RS. *Chemical Engineering and Processing: Process Intensification*, **99**, 97–106.

41 Zhou, T., Wang, J., McBride, K., and Sundmacher, K. (2016) Optimal design of solvents for extractive reaction processes. *AIChE Journal*, **62**, 3238–3249.

42 Adjiman, C.S., Galindo, A., and Jackson, G. (2014) Molecules matter: the expanding envelope of process design, in *8th International Conference on Foundations of Computer-Aided Process Design – FOCAPD* (eds M. Eden, J.D. Siirola, and G.P. Towler), Elsevier B.V., Washington, pp. 55–64.

43 Papadopoulos, A.I., Badr, S., Chremos, A., Forte, E., Zarogiannis, T., Seferlis, P., Papadokonstantakis, S., Galindo, A., Jackson, G., and Adjiman, C.S. (2016) Computer-aided molecular design and selection of CO_2 capture solvents based on thermodynamics, reactivity and sustainability. *Molecular Systems Design and Engineering*, **1**, 313–334.

44 Limleamthong, P., Gonzalez-Miquel, M., Papadokonstantakis, S., Papadopoulos, A.I., Seferlis, P., and Guillen-Gosalbez, G. (2016) Multi-criteria screening of chemicals considering thermodynamic and life cycle assessment metrics via data envelopment analysis: application to CO_2 capture. *Green Chemistry*, **18**, 6468–6481.

45 Gani, R. and Brignole, E.A. (1983) Molecular design of solvents for liquid extraction based on UNIFAC. *Fluid Phase Equilibria*, **13**, 331–340.

46 Achenie, L.E.K., Gani, R., and Venkatasubramanian, V. (eds) (2003) *Computer Aided Molecular Design: Theory and Practice*, Elsevier, Amsterdam, The Netherlands.

47 Gani, R. (2004) Chemical product design: challenges and opportunities. *Computers and Chemical Engineering*, **28**, 2441–2457.

48 Gani, R. and Ng, K.M. (2015) Product design: molecules, devices, functional products, and formulated products. *Computers & Chemical Engineering*, **81**, 70–79.

49 Ng, L.Y., Chong, F.K., and Chemmangattuvalappil, N.G. (2015) Challenges and opportunities in computer-aided molecular design. *Computers & Chemical Engineering*, **81**, 115–129.

50 Austin, N.D., Sahinidis, N.V., and Trahan, D.W. (2016) Computer-aided molecular design: an introduction and review of tools, applications, and solution techniques. *Chemical Engineering Research and Design*, **116**, 2–26.

51 Maranas, C.D. (1996) Optimal computer-aided molecular design: a polymer design case study. *Industrial and Engineering Chemistry Research*, **35**, 3403–3414.

52 Sahinidis, N.V., Tawarmalani, M., and Yu, M. (2003) Design of alternative refrigerants via global optimization. *AIChE Journal*, 49, 1761–1775.

53 Apostolakou, A. and Adjiman, C.S. (2003) Refrigerant design case study, in *Computer Aided Molecular Design: Theory and Practice*, vol. **12** (eds L.E.K. Achenie, R. Gani, and V. Venkatasubramanian), Elsevier, Amsterdam, The Netherlands, pp. 289–301.

54 Sheldon, T.J., Folić, M., and Adjiman, C.S. (2006) Solvent design using a quantum mechanical continuum solvation model. *Industrial and*

Engineering Chemical Research, **45**, 1128–1140.

55 Weis, D.C. and Visco, D.P. (2010) Computer-aided molecular design using the Signature molecular descriptor: application to solvent selection. *Computers and Chemical Engineering*, **34**, 1018–1029.

56 Mattei, M., Kontogeorgis, G.M., and Gani, R. (2014) A comprehensive framework for surfactant selection and design for emulsion based chemical product design. *Fluid Phase Equilibria*, **362**, 288–299.

57 Giovanoglou, A., Barlatier, J., Adjiman, C.S., Pistikopoulos, E.N., and Cordiner, J.L. (2003) Optimal solvent design for batch separation based on economic performance. *AIChE Journal*, **49**, 3095–3109.

58 Eden, M.R., Jørgensen, S.B., Gani, R., and El-Halwagi, M.M. (2004) A novel framework for simultaneous separation process and product design. *Chemical Engineering and Processing*, **43**, 595–608.

59 Burger, J., Papaioannou, V., Gopinath, S., Jackson, G., Galindo, A., and Adjiman, C.S. (2015) A hierarchical method to integrated solvent and process design of physical CO_2 absorption using the SAFT-γ Mie approach. *AIChE Journal*, **610** (10), 3249–3269.

60 Papadopoulos, A.I., Stijepovic, M., Linke, P., Seferlis, P., and Voutetakis, S. (2013) Toward optimum working fluid mixtures for organic Rankine cycles using molecular design and sensitivity analysis. *Industrial and Engineering Chemistry Research*, **52**, 12116–12133.

61 Vaidyanathan, R. and El-Halwagi, M. (1996) Computer-aided synthesis of polymers and blends with target properties. *Industrial & Engineering Chemistry Research*, **35**, 627–634.

62 Buxton, A., Livingston, A.G., and Pistikopoulos, E.N. (1999) Optimal design of solvent blends for environmental impact minimization. *AIChE Journal*, **45**, 817–843.

63 Karunanithi, A.T., Achenie, L.E.K., and Gani, R. (2005) A new decomposition-based computer-aided molecular/mixture design methodology for the design of optimal solvents and solvent mixtures. *Industrial and Engineering Chemistry Research*, **44**, 4785–4797.

64 Solvason, C.C., Chemmangattuvalappil, N.G., Eljack, F.T., and Eden, M.R. (2009) Efficient visual mixture design of experiments using property clustering techniques. *Industrial & Engineering Chemistry Research*, **48**, 2245–2256.

65 Nunes, S., Matos, A.M., Duarte, T., Figueiras, H., and Sousa-Coutinho, J. (2013) Mixture design of self-compacting glass mortar. *Cement and Concrete Composites*, **43**, 1–11.

66 Yunus, N.A., Gernaey, K.V., Woodley, J.M., and Gani, R. (2014) A systematic methodology for design of tailor-made blended products. *Computers and Chemical Engineering*, **66**, 201–213.

67 Cignitti, S., Zhang, L., and Gani, R. (2015) Computer-aided framework for design of pure, mixed and blended products. *Computers Aided Chemical Engineering*, **37**, 2093–2098.

68 Jonuzaj, S., Akula, P.T., Kleniati, P.-M., and Adjiman, C.S. (2016) The formulation of optimal mixtures with generalized disjunctive programming: a solvent design case study. *AIChE Journal*, **62**, 1616–1633.

69 Jonuzaj, S. and Adjiman, C.S. (2017) Designing optimal mixtures using generalized disjunctive programming: Hull relaxations. *Chemical Engineering Science*, **159**, 106–130.

70 Harper, P.M., Hostrup, M., and Gani, R. (2003) A hybrid CAMD method, in *Computer Aided Molecular Design: Theory and Practice*, vol. 12 (eds L.E.K. Achenie, R. Gani, and V. Venkatasubramanian), Elsevier, Amsterdam, The Netherlands, pp. 139–165.

71 Odele, O. and Macchietto, S. (1993) Computer-aided molecular design: a novel method for optimal solvent selection. *Fluid Phase Equilibria*, **82**, 47–54.

72 Chipperfield, J.R. (ed.) (1999) *Non-Aqueous Solvents*, Oxford University Press, Oxford.

73 Berthelot, M. and Pean de Saint-Gilles, L. (1862) Recherches sur les affinités. De la

formation et de la décomposition des éthers. *Annales de Chimie et de Physique, 3e série*, **65**, 385–422.

74 Menschutkin, N. (1890) Beiträge zur kenntnis der affinitätskoeffizienten der alkylhaloide und der organischen amine. *Zeitschrift für Physikalische Chemie*, **5**, 589–600.

75 Menschutkin, N. (1890) Über die affinitätskoeffizienten der alkylhaloide und der amine. *Zeitschrift für Physikalische Chemie*, **6**, 41–57.

76 Guillaume, A., Bruylants, A. (1976) 4. Kinetic study of carbocyanine synthesis. II. Catalysis of the aminolysis of 2-[β-(ethylthio)vinyl]cycloammonium salts by anions and solvents. *Bulletin de la Classe des Sciences, Academie Royale de Belgique*, **62**, 89–106.

77 Reichardt, C. and Welton, T. (2011) *Solvents and Solvents Effects in Organic Chemistry*, Wiley-VCH Verlag GmbH, Weinheim, Germany.

78 Hintermann, L., Masuo, R., and Suzuki, K. (2008) Solvent-controlled leaving-group selectivity in aromatic nucleophilic substitution. *Organic Letters*, **10**, 4859–4862.

79 Scholl, M., Ding, S., Lee, C.W., and Grubbs, R.H. (1999) Synthesis and activity of a new generation of ruthenium-based olefin metathesis catalysts coordinated with 1,3-dimesityl-4,5-dihydroimidazol-2-ylidene ligands. *Organic Letters*, **1**, 953–956.

80 Adjiman, C.S., Clarke, A.J., Cooper, G., and Taylor, P.C. (2008) Solvents for ring-closing metathesis reactions. *Chemical Communications*, 2806–2808.

81 Ashworth, I.W., Nelson, D.J., and Percy, J.M. (2013) Solvent effects on Grubbs' pre-catalyst initiation rates. *Dalton Transactions*, **42**, 4110–4113.

82 Ganase, Z. (2015) Ph.D. thesis. An Experimental Study on the Effects of Solvents on the Rate and Selectivity of Organic Reactions. Imperial College London.

83 Eyring, H. (1935) The activated complex in chemical reactions. *Journal of Chemical Physics*, **3**, 107–115.

84 Evans, M.G. and Polanyi, M. (1935) Some applications of the transition state method to the calculation of reaction velocities, especially in solution. *Transactions of the Faraday Society*, **31**, 875.

85 Fredenslund, A., Jones, R.L., and Prausnitz, J.M. (1975) Group-contribution estimation of activity coefficients in nonideal liquid mixtures. *AIChE Journal*, **21**, 1086–1099.

86 Lymperiadis, A., Adjiman, C.S., Galindo, A., and Jackson, G. (2007) A group contribution method for associating chain molecules based on the statistical associating fluid theory (SAFT-γ). *Journal of Chemical Physics*, **127**, 234903.

87 Papaioannou, V., Lafitte, T., Avendaño, C., Adjiman, C.S., Jackson, G., Müller, E.A., and Galindo, A. (2014) Group contribution methodology based on the statistical associating fluid theory for heteronuclear molecules formed from Mie segments. *Journal of Chemical Physics*, **140** (5).

88 Mennucci, B., Cancès, E., and Tomasi, J. (1997) Evaluation of solvent effects in isotropic and anisotropic dielectrics and in ionic solutions with a unified integral equation method: theoretical bases, computational implementation, and numerical applications. *The Journal of Physical Chemistry B*, **1010** (49), 10506–10517.

89 Marenich, A., Cramer, C.J., and Truhlar, D.G. (2009) Universal solvation model based on solute electron density and on a continuum model of the solvent defined by bulk dielectric constant and atomic surface tensions. *Journal of Physical Chemistry B*, **113**, 6378–6396.

90 Eckert, F. and Klamt, A. (2002) Fast solvent screening via quantum chemistry: COSMO-RS approach. *AIChE Journal*, **48**, 369–385.

91 Lin, S.-T. and Sandler, S.I. (2002) *A priori* phase equilibrium prediction from a segment contribution solvation model. *Industrial & Engineering Chemistry Research*, **410** (5), 899–913.

92 Struebing, H., Obermeier, S., Siougkrou, E., Adjiman, C.S., and Galindo, A. (2017) A QM-CAMD approach to solvent design for optimal reaction rates. *Chemical Engineering Science*, **159**, 69–83.

93 Constantinou, L. and Gani, R. (1994) New group-contribution method for estimating properties of pure compounds. *AIChE Journal*, **40**, 1697–1710.

94 Marrero, J. and Gani, R. (2001) Group-contribution based estimation of pure compound properties. *Fluid Phase Equilibria*, **183**, 183–208.

95 Sheldon, T.J., Adjiman, C.S., and Cordiner, J.L. (2005) Pure component properties from group contribution: hydrogen-bond basicity, hydrogen-bond acidity, Hildebrand solubility parameter, macroscopic surface tension, dipole moment, refractive index and dielectric constant. *Fluid Phase Equilibria*, **231**, 27–37.

96 Branduardi, D., Gervasio, F.L., and Parrinello, M. (2007) From A to B in free energy space. *J. Chem. Phys.*, **126**, 054103.

97 Kamlet, M.J. and Taft, R.W. (1976) Solvatochromic comparison method.1. Beta-scale of solvent hydrogen-bond acceptor (HBA) basicities. *Journal of American Chemical Society*, **98**, 377–383.

98 Taft, R.W. and Kamlet, M.J. (1976) Solvatochromic comparison method.2. Alpha-scale of solvent hydrogen-bond donor (HBD) acidities. *Journal of American Chemical Society*, **98**, 2886–2894.

99 Abraham, M.H., Doherty, R.M., Kamlet, M.J., Harris, J.M., and Taft, R.W. (1987) Linear solvation energy relationships. Part 38. An analysis of the use of solvent parameters in the correlation of rate constants, with special reference to the solvolysis of *t*-butyl chloride. *Journal of Chemical Society Perkin Transactions 2*, 1097–1101.

100 Abraham, M.H. (1993) Application of solvation equations to chemical and biochemical processes. *Pure and Applied Chemistry*, **65**, 2503–2512.

101 Cativiela, C., García, J.I., Gil, J., Martínez, R.M., Mayoral, J.A., Salvatella, L., Urieta, J.S., Mainar, A.M., and Abraham, M.H. (1997) Solvent effects on Diels–Alder reactions. The use of aqueous mixtures on fluorinated alcohols and the study of reactions of acrylonitrile. *Journal of Chemical Society Perkin Transactions 2*, 653–660.

102 Klamt, A. (1995) Conductor-like screening model for real solvents: a new approach to the quantitative calculation of solvation phenomena. *Journal of Physical Chemistry*, **99**, 2224–2235.

103 Eliasson, B., Johnels, D., Wold, S., Edlund, U., and Sjöström, M. (1982) On the correlation between solvent scales and solvent-induced C-13 NMR chemical-shifts of a planar lithium carbanion: a multivariate data-analysis using a principal component-multiple-regression-like formalism. *Acta Chemica Scandinavica Series B: Organic Chemistry and Biochemistry*, **36**, 155–164.

104 Faria, R.P.V., Pereira, C.S.M., Silva, V.M.T.M., Loureiro, J.M., and Rodrigues, A.E. (2013) Glycerol valorisation as biofuels: selection of a suitable solvent for an innovative process for the synthesis of GEA. *Chemical Engineering Journal*, **233**, 159–167.

105 National Fire Protection Association (2012) Standard System for the Identification of the Hazards of Materials for Emergency Response.

106 Brignole, E.A., Bottini, S., and Gani, R. (1986) A strategy for the design and selection of solvents for separation processes. *Fluid Phase Equilibria*, **29**, 125–132.

107 ICAS documentation. Internal report PEc2-14. Computer Aided Process Engineering Centre. Department of Chemical Engineering. Technical University of Denmark, Lyngby, Denmark, 1994–004.

108 Gani, R., Gómez, P.A., Folić, M., Jiménez-González, C., and Constable, D.J.C. (2008) Solvents in organic synthesis: replacement and multi-step reaction systems. *Computers and Chemical Engineering*, **32**, 2420–2444.

109 Abildskov, J., van Leeuwen, M.B., Boeriu, C.G., and L. A. M., vandenBroek. (2013) Computer-aided solvent screening for biocatalysis. *Journal of Molecular Catalysis B: Enzymatic*, **85–86**, 200–213.

110 Martin, T.M. and Young, D.M. (2001) Prediction of the acute toxicity (96-h

110 LC$_{50}$) of organic compounds to the fathead minnow (*Pimephales promelas*) using group contribution method. *Chemical Research in Toxicology*, **14**, 1378–1385.
111 CAPEC (2012) ProCAMD Documentation. Denmark Technical University, Department of Chemical Engineering.
112 Mu, T., Rarey, J., and Gmehling, J. (2007) Group contribution prediction of surface charge density profiles for COSMO-RS(Ol). *AIChE Journal*, **53**, 3231–3240.
113 Diamanti, A., Adjiman, C.S., Piccione, P.M., Rea, A.M., and Galindo, A. (2017) Development of predictive models of the kinetics of a hydrogen abstraction reaction combining quantum-mechanical calculations and experimental data. *Industrial & Engineering Chemistry Research*, **56**, 815–831.
114 Rosés, M., Buhvestov, U., Ráfols, C., Rived, F., and Bosch, E. (1997) Solute–solvent and solvent–solvent interactions in binary solvent mixtures. Part 6. A quantitative measurement of the enhancement of the water structure in 2-methylpropan-2-ol-water and propan-2-ol-water mixtures by solvatochromic indicators. *Journal of Chemical Society Perkin Transactions 2*, 1341–1348.
115 Buhvestov, U., Rived, F., Ráfols, C., Bosch, E., and Rosés, M. (1998) Solute–solvent and solvent–solvent interactions in binary solvent mixtures. Part 7. Comparison of the enhancement of the water structure in alcohol–water mixtures measured by solvatochromic indicators. *Journal of Physical Organic Chemistry*, **11**, 185–192.
116 Ahlers, J., Yamaguchi, T., and Gmehling, J. (2004) Development of a universal group contribution equation of state. 5. Prediction of the solubility of high-boiling compounds in supercritical gases with the group contribution equation of state Volume-Translated Peng-Robinson. *Industrial and Engineering Chemistry Research*, **43**, 6569–6576.
117 Eckert, C.A., Liotta, C.L., Bush, D., Brown, J.S., and Hallett, J.P. (2004) Sustainable reactions in tunable solvents. *The Journal of Physical Chemistry B*, **1080** (47), 18108–18118.
118 Pearson, R.G., Langer, S.H., Williams, F.V., and McGuire, W.J. (1952) Mechanism of the reaction of α-Haloketones with weakly basic nucleophilic reagents. *Journal of the American Chemical Society*, **74**, 5130–5132.
119 Halvorsen, A. and Songstad, J. (1978) The reactivity of 2-bromo-1-phenylethanone (phenacyl bromide) toward nucleophilic species. *Journal of Chemical Society, Chemical Communications*, 327–328.
120 Barnard, P.W.C. and Smith, B.V. (1981) The Menschutkin reaction: a group experiment in a kinetic study. *Journal of Chemical Education*, **580** (3), 282–285.
121 Yoh, S.D., Shim, K.T., and Lee, K.A. (1981) Studies on the quaternization of tertiary amines (II). Kinetics and mechanism for the reaction of substituted phenacyl bromides with substituted pyridines. *Journal of the Korean Chemical Society*, **25**, 110–118.
122 Forster, W. and Laird, R.M. (1982) The mechanism of alkylation reactions. Part 1. The effect of substituents on the reaction of phenacyl bromide with pyridine in methanol. *Journal of the Chemical Society Perkin Transactions 2*, 135–138.
123 Hwang, J.U., Chung, J.J., Yoh, S.D., and Jee, J.G. (1983) Kinetics for the reaction of phenacyl bromide with pyridine in acetone under high pressure. *Bulletin of the Korean Chemical Society*, **4**, 273–240.
124 Shunmugasundaram, A. and Balakumar, S. (1985) Kinetics of reaction of phenacyl bromide with 3- & 4-substituted pyridines and 4-substituted 4-styrylpyridines. *Indian Journal of Chemistry*, **24A**, 775–777.
125 Winston, S.J., Rao, P.J., Sethuram, B., and Rao, T.N. (1996) The LFER correlations for S_N2 reactions of phenacyl bromide. *Indian Journal of Chemistry*, **35A**, 979–982.
126 Koh, H.J., Han, K.L., Lee, H.W., and Lee, I. (2000) Kinetics and mechanism of the pyridinolysis of phenacyl bromides in acetonitrile. *Journal of Organic Chemistry*, **65**, 4706–4711.

3
Hierarchically Structured Pt and Non-Pt-Based Electrocatalysts for PEM Fuel Cells

Panagiotis Trogadas and Marc-Olivier Coppens

3.1
Introduction

Electrocatalysis is the field of chemistry that deals with the catalysis of redox reactions and links electrochemistry to catalysis [1, 2]. Redox reactions play a key role in electrochemical conversion devices, such as fuel cells, batteries, electrolyzers, and sensors, as well as in naturally occurring energy conversion processes in living systems [1].

Electrocatalysis started to see significant developments as interest mounted in experimental research toward the hydrogen evolution reaction [3]; research in this field flourished during the development of fuel cells in the 1960s and the investigation of oxygen reduction (ORR) and fuel oxidation to carbon dioxide reactions [3]. Thus, electrocatalysis plays an important role in the context of renewable energy and "green chemistry" more generally.

Electrocatalytic reactions are similar to heterogeneous catalytic reactions, but electrocatalysis has certain specific characteristics [3]:

- The rate of an electrochemical reaction depends not only on the experimental conditions that are relevant to heterogeneous catalysis in general (such as temperature, electrolyte composition, catalyst composition, and surface structure), but also on the electrode potential, which significantly affects the reaction rate.
- The catalyst surface is in contact with reacting species and electrolyte ions, influencing the surface properties.
- Electrons and reactant species have to be supplied to or withdrawn from the catalyst, hence electrocatalysts must have electrical conductivity.

Transition metals (containing atoms with empty d-electron shells) are commonly used as catalysts in electrochemical energy conversion devices. These include platinum, iron, manganese, vanadium, and so on. Despite its high cost, platinum (Pt) is widely used as electrocatalyst in fuel cells, due to its high activity in acid and alkaline solutions over a wide range of applied potentials [3].

Handbook of Green Chemistry Volume 12: Green Chemical Engineering, First Edition. Edited by Alexei A. Lapkin.
© 2018 Wiley-VCH Verlag GmbH & Co. KGaA. Published 2018 by Wiley-VCH Verlag GmbH & Co. KGaA.

The understanding of particle size effect on the activity of metal nanoparticles is one of the most important objectives in electrocatalysis [4, 5]. ORR is among the dominant reactions in electrochemistry [6], along with H_2 evolution and methanol oxidation reactions, and its slow kinetics hinders the performance of low-temperature fuel cells [7]. The overall activity of a Pt electrocatalyst depends on structural (particle size, facets, etc.) and electronic effects (surface electronic structure and energetics) [1]. However, the particle size effect on the activity of Pt is still controversial [8, 9]: There are reports in the literature demonstrating a decrease in Pt activity as particle size decreases, whereas other reports suggest that Pt activity depends on the interparticle distance and not on the particle size [10–12]. Different explanations for the relation between particle size and Pt activity have been presented, including a lower ratio of preferable crystal facets [9, 13, 14], stronger interaction between oxygen containing species and Pt atoms [15, 16], and a lower potential of total zero charge [17]. A recent study demonstrated that there is indeed a particle size effect on Pt activity, which increases with increasing Pt diameter [8]. However, the particle size effect loses its strength as the catalyst ages, especially at small particle sizes where Pt dissolution causes agglomeration of nanoparticles and, hence, wider particle size distributions and particle shape alterations [8]. Thus, particle stability is still a significant issue to be resolved that could provide ways to stabilize a particular shape or facet associated with a desired architecture of nanoparticles providing the highest activity [8].

Many electrocatalysts are porous, or are supported on a porous support, to achieve a high, accessible, specific surface area for catalysis and for transport of both neutral and charged species, including electrons. For heterogeneous catalysts in general, their overall performance depends on their architecture at multiple scales. There has been significant progress in the synthesis of porous materials with a controlled hierarchical structure, including for electrocatalytic devices. The majority of these synthesis methods is based on templating (soft, hard, or dual templating), scaffolding, galvanic replacement, and the Kirkendall effect. Soft (macro-) molecular and hard templating approaches are effective to produce structured hierarchical materials with controlled pore sizes, even though they usually do not allow for accurate control over pore sizes larger than 10 µm [18–23]. The creation of such large macropores is attainable via the utilization of gas bubbles or droplets as templates; however, foaming or emulsification methods result in broad pore size distributions and are prone to Ostwald ripening and coalescence [24, 25]. Finally, scaffolding is a template-free method producing meso/macroporous hierarchical materials, even though precise control over shape and size is difficult to achieve [26].

At smaller length scales, hollow nanostructures can be obtained via the galvanic replacement reaction and the Kirkendall effect [26]. The latter refers to preferred outward elemental diffusion, leading to a net material flux across the spherical interface and the subsequent formation of a single void at the center of the structure, while the former is based on the electrochemical potential difference between the metals participating in the reaction.

Detailed reviews on the synthesis, properties, and applications of hierarchically structured materials for electrochemical energy conversion are available in the literature [26–31]. The unique characteristics (large surface area, high meso/macroporosity) of the resulting hierarchical nanomaterials transform them into ideal materials for electrocatalysis and energy storage. The large surface area of the mesopores provides short diffusion lengths and a large contact area, while the macropores enhance their transport properties and can store electrochemically active ions, improving device performance [26, 27].

This chapter focuses on structuring electrocatalysts at the nanoscale. As mentioned before, Pt is typically used as catalytically active phase. However, the kinetics and corrosion stability of Pt, as well as its tolerance to CO poisoning (for example, if hydrogen as a fuel is derived from steam reforming or partial oxidation) have to be improved significantly to transform PEMFCs into cost-competitive technology. This could be achieved by the utilization of hollow Pt or Pt-alloys as electrocatalysts, since hollow nanostructures possess structurally tunable features such as shell thickness, interior cavity size, and composition [32], leading to a higher surface area and utilization efficiency over pristine Pt [33–38], as well as increased catalytic activity over monometallic Pt, due to compositional effects [39–61]. Specifically, Pt-based nanoporous alloys with transition metals (PtM, where M = Au, Ag, Ni, Co, Pd, Cu) are regarded as the most promising catalytic materials for PEMFCs, as they offer comparable catalytic activities to pristine Pt at reduced overall catalyst loading cost, even though extensive research is still needed to determine the optimum porosity and composition [32, 62, 63]. Another issue to be considered is electrocatalyst support corrosion, especially under automotive drive cycles [64]. This can be mitigated by the synthesis and use of corrosion resistant metal oxide supports and derivative electrocatalysts [65–70]. We now discuss some of these manifestations in greater detail.

3.2
Pure Hollow Pt Nanoparticles

Pt hollow nanospheres and solid nanoclusters have been synthesized via a modified galvanic replacement reaction using Co nanoparticles as the sacrificial template and the seeding method, respectively [36, 71]. Each Pt nanosphere (~24 nm) consisted of an empty core with a shell of ~2 nm thickness, while each nanocluster (~29 nm) was composed of small nanoparticles (~4 nm) [36]. Cyclic voltammogram (CV) measurements in sulfuric acid (0.5 M H_2SO_4) and methanol (0.6 M) solution revealed that the Pt hollow nanospheres have a significantly higher activity than Pt nanoclusters (54 µA and 19 µA peak current, respectively) at the same metal loading (~0.0175 mg Pt cm^{-2}) [36]. This difference in activity is directly related to the available surface area; because Pt hollow nanospheres are coreless, a larger number of hollow Pt nanospheres is obtained compared to the synthesized solid Pt nanoclusters with the same loading. Taking into account that the surface area of a single hollow nanosphere is approximately 1.16 times the surface area of a single solid nanocluster and that both the inner and outer surface of Pt hollow nanospheres participate in the catalytic reaction, hollow Pt nanospheres result in enhanced catalytic activity [36, 72].

Similar results are observed for hollow Pt nanospheres synthesized via galvanic replacement using Ag nanoparticles as the sacrificial template [73]. Methanol oxidation on hollow Pt nanospheres is initiated at ∼0.3–0.4 V (versus SCE) and a fully developed oxidation peak is formed at ∼0.7 V (versus SCE). The peak current density of hollow Pt nanospheres is −80 and −68 μA mg^{-1} in comparison to −26 and −11 μA mg^{-1} of the core–shell PtAg nanoparticles in the forward and reverse scan, respectively (Figure 3.1) [73]. The measured current densities are normalized by the accessible surface areas in hollow and core–shell particles, showing that the higher catalytic activity of hollow Pt nanospheres is, again, a surface area effect. In this case, the surface area per hollow Pt nanosphere is approximately 1.52 times the area of core–shell PtAg nanoparticles [73].

The effect of the bimetallic composition of PtAg nanocubes on the catalytic activity of the hydrogen evolution reaction (HER) has also been investigated [74]. Pure Ag nanocubes do not exhibit any activity toward the HER. As the amount

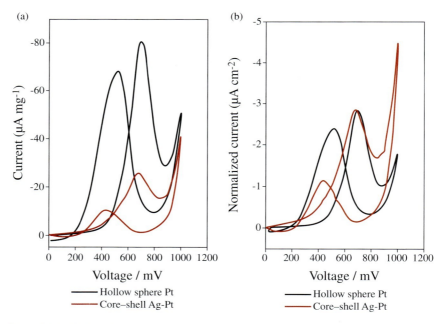

Figure 3.1 (a) Cyclic voltammograms of hollow Pt nanospheres (black line) and core–shell PtAg nanoparticles (red line) in electrolyte solution (0.5 M H$_2$SO$_4$ and 0.6 M methanol). (b) Renormalization of cyclic voltammograms by the available surface areas of the catalysts (experimental conditions – catalyst loading: 0.07 mg cm^{-2} for Pt hollow nanospheres and 0.11 mg cm^{-2} for the core–shell nanoparticles, 50 mV s^{-1} scan rate). (Reprinted with permission from Ref. [73]. Copyright 2005, American Chemical Society.)

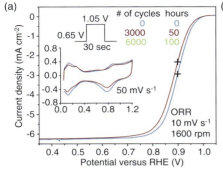 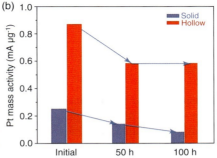

Figure 3.2 (a) Sequential ORR polarization curves and voltammograms (inset) for hollow Pt nanospheres. (b) Pt mass activity as a function of potential cycling time for hollow and solid Pt nanoparticles. (Reprinted with permission from Ref. [75]. Copyright 2011, American Chemical Society.)

of Pt is increased in the Ag nanocubes, there is a shift to less negative potentials for the HER [74]. Upon data normalization with Pt mass, a significant increase in the catalytic activity is observed once the percentage of Pt in Ag nanocubes reaches 4 wt%. However, there is a marginal increase in HER activity as Pt loading reaches 20 wt% due to the breakdown of the PtAg nanocubes [74].

Additionally, it has been revealed that the enhanced Pt mass activity of hollow Pt nanospheres is stable during potential cycling [75]. No electrochemical area (ECA) and ORR losses were observed after 10 000 potential cycles between 0.65 and 1.05 V at 50 mV s^{-1} (versus RHE) [75]. However, when pulsed potential cycling is used between 0.65 and 1.05 V (30 s period), an approximate 33% loss in the ORR activity was observed after 3000 pulse potential cycles, with no further loss for the whole duration of the cycling (7 days). Pulsed potential cycling between two limiting potentials is a more severe test than a linear sweep, due to the dissolution of the low coordinate atoms at 0.65 V and defect generation at the high potential limit [75, 76]. In comparison, the ORR activity of Pt nanoparticles decreases after 3000 potential cycles and continues to do so during 3000 additional cycles (Figure 3.2) [75]. The Pt mass activity of hollow Pt nanoparticles is six-fold higher than that of solid Pt nanoparticles after 100 h of durability testing (0.6 mA µg^{-1} and 0.1 mA µg^{-1}, respectively) [75].

3.3 Hollow Pt Metal Alloys

Apart from hollow Pt nanoparticles, hollow Pt alloys (PtM, M = Au, Ag, Co, Ni, Ru, Pd, Cu) are also used as electrocatalysts in PEMFCs, due to their enhanced activity and stability in PEMFCs compared to pure Pt [40–60]. The galvanic replacement reaction is the most common method to synthesize these hollow nanoparticles, since it allows for easy control of the shape and size of the product by varying the shape of the sacrificial template and the ratio of surfactant/reductant [32, 77, 78].

3.3.1
PtAu

Raspberry-like hierarchical PtAu hollow spheres (RHAHS) are synthesized via the sacrificial template approach without any post-treatment, where TiO_2 is the template. Titania precursor spheres are functionalized with $-NH_2$ groups and the amino-functionalized TiO_2 spheres are mixed with gold nanoparticles to obtain TiO_2/Au intermediates. The raspberry-like, hierarchical PtAu hollow spheres (Figure 3.3) are obtained through heating of TiO_2/Au and H_2PtCl_6 solution in the presence of a reducing agent [79, 80].

Cyclic voltammograms in acidic solution (0.5 M H_2SO_4; N_2 purge) of RHAHS (Figure 3.4a; line b) and commercial Pt black (Figure 3.4a; line a) demonstrate that RHAHS have a higher electrochemically active surface area, and gold nanoparticles are in the interior of Pt shell, as there is no reduction peak of gold oxide present in the voltammograms [79]. It is also evident that RHAHS have a higher

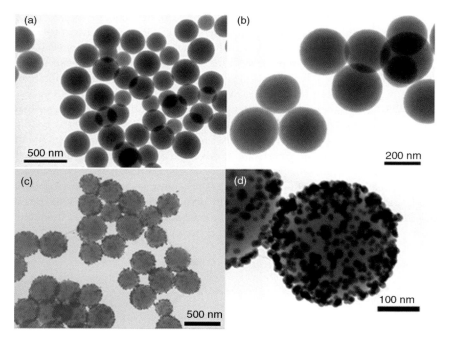

Figure 3.3 (a and b) TEM image of TiO_2 precursor spheres at different magnifications. (c and d) TEM images of TiO_2 (precursor sphere)/Au hybrid spheres at different magnifications. (Reprinted with permission from Ref. [79]. Copyright 2009, American Chemical Society.

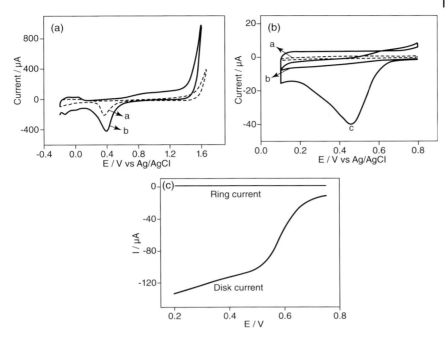

Figure 3.4 (a) CVs of CPB (line a)- and RHAHS (line b)-modified GC electrode (5 mm) in a N_2-sparged 0.5 M H_2SO_4 solution. (b) CVs of O_2 reduction at RHAHS-modified gold electrode (curves b and c) and bare gold electrode (curve a) in air-saturated (curves a and c) and N_2-saturated (curve b) 0.5 M H_2SO_4 solution. (c) Current–potential curves for the reduction of O_2 at a rotating platinum ring-GC (5 mm) disk electrode with RHAHS adsorbed on the disk electrode in the presence of air (experimental conditions: 1 V constant ring potential, 500 rpm rotation rate, 50 mVs^{-1} scan rate, 0.5 M H_2SO_4 solution). (Reprinted with permission from Ref. [79]. Copyright 2009, American Chemical Society.)

oxygen reduction peak potential (0.46 V compared to 0.1 and 0.25 V versus Ag/AgCl) [81, 82] (Figure 3.4b) and probe the 4-electron reduction of water (based on the ring and disk current values of RHAHS; Figure 3.4c). Specific activity of RHAHS is 6.3 and 2.7 times higher than Pt/C at 0.595 and 0.465 V, and their mass activity is 8 and 3.5 times higher than Pt/C at 0.595 and 0.465 V, respectively [79].

Hollow PtAu spheres are also used in direct formic acid fuel cells, increasing the catalytic activity and the tolerance to CO poisoning through ensemble effects [39, 83].

3.3.2
PtAg

Bimetallic hollow PtAg nanoparticles can be synthesized via either a successive reduction of $AgNO_3$ and H_2PtCl_6 solutions with hydrazine [84] or a galvanic replacement reaction [32]. When $AgNO_3$ solution is reduced by hydrazine, the formed Ag particles are used as templates for the reduction of $PtCl_6^{2-}$ and deposition of Pt, as well as hydrazine reduction [85].

$$PtCl_{6(aq)}^{2-} + 4Ag_{(s)} \rightarrow Pt_{(s)} + 4Ag_{(aq)}^+ + 6Cl_{(aq)}^-$$
$$Ag_{(s)} + Pt_{(s)} \rightarrow Ag_{(s)}Pt_{(s)}$$
(3.1)

where (aq) indicates aqueous solution and (s) indicates solid.

The oxidized Ag^+ reacts with the excess hydrazine and is reduced back to Ag. At the same time, polyvinylpyrrolidone (PVP), which acts as stabilizer, entraps the Ag^+ and reacts with $PtCl_6^{2-}$ to co-reduce near the surface of the template particles [77] and form composite nanoparticles [84]. Ag atoms migrate outward from the core during these reactions, producing bimetallic PtAg with a hollow interior [84].

The synthesis of distinct hollow PtAg nanostructures with a tailored number of void spaces can be achieved by tuning the galvanic replacement reaction between Ag nanocubes and K_2PtCl_4 in the presence of HCl and PVP [86]. PtAg nanoboxes are obtained when both HCl and PVP are used (Figure 3.5a), whereas hollow PtAg heterodimers (Figure 3.5b) are produced when only HCl is used. The presence of HCl causes rapid precipitation of AgCl that grows on the surface of Ag nanocubes and acts as a removable secondary template for the

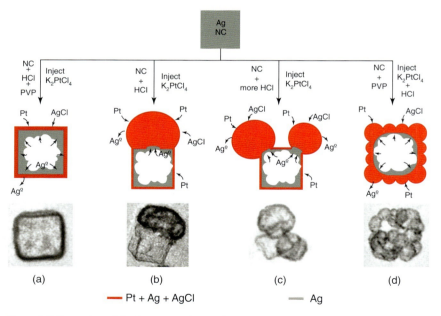

Figure 3.5 Illustration of the formation of PtAg. (a) Nanobox; (b) heterodimer; (c) multimer; and (d) popcorn-shaped nanoparticle from the galvanic replacement reaction between Ag NCs and K_2PtCl_4 in the presence of HCl. (Reprinted with permission from Ref. [87]. Copyright 2012, American Chemical Society.)

deposition of Pt [86]. At high HCl concentrations, the number of attached particles on each hollow nanobox increases, leading to the formation of multimeric structures (Figure 3.5c). However, if HCl is mixed with K_2PtCl_4 solution before it reacts with Ag nanocubes, hollow PtAg popcorn-shaped particles are formed (Figure 3.5d) [80, 86]. Subsequent washing with sodium chloride (NaCl) and iron(III) nitrate ($Fe(NO_3)_3$) solutions can further convert the attached solid particles into hollow particles [86].

The presence of rough hollow interiors and porous walls in PtAg dimers and popcorns results in a high specific surface area, favoring electrocatalysis; the ECA of PtAg dimers and popcorns is ~82 and 91 $m^2 g^{-1}$, respectively, much higher than commercial Pt/C (52 $m^2 g^{-1}$). As a result, these hollow PtAg nanostructures exhibit improved activity toward MOR (51 and 28% higher activity for PtAg popcorns and dimers, respectively, compared to Pt/C) [86]. The measured ECA values and MOR activity of these hollow PtAg nanostructures can only be compared to existing Pt nanostructures such as ultrathin Pt nanowires with 18% higher MOR activity than commercial Pt/C (Figure 3.6) [86]. Chronoamperometry measurements at constant potential (−0.4 V versus SCE) show that hollow PtAg nanostructures are more stable than commercial Pt/C [86].

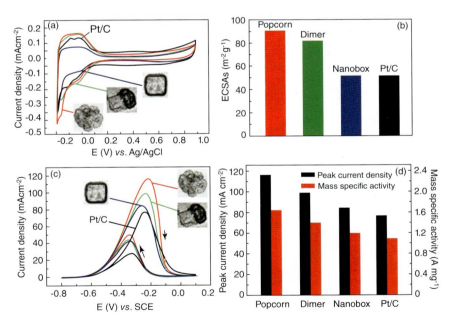

Figure 3.6 (a) Cyclic voltammograms of Pt/C, hollow PtAg nanoboxes, dimers, and popcorns in 1 M $HClO_4$. (b) Electrochemical surface area (ECSA) of each catalyst. (c) CVs for MOR in a solution containing 1 M CH_3OH and 1 M KOH. (d) Peak current density and mass specific activity of each catalyst. (Reprinted with permission from Ref. [87]. Copyright 2012, American Chemical Society.)

3.3.3
PtCo

Hollow PtCo nanospheres are synthesized via a novel galvanic replacement reaction [77]. The Co core is selectively dissolved during Pt shell formation/deposition during the galvanic reaction. The use of reducing conditions in conjunction with a polymer stabilizer (PVP) allows the Pt and Co to co-reduce and form PtCo face centered cubic (fcc) alloys with the morphology of the sacrificial Co template [77].

TEM images reveal that the products of the galvanic reaction are PtCo nanospheres with an average diameter of 10–50 nm (Figure 3.7a). Each nanosphere is made of randomly oriented Pt and Co nanocrystals in the shell (Figure 3.7b, c).

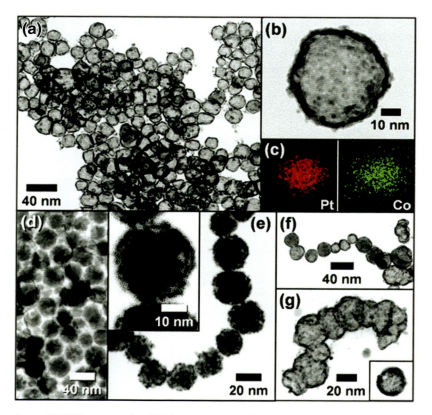

Figure 3.7 TEM micrographs of (a) hollow PtCo nanospheres and (b) a single hollow nanosphere showing that it is comprised of smaller particles. (c) Elemental mapping data (Pt and Co) for the nanosphere in (b). TEM micrographs of (d) the Co nanoparticle template generated *in situ* and PtCo samples taken after reaction times of (e) 1 min and (f) 5 min. The hollow structures remain stable after heating at 300 °C on a TEM grid (g). (Reprinted with permission from Ref. [77]. Copyright 2005, American Chemical Society.)

The formation of PtCo nanospheres is directly related to the excess sodium boron hydride (NaBH$_4$) and polymer stabilizer (PVP); the excess BH$_4^-$ present in the solution reduces Co^{2+} formed during Pt deposition back to Co, while PVP encapsulates the produced Co^{2+} during the replacement reaction, allowing it to combine with Pt^{4+} and co-reduce near the surface of the nanosphere (Figure 3.7d–f) [77].

The same experimental method is used to produce hollow PtCo nanospheres deposited on indium tin oxide (ITO) [63]. The ORR follows a four-electron pathway to H$_2$O, based on the ring and disk currents measured during rotating ring-disk electrode (RRDE) experiments (Figure 3.8), and a ~2–3-fold increase in mass and specific activity was observed in acidic and alkaline solutions [88]. However, the exact rate constant of the oxygen reduction reaction cannot be obtained via Koutecky–Levich plots due to the different oxygen concentration on the inner and outer surface of hollow PtCo [63].

The catalytic activity of PtCo toward the methanol oxidation reaction (MOR) is tested by CV in alkaline solution (0.1 M CH$_3$OH and 0.1 M NaOH) and compared with the performance of pristine Pt and Co [89, 90]. PtCo shifts the onset potential of MOR by 80 mV (−0.67 V onset potential) with a maximum current density of 1.8 mA cm^{-2}; on the contrary, pristine Pt has an onset potential of −0.59 V with 0.95 mA cm^{-2} maximum current density (Figure 3.9) [90]. The improved catalytic activity of PtCo is attributed to the porous core–shell structure of PtCo, providing more active sites for catalysis, as well as to the improved CO tolerance, as Co is in the inner layer of PtCo [62, 90].

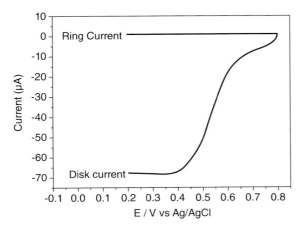

Figure 3.8 Current–potential curve for hollow PtCo nanospheres (experimental conditions: air saturated 0.1 M HClO$_4$ solution, 100 rpm rotating speed, 50 mVs^{-1} scan rate). (Reproduced with permission from Ref. [63] Copyright 2008, the Royal Society of Chemistry.)

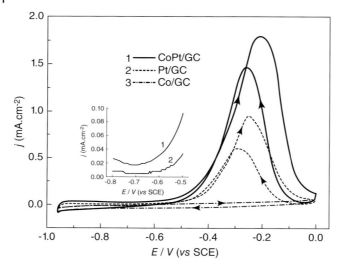

Figure 3.9 CVs of Co/Graphitic Carbon (GC), Pt/GC, and PtCo/GC in 0.1 M NaOH and 0.1 M CH$_3$OH solution (50 mVs^{-1} scan rate). The inset shows the onset region of methanol oxidation on PtCo/GC and Pt/GC, respectively. (Reproduced from Ref. [90] with permission from the Royal Society of Chemistry.)

3.3.4
PtNi

Nest-like, hollow Pt$_x$Ni$_{1-x}$ spheres of sub-micrometer size are prepared through a combination of a (polystyrene-co-methacrylic acid) PSA-assisted templating method and galvanic replacement reaction [62]. The formation of Pt$_x$Ni$_{1-x}$ is attained through the replacement reaction between K$_2$PtCl$_6$ and Ni, which is first deposited on the surface of the PSA [62]. After the partial consumption of Ni, the resultant Pt$_x$Ni$_{1-x}$-PSA core–shell structures are dispersed in a toluene solution to dissolve the PSA template cores, resulting in the formation of Pt$_x$Ni$_{1-x}$ (further denoted "PtNi") hollow spheres [62].

These hollow PtNi nanospheres (∼32 nm) exhibit higher catalytic activity toward methanol oxidation than commercial Pt/C and solid PtNi nanoparticles [91]. The ratio of forward to backward peak currents for PtNi nanostructures is significantly higher than that for Pt/C (∼1.04 and 0.91, respectively), indicating that PtNi has a higher tolerance to poisoning from methanol oxidation intermediates than Pt/C [91]. Chronoamperometry measurements at constant potential (0.45 V versus SCE) reveal the high activity of hollow PtNi nanospheres toward MOR, since their oxidation current density is significantly higher than Pt/C (∼200 and 50 mA mg^{-1}, respectively) due to the high surface area of the hollow structure and the change in Pt electronic structure after the addition of Ni [91]. Ni(0) occupies the Pt lattice and the metallic grains are intermixed with amorphous Ni hydroxides such as Ni(OH)$_2$ and NiOOH [91]. The presence of these Ni oxides results in the increase of Pt(0) and decrease of Pt(IV) content, due to the electronic effect of Ni on Pt and, thus, a higher catalytic activity of PtNi [91]. Moreover, Ni hydroxides can offer OH species to remove the intermediate CO that is strongly adsorbed on the Pt surface, regenerate the Pt active sites for methanol adsorption [86], and thus improve the catalytic activity of the nanospheres.

Hollow PtNi nanospheres supported on graphene demonstrate, for methanol oxidation, higher activity and current per unit Pt mass, as well as a higher specific activity when current is normalized by ECSA (electrochemical surface area), compared to solid PtNi on graphene and commercial Pt/C (∼2 and 3 times higher, and ∼1.7 and 2.2 times higher, respectively) [86]. The observed catalytic activity is similar to mesoporous Pt (460 mA mg$_{Pt}^{-1}$, for 1 M methanol) [92], Pt-Pd (376 mA mg$_{Pt}^{-1}$) [93], and hollow PtNi nanospheres (∼380 mA mg$_{Pt}^{-1}$) [94].

The ratio of forward to backward peak currents for hollow PtNi nanospheres (∼1.33) is higher than for solid PtNi nanoparticles (∼1.1) and commercial Pt/C (∼1.1), showing that hollow PtNi has good tolerance to poisoning from methanol oxidation intermediates [86]. Chronoamperometry measurements at constant 0.62 V (versus SCE) reveal that hollow PtNi nanospheres supported on graphene are more stable than solid PtNi and commercial Pt/C; however, longer stability tests are required to fully evaluate the stability of this material, as Ni might be unstable in acidic solution [86].

Hollow PtNi nanospheres supported on carbon exhibit higher catalytic activity toward ORR [95]; the mass and specific activity is 0.5 A mg$_{Pt}^{-1}$ and 1530 μA cm^{-2} at 0.9 V, respectively, which are 3.3- and 7.8-fold enhanced activities relative to those of Pt/C (0.153 A mg$_{Pt}^{-1}$ and 197 μA cm^{-2} based on Pt mass and ECSA, respectively) [95].

The significant increase in ORR activity is attributed to the hollow structure of the material, as there are more surface sites, which is favorable for the ORR [75]. TEM images demonstrate that the shell consists of a Pt-enriched surface layer and an inner layer of PtNi alloy. Due to the perturbation caused by the alloy core, the Pt surface layer has reduced interatomic distance and modified electronic structure, which weakens the adsorption of nonreactive oxygenated species on Pt surface atoms [96, 97] and lead to enhancement of the ORR activity [75]. The loss in ECSA after 10 000 potential cycles (0.65–1.05 versus RHE at 50 mV s^{-1} in O$_2$ saturated HClO$_4$ solution) was 11 and 33% for PtNi/C and commercial Pt/C, respectively, demonstrating the improved durability of hollow PtNi nanostructures [75].

3.3.5
PtRu

Hollow PtRu nanotubes, prepared via chemical etching (ZnO as sacrificial template), have been employed as catalysts for methanol oxidation [98]. A negative shift in the onset potential for methanol oxidation is observed in the case of PtRu (0.62 and 0.52 V versus SHE for Pt and PtRu nanotubes, respectively) due to the bifunctional mechanism of PtRu alloys [99]. The bifunctional mechanism involves the adsorption and initial dehydrogenation of methanol on Pt surface atoms, followed by the oxidative removal of methanol dehydrogenation products (CO_{ads}) via oxygen-containing species (OH_{ads}) on adjacent Ru atoms (Eq. (3.2)) [99].

$$OH_{ads} + CO_{ads} \longleftrightarrow CO_2 + H^+ + e^- \qquad (3.2)$$

Moreover, hollow PtRu nanospheres with different Pt/Ru ratios have been synthesized via mixing of negatively charged hollow Pt nanospheres with positively charged ultrafine Ru nanoparticles [98, 100]. Hollow Pt nanospheres are produced via a galvanic replacement reaction, using Ag as the sacrificial template [100].

Hollow PtRu (2:1) nanospheres (synthesized via galvanic replacement reaction) [100] have the highest catalytic activity toward methanol oxidation compared to commercial PtRu/C and hollow PtRu nanospheres with different Pt:Ru ratios (Table 3.1). This phenomenon is attributed to the hollow surface structure of these materials. In the case of commercial PtRu/C, the atomic mixing and uniform distribution of Pt and Ru on the surface inhibits the adsorption of methanol along the directions 1 and 3 (Figure 3.10a), resulting in a decreased activity for methanol oxidation [100]. However, in the case of hollow PtRu, several neighboring Pt atoms share an ultrafine Ru cluster on the surface, allowing methanol to adsorb in all directions and utilizing the oxygen-containing species on a Ru cluster (Figure 3.10b).

The CO tolerance of these catalysts is examined via chronoamperometry at 0.45 V (versus Ag/AgCl) constant potential [100]; at that potential, methanol is continuously oxidized on the catalyst surface, and reaction intermediates (CO_{ads}) would begin to accumulate if the kinetics of CO removal are slow [100]. Hence, a more gradual decay of current density with time is an indication of improved CO tolerance. Hollow PtRu nanospheres demonstrate the slowest decay rate in current density, indicating superior CO tolerance over commercial PtRu/C and pure hollow Pt nanospheres [100].

Table 3.1 Electrochemical measurements of MOR on commercial PtRu/C catalysts, hollow Pt nanospheres, and hollow PtRu assemblies at different Pt/Ru ratios [100].

	Forward scan peak potential (V versus Ag/AgCl)	Forward scan peak current density (mAcm^{-2})	Backward scan peak potential (V versus Ag/AgCl)	Backward scan peak current density (mAcm^{-2})
PtRu/C	0.66	7.04	0.46	3.36
Hollow Pt	0.68	6.86	0.52	5.45
Hollow PtRu (2:1)	0.70	14.5	0.54	10.6
Hollow PtRu (1:1)	0.70	10.9	0.54	8.69
Hollow PtRu (1:2)	0.66	8.65	0.50	5.30

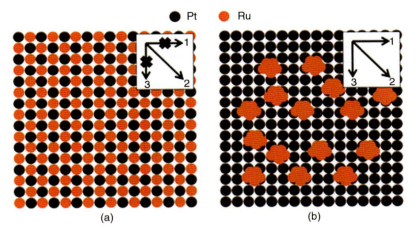

Figure 3.10 Schematic illustrations of the surface conditions of PtRu catalysts (a) and hollow PtRu assemblies (b). Insets show the possible direction for methanol adsorption on the surface of each catalyst. (Reproduced from Ref. [100] with permission from the Royal Society of Chemistry.)

3.3.6
PtPd

Hollow and dendritic PtPd nanocrystals (NCs) can be synthesized by a galvanic replacement reaction, using uniform cubic and octahedral Pd nanocrystals as sacrificial templates and cetyltrimethylammonium chloride (CTAC) as surfactant [78]. Four different types of hollow PtPd nanocrystals have been prepared, namely, octahedral (ONC) and cubic nanocages (CNC), as well as dendritic hollow nanocrystals (ODH and CDH) [78]. The synthesized hollow and dendritic nanocages have porous walls of 15 nm average thickness and various lengths of dendritic branches ranging from 7.9 to 8.5 nm (Figure 3.11). The d-spacing between adjacent lattice fringes of octahedral nanocages (ONCs), octahedral dendritic hollow nanocrystals (ODH), and octahedral dendritic nanocrystals (OD NCs) is 2.24 Å, corresponding to the (1,1,1) planes of fcc Pt, Pd, or PtPd [101]. On the contrary, the d-spacing between adjacent lattice fringes of CNCs, CDH, and CD NCs is 2 Å, corresponding to the (1,0,0) planes [78, 102].

All PtPd NCs produce very low yields of hydrogen peroxide (less than 0.15%), indicating a four-electron pathway toward ORR [103, 104], and have high electrochemical stability [78]. The ORR activities of PtPd nanocubes follow the order ONCs > ODH NCs > CDH NCs > OD NCs > Pt/C > CNCs > CD NCs [78]. The mass activity of ONCs is ~765 mA mg_{PtPd}^{-1} and is ~2 and 4 times higher than ODH NCs (394 mA mg_{PtPd}^{-1}) and Pt/C (201 mA mg_{PtPd}^{-1}), respectively [78]. Area-specific activities show a similar trend [78]. The enhanced activity of NCs is attributed to their surface facets, as (1,1,1) planes are more active than (1,0,0) planes [97, 105, 106]. Thus, the type of surface facet is the most important parameter to determine the ORR activity of PtPd NCs.

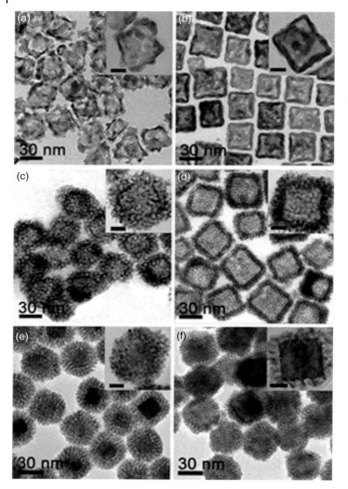

Figure 3.11 TEM images of (a) ONCs; (b) CNCs; (c) ODH NCs; (d) CDH NCs; (e) OD NCs; and (f) CD NCs. High-magnification TEM images are shown in each inset. Scale bars in the insets indicate 10 nm. (Reprinted with permission from Ref. [78]. Copyright 2012, American Chemical Society.)

Hollow spherical PtPd/C are also effective catalysts toward MOR, exhibiting twice the activity of commercial Pt/C [107]. Pd can directly oxidize the poisoning intermediates of methanol oxidation, such as HCOOH and CO_{ads} to CO_2, accelerating the reaction [107].

3.3.7
PtCu

Hollow PtCu nanospheres can be synthesized via a galvanic replacement reaction between Cu and K_2PtCl_6 in the presence of oleylamine (OAm) as surfactant/reductant and cetyltrimethylammonium bromide (CTAB) as co-surfactant [108]. The average edge length of the hollow nanospheres is ~11.5 nm with inner fringe distance of 0.2 nm [108].

Hollow PtCu nanospheres demonstrate ~2 and 3 times higher activity than Pt black and Pt/C (2.1 mA cm^{-2}, 1.1 mA cm^{-2}, and 0.7 mA cm^{-2}, respectively) due to the high active surface area of the hollow structure (53 m^2 g^{-1} compared to 41.9 m^2 g^{-1} of Pt black and 35 m^2 g^{-1} of Pt/C) [108].

Similar catalytic behavior is observed for hollow PtCu$_3$ nanocages [109]. PtCu$_3$ nanocages have a higher oxidation density (~14 mA cm^{-2}) than PtCu nanoparticles (~8 mA cm^{-2}) and commercial Pt/C (~13 mA cm^{-2}), due to their high surface area and synergetic effects between Pt and Cu, namely, the oxidation of CO species by Cu [110]. Additionally, PtCu$_3$ nanocages exhibit the highest forward to backward current density ratio (~2), demonstrating that methanol is more effectively oxidized on nanocages, generating less poisoning species [109, 111].

3.4 Non-Pt Alloy Nanostructures

Hollow non-Pt alloys have been mainly used for the electrochemical oxidation of formic acid [112, 113]. Pd catalysts possess higher activity in formic acid oxidation than Pt-based catalysts [112–114]. Small Pd particles (~9–11 nm) exhibit the highest binding energy shift and valence band center downshift with respect to the Fermi level, resulting in a decrease in the bond strength of the adsorbents [115–117] and, thus, higher formic acid reactivity [114]. The low d-band center of small Pd nanoparticles allows them to bind less strongly with the COOH intermediate than larger Pd nanoparticles, thus reducing the surface (COOH)$_{ads}$ coverage (Eqs (3.4) and (3.5)) [114]. As a result, it can be avoided that desorption is the rate determining step (Eq. (3.5)) and a higher rate of HCOOH decomposition to CO$_2$ is achieved (Eq. (3.3)) [114].

$$HCOOH_{bulk} \rightarrow 2H^+ + 2e^- + CO_2 \tag{3.3}$$

$$HCOOH_{ads} \rightarrow (COOH)_{ads} + H^+ + e^- \tag{3.4}$$

$$(COOH)_{ads} \xrightarrow{rds} CO_2 + H^+ + e^- \tag{3.5}$$

To further improve the catalytic activity of Pd, Pd-M alloys (M = Au, Rh, Cu, Ag) with hollow/core–shell structure are used due to their high surface area and the modulated electron structure [80, 108, 118–122].

Hollow PdAu nanoparticles are synthesized via galvanic replacement reaction, using Co as sacrificial template [80]. Hollow Au nanospheres are produced first, due to the large reduction potential difference between AuCl$_4^-$/Au and Co^{2+}/Co (0.935 and −0.377 V versus SHE, respectively); palladium chloride in the form of H$_2$PdCl$_4$ is then added to the solution in the presence of ascorbic acid and is gradually reduced to form metallic Pd [80]. The interfacial energy between Au and Pd lattices is low and, as a result, metallic Pd is deposited on the surface of hollow Au nanospheres [80].

Hollow PdAu nanospheres demonstrate higher activity toward formic acid oxidation than pristine Pd and Au nanoparticles, due to their increased surface-to-volume ratio [80]; the raspberry surface of Au nanospheres consists of irregular Pd grains, increasing the surface-to-volume ratio and surface activity of the PdAu catalyst [80].

Similar observations were made in the case of hollow PdCu nanospheres [118, 123]. Hollow PdCu nanospheres supported on multiwalled carbon nanotubes (MWCNTs) exhibit the highest activity toward formic acid oxidation. Hollow PdCu/MWCNT has the highest peak current density compared to solid PdCu/MWCNT and pristine Pd/MWCNT (~173, 143, and 107 mA cm^{-2} respectively), as well as a significantly lower onset negative potential of formic acid oxidation (−0.1 V compared to −0.01 and −0.05 V SHE for solid PdCu and Pd/MWCNTs, respectively) [123].

Hollow PdAg is synthesized via the galvanic replacement reaction using Ag nanowire (~80 nm diameter and 5 μm length) as sacrificial template [119]. Once the Pt salt is reduced and deposited on Ag, the structure changes into hollow nanotubes with a coarse surface. High resolution TEM images show that hollow PdAg nanotubes have a single-layer crystalline structure comprised of Ag(1,1,1) and Pd(1,1,1) planes with sixfold rotational symmetry. Hollow PdAg nanotubes exhibit double the activity of pristine Pd with the same Pd loading and a high tolerance to CO poisoning, as evidenced by the shift of the onset potential for formic acid oxidation toward more positive values [119].

3.5
Conclusions and Outlook

In conclusion, the utilization of hierarchically structured nanomaterials in electrochemical energy conversion to optimize the electronic, ionic mobility, and kinetics at multiphase boundaries is an area of growing interest. Controlling the architecture of these materials at multiple length scales, and not only at the nanoscale, is critical to further improve their properties. In particular, the mesopores of these materials lead to a large surface area, providing short diffusion lengths and excellent multiphase contact, while the macropores improve the overall transport properties and maximize effective performance; this was recently reviewed elsewhere [26, 124].

Complementary to the hierarchical structuring of the pore space, in this review, several experimental methods have been discussed for the synthesis of hierarchical nanomaterials for electrocatalysis, including templating, self-formation, and galvanic replacement. Hollow, Pt-based catalysts can be used as electrocatalysts in PEMFCs toward oxygen and alcohol oxidation reactions, due to their increased surface area and high Pt utilization. The galvanic replacement method is commonly used for the synthesis of these hollow nanostructures, since it is easy to control the shape and size of the product by varying the shape of the sacrificial template and the ratio of surfactant/reductant. The hollow Pt or Pt alloys produced exhibit significantly higher catalytic activity toward MOR and AOR, as well as high tolerance to CO poisoning intermediates. To further improve the design of these hierarchical nanomaterials, in-depth research on the effect of their architecture on reaction and transport kinetics is necessary. This requires multiscale models that properly account for the geometry as well as the physical and chemical phenomena at all scales [26, 124].

Another avenue is to use techniques inspired by biological systems. Nature can be an excellent guide to rational design, as it is full of hierarchical structures that are intrinsically scaling, efficient, and robust. However, the biological example needs to be properly chosen and the actual physical processes governing the system need to be taken into account; the imitation of isolated features of biological structures to synthesize new catalysts for PEMFCs lead to suboptimal results. Hence, we suggest focusing research on the fundamental understanding of hierarchical structure/function relationships in biological systems, and to use this understanding in guiding the computationally assisted development of novel, nature-inspired hierarchical catalytic nanomaterials and electrochemical devices [26, 124].

Acknowledgment

MOC and PT gratefully acknowledge the EPSRC for funding via a "Frontier Engineering" Award, EP/K038656/1.

References

1 Bandarenka, A.S. and Koper, M.T.M. (2013) Structural and electronic effects in heterogeneous electrocatalysis: toward a rational design of electrocatalysts. *Journal of Catalysis*, **308**, 11–24.
2 Koper, M.T.M. and Iwasawa, Y. (2014) Electrocatalysis. *Physical Chemistry Chemical Physics*, **16** (27), 13567.
3 Bagotsky, V.S. (2012) Electrocatalysis. *Fuel Cells*, John Wiley & Sons, Inc., New York, p. 207–231.
4 Hayden, B.E. and Suchsland, J.-P. (2008) Support and particle size effects in electrocatalysis. *Fuel Cell Catalysis*, John Wiley & Sons, Inc., New York, pp. 567–592.
5 Maillard, F., Pronkin, S., and Savinova, E.R. (2008) Size effects in electrocatalysis of fuel cell reactions on supported metal nanoparticles. *Fuel Cell Catalysis*, John Wiley & Sons, Inc., New York, pp. 507–566.
6 Tarasevich, M.R., Sadkowski, A., and Yeager, E. (1983) Oxygen electrochemistry, in *Comprehensive Treatise of Electrochemistry: Volume 7 Kinetics and Mechanisms of Electrode Processes* (eds B.E. Conway, J.O.M. Bockris, E. Yeager, S.U.M. Khan, and R.E. White), Springer US, Boston, MA, pp. 301–398.
7 Gasteiger, H.A., Kocha, S.S., Sompalli, B., and Wagner, F.T. (2005) Activity benchmarks and requirements for Pt, Pt-alloy, and non-Pt oxygen reduction catalysts for PEMFCs. *Applied Catalysis B: Environmental*, **56** (1–2), 9–35.

8 Li, D., Wang, C., Strmcnik, D.S., Tripkovic, D.V., Sun, X., Kang, Y. et al. (2014) Functional links between Pt single crystal morphology and nanoparticles with different size and shape: the oxygen reduction reaction case. *Energy & Environmental Science*, **7** (12), 4061–4069.

9 Shao, M., Peles, A., and Shoemaker, K. (2011) Electrocatalysis on platinum nanoparticles: particle size effect on oxygen reduction reaction activity. *Nano Letters*, **11** (9), 3714–3719.

10 Masahiro, W., Sigeru, S., and Paul, S. (1988) Electro-catalytic activity on supported platinum crystallites for oxygen reduction in sulphuric acid. *Chemistry Letters*, **17** (9), 1487–1490.

11 Watanabe, M., Sei, H., and Stonehart, P. (1989) The influence of platinum crystallite size on the electroreduction of oxygen. *Journal of Electroanalytical Chemistry and Interfacial Electrochemistry*, **261** (2), 375–387.

12 Yano, H., Inukai, J., Uchida, H., Watanabe, M., Babu, P.K., Kobayashi, T. et al. (2006) Particle-size effect of nanoscale platinum catalysts in oxygen reduction reaction: an electrochemical and 195Pt EC-NMR study. *Physical Chemistry Chemical Physics*, **8** (42), 4932–4939.

13 Kinoshita, K. (1990) Particle size effects for oxygen reduction on highly dispersed platinum in acid electrolytes. *Journal of The Electrochemical Society*, **137** (3), 845–848.

14 Markovic, N., Gasteiger, H., and Ross, P.N. (1997) Kinetics of oxygen reduction on Pt(hkl) electrodes: implications for the crystallite size effect with supported Pt electrocatalysts. *Journal of The Electrochemical Society*, **144** (5), 1591–1597.

15 Mayrhofer, K.J.J., Blizanac, B.B., Arenz, M., Stamenkovic, V.R., Ross, P.N., and Markovic, N.M. (2005) The impact of geometric and surface electronic properties of Pt-catalysts on the particle size effect in electrocatalysis. *The Journal of Physical Chemistry B*, **109** (30), 14433–14440.

16 Mukerjee, S. and McBreen, J. (1998) Effect of particle size on the electrocatalysis by carbon-supported Pt electrocatalysts: an *in situ* XAS investigation1. *Journal of Electroanalytical Chemistry*, **448** (2), 163–171.

17 Stamenkovic, V., Mun, B.S., Mayrhofer, K.J.J., Ross, P.N., Markovic, N.M., Rossmeisl, J. et al. (2006) Changing the activity of electrocatalysts for oxygen reduction by tuning the surface electronic structure. *Angewandte Chemie, International Edition*, **45** (18), 2897–2901.

18 Rhodes, K.H., Davis, S.A., Caruso, F., Zhang, B., and Mann, S. (2000) Hierarchical assembly of zeolite nanoparticles into ordered macroporous monoliths using core–shell building blocks. *Chemistry of Materials*, **12** (10), 2832–2834.

19 Soler-Illia, G.J.d.A.A., Sanchez, C., Lebeau, B., and Patarin, J. (2002) Chemical strategies to design textured materials: from microporous and mesoporous oxides to nanonetworks and hierarchical structures. *Chemical Reviews*, **102** (11), 4093–4138.

20 Studart, A.R., Studer, J., Xu, L., Yoon, K., Shum, H.C., and Weitz, D.A. (2011) Hierarchical porous materials made by drying complex suspensions. *Langmuir*, **27** (3), 955–964.

21 Yang, P., Deng, T., Zhao, D., Feng, P., Pine, D., Chmelka, B.F. et al. (1998) Hierarchically ordered oxides. *Science*, **282** (5397), 2244–2246.

22 Yang, X.-Y., Leonard, A., Lemaire, A., Tian, G., and Su, B.-L. (2011) Self-formation phenomenon to hierarchically structured porous materials: design, synthesis, formation mechanism and applications. *Chemical Communications*, **47** (10), 2763–2786.

23 Zhao, D., Feng, J., Huo, Q., Melosh, N., Fredrickson, G.H., Chmelka, B.F. et al. (1998) Triblock copolymer syntheses of mesoporous silica with periodic 50 to 300 angstrom pores. *Science*, **279** (5350), 548–552.

24 Imhof, A. and Pine, D.J. (1997) Ordered macroporous materials by emulsion templating. *Nature*, **389** (6654), 948–951.

25 Choi, S.-W., Xie, J., and Xia, Y. (2009) Chitosan-based inverse opals: three-dimensional scaffolds with uniform pore structures for cell culture. *Advanced Materials*, **21** (29), 2997–3001.

26 Trogadas, P., Ramani, V., Strasser, P., Fuller, T.F., and Coppens, M.-O. (2016) Hierarchically structured nanomaterials for electrochemical energy conversion. *Angewandte Chemie, International Edition*, **55**, 122–148.

27 Trogadas, P., Fuller, T.F., and Strasser, P. (2014) Carbon as catalyst and support for electrochemical energy conversion. *Carbon*, **75** (0), 5–42.

28 Long, N.V., Yang, Y., Minh Thi, C., Minh, N.V., Cao, Y., and Nogami, M. (2013) The development of mixture, alloy, and core-shell nanocatalysts with nanomaterial supports for energy conversion in low-temperature fuel cells. *Nano Energy*, **2** (5), 636–676.

29 Parlett, C.M.A., Wilson, K., and Lee, A.F. (2013) Hierarchical porous materials: catalytic applications. *Chemical Society Reviews*, **42** (9), 3876–3893.

30 Qiao, Y. and Li, C.M. (2011) Nanostructured catalysts in fuel cells. *Journal of Materials Chemistry*, **21** (12), 4027–4036.

31 Zhang, S., Shao, Y., Yin, G., and Lin, Y. (2013) Recent progress in nanostructured electrocatalysts for PEM fuel cells. *Journal of Materials Chemistry A*, **1** (15), 4631–4641.

32 Kim, M.R., Lee, D.K., and Jang, D.-J. (2011) Facile fabrication of hollow Pt/Ag nanocomposites having enhanced catalytic properties. *Applied Catalysis B: Environmental*, **103** (1–2), 253–260.

33 Bell, A.T. (2003) The impact of nanoscience on heterogeneous catalysis. *Science*, **299** (5613), 1688–1691.

34 Chen, M., Pica, T., Jiang, Y.-B., Li, P., Yano, K., Liu, J.P. et al. (2007) Synthesis and self-assembly of fcc phase FePt nanorods. *Journal of the American Chemical Society*, **129** (20), 6348–6349.

35 Guo, S., Fang, Y., Dong, S., and Wang, E. (2007) High-efficiency and low-cost hybrid nanomaterial as enhancing electrocatalyst: spongelike Au/Pt core/shell nanomaterial with hollow cavity. *The Journal of Physical Chemistry C*, **111** (45), 17104–17109.

36 Liang, H.-P., Zhang, H.-M., Hu, J.-S., Guo, Y.-G., Wan, L.-J., and Bai, C.-L. (2004) Pt hollow nanospheres: facile synthesis and enhanced electrocatalysts. *Angewandte Chemie, International Edition*, **43** (12), 1540–1543.

37 Rolison, D.R. (2003) Catalytic nanoarchitectures – the importance of nothing and the unimportance of periodicity. *Science*, **299** (5613), 1698–1701.

38 Steele, B.C.H. and Heinzel, A. (2001) Materials for fuel-cell technologies. *Nature*, **414** (6861), 345–352.

39 Kim, Y., Kim, H.J., Kim, Y.S., Choi, S.M., Seo, M.H., and Kim, W.B. (2012) Shape- and composition-sensitive activity of Pt and PtAu catalysts for formic acid electrooxidation. *Journal of Physical Chemistry C*, **116** (34), 18093–18100.

40 Petrii, O. (2008) Pt–Ru electrocatalysts for fuel cells: a representative review. *Journal of Solid State Electrochemistry*, **12** (5), 609–642.

41 Cui, C., Gan, L., Heggen, M., Rudi, S., and Strasser, P. (2013) Compositional segregation in shaped Pt alloy nanoparticles and their structural behaviour during electrocatalysis. *Nature Materials*, **12** (8), 765–771.

42 Cui, C., Gan, L., Li, H.-H., Yu, S.-H., Heggen, M., and Strasser, P. (2012) Octahedral PtNi nanoparticle catalysts: exceptional oxygen reduction activity by tuning the alloy particle surface composition. *Nano Letters*, **12** (11), 5885–5889.

43 Cui, C., Gan, L., Neumann, M., Heggen, M., Roldan Cuenya, B., and Strasser, P. (2014) Carbon monoxide-assisted size confinement of bimetallic alloy nanoparticles. *Journal of the American Chemical Society*, **136** (13), 4813–4816.

44 Gan, L., Cui, C., Rudi, S., and Strasser, P. (2014) Core–shell and nanoporous particle architectures and their effect on the activity and stability of Pt ORR electrocatalysts. *Topics in Catalysis*, **57** (1–4), 236–244.

45 Gan, L., Heggen, M., Rudi, S., and Strasser, P. (2012) Core–shell

compositional fine structures of dealloyed Pt$_x$Ni$_{1-x}$ nanoparticles and their impact on oxygen reduction catalysis. *Nano Letters*, **12** (10), 5423–5430.

46 Hasché, F., Fellinger, T.-P., Oezaslan, M., Paraknowitsch, J.P., Antonietti, M., and Strasser, P. (2012) Mesoporous nitrogen doped carbon supported platinum PEM fuel cell electrocatalyst made from ionic liquids. *ChemCatChem*, **4** (4), 479–483.

47 Hasche, F., Oezaslan, M., and Strasser, P. (2010) Activity, stability and degradation of multi walled carbon nanotube (MWCNT) supported Pt fuel cell electrocatalysts. *Physical Chemistry Chemical Physics*, **12** (46), 15251–15258.

48 Hasché, F., Oezaslan, M., and Strasser, P. (2011) Activity, stability, and degradation mechanisms of dealloyed PtCu3 and PtCo3 nanoparticle fuel cell catalysts. *ChemCatChem*, **3** (11), 1805–1813.

49 Koh, S., Leisch, J., Toney, M.F., and Strasser, P. (2007) Structure-activity-stability relationships of Pt–Co alloy electrocatalysts in gas-diffusion electrode layers. *The Journal of Physical Chemistry C*, **111** (9), 3744–3752.

50 Koh, S. and Strasser, P. (2007) Electrocatalysis on bimetallic surfaces: modifying catalytic reactivity for oxygen reduction by voltammetric surface dealloying. *Journal of the American Chemical Society*, **129** (42), 12624–12625.

51 Koh, S., Yu, C., Mani, P., Srivastava, R., and Strasser, P. (2007) Activity of ordered and disordered Pt–Co alloy phases for the electroreduction of oxygen in catalysts with multiple coexisting phases. *Journal of Power Sources*, **172** (1), 50–56.

52 Liu, Z., Yu, C., Rusakova, I., Huang, D., and Strasser, P. (2008) Synthesis of Pt3Co alloy nanocatalyst via reverse micelle for oxygen reduction reaction in PEMFCs. *Topics in Catalysis*, **49** (3–4), 241–250.

53 Mani, P., Srivastava, R., and Strasser, P. (2008) Dealloyed Pt–Cu core–shell nanoparticle electrocatalysts for use in PEM fuel cell cathodes. *The Journal of Physical Chemistry C*, **112** (7), 2770–2778.

54 Oezaslan, M., Hasché, F., and Strasser, P. (2012) Oxygen electroreduction on PtCo3, PtCo and Pt3Co alloy nanoparticles for alkaline and acidic PEM fuel cells. *Journal of The Electrochemical Society*, **159** (4), B394–B395.

55 Oezaslan, M., Hasché, F., and Strasser, P. (2012) PtCu3, PtCu and Pt3Cu alloy nanoparticle electrocatalysts for oxygen reduction reaction in alkaline and acidic media. *Journal of The Electrochemical Society*, **159** (4), B444–B454.

56 Oezaslan, M., Heggen, M., and Strasser, P. (2011) Size-dependent morphology of dealloyed bimetallic catalysts: linking the nano to the macro scale. *Journal of the American Chemical Society*, **134** (1), 514–524.

57 Srivastava, R., Mani, P., Hahn, N., and Strasser, P. (2007) Efficient oxygen reduction fuel cell electrocatalysis on voltammetrically dealloyed Pt–Cu–Co nanoparticles. *Angewandte Chemie, International Edition*, **46** (47), 8988–8991.

58 Strasser, P., Koh, S., Anniyev, T., Greeley, J., More, K., Yu, C. et al. (2010) Lattice-strain control of the activity in dealloyed core–shell fuel cell catalysts. *Nature Chemistry*, **2** (6), 454–460.

59 Strasser, P., Koh, S., and Greeley, J. (2008) Voltammetric surface dealloying of Pt bimetallic nanoparticles: an experimental and DFT computational analysis. *Physical Chemistry Chemical Physics*, **10** (25), 3670–3683.

60 Yang, R., Leisch, J., Strasser, P., and Toney, M.F. (2010) Structure of dealloyed PtCu$_3$ thin films and catalytic activity for oxygen reduction. *Chemistry of Materials*, **22** (16), 4712–4720.

61 Menzel, N., Ortel, E., Kraehnert, R., and Strasser, P. (2012) Electrocatalysis using porous nanostructured materials. *ChemPhysChem*, **13** (6), 1385–1394.

62 Cheng, F., Ma, H., Li, Y., and Chen, J. (2007) Ni1-xPtx ($x = 0$–0.12) hollow spheres as catalysts for hydrogen generation from ammonia borane. *Inorganic Chemistry*, **46** (3), 788–794.

63 Zhai, J., Huang, M., Zhai, Y., and Dong, S. (2008) Magnet-assisted assembly of 1-dimensional hollow PtCo nanomaterials on an electrode surface. *Journal of Materials Chemistry*, **18** (8), 923–928.

64 Reiser, C.A., Bregoli, L., Patterson, T.W., Yi, J.S., Yang, J.D., Perry, M.L. et al. (2005) A reverse-current decay mechanism for fuel cells. *Electrochemical and Solid-State Letters*, **8** (6), A273–A276.

65 Kumar, A. and Ramani, V. (2013) Ta0.3Ti0.7O2 electrocatalyst supports exhibit exceptional electrochemical stability. *Journal of The Electrochemical Society*, **160** (11), F1207–F1215.

66 Kumar, A. and Ramani, V. (2014) Strong metal–support interactions enhance the activity and durability of platinum supported on tantalum-modified titanium dioxide electrocatalysts. *ACS Catalysis*, **4** (5), 1516–1525.

67 Kumar, A. and Ramani, V.K. (2013) RuO$_2$–SiO$_2$ mixed oxides as corrosion-resistant catalyst supports for polymer electrolyte fuel cells. *Applied Catalysis B: Environmental*, **138–139** (0), 43–50.

68 Lo, C.-P. and Ramani, V. (2012) SiO$_2$–RuO$_2$: a stable electrocatalyst support. *ACS Applied Materials & Interfaces*, **4** (11), 6109–6116.

69 Lo, C.-P., Wang, G., Kumar, A., and Ramani, V. (2013) TiO$_2$–RuO$_2$ electrocatalyst supports exhibit exceptional electrochemical stability. *Applied Catalysis B: Environmental*, **140–141** (0), 133–140.

70 Parrondo, J., Han, T., Niangar, E., Wang, C., Dale, N., Adjemian, K. et al. (2014) Platinum supported on titanium–ruthenium oxide is a remarkably stable electrocatalyst for hydrogen fuel cell vehicles. *Proceedings of the National Academy of Sciences of the United States of America*, **111** (1), 45–50.

71 Zhou, X., Gan, Y., Du, J., Tian, D., Zhang, R., Yang, C. et al. (2013) A review of hollow Pt-based nanocatalysts applied in proton exchange membrane fuel cells. *Journal of Power Sources*, **232** (0), 310–322.

72 Zhao, J., Chen, W., Zheng, Y., and Li, X. (2006) Novel carbon supported hollow Pt nanospheres for methanol electrooxidation. *Journal of Power Sources*, **162** (1), 168–172.

73 Yang, J., Lee, J.Y., Too, H.-P., and Valiyaveettil, S. (2005) A bis(p-sulfonatophenyl)phenylphosphine-based synthesis of hollow Pt nanospheres. *The Journal of Physical Chemistry B*, **110** (1), 125–129.

74 Bansal, V., O'Mullane, A.P., and Bhargava, S.K. (2009) Galvanic replacement mediated synthesis of hollow Pt nanocatalysts: significance of residual Ag for the H$_2$ evolution reaction. *Electrochemistry Communications*, **11** (8), 1639–1642.

75 Wang, J.X., Ma, C., Choi, Y., Su, D., Zhu, Y., Liu, P. et al. (2011) Kirkendall effect and lattice contraction in nanocatalysts: a new strategy to enhance sustainable activity. *Journal of the American Chemical Society*, **133** (34), 13551–13557.

76 Komanicky, V., Chang, K.C., Menzel, A., Markovic, N.M., You, H., Wang, X. et al. (2006) Stability and dissolution of platinum surfaces in perchloric acid. *Journal of The Electrochemical Society*, **153** (10), B446–B451.

77 Vasquez, Y., Sra, A.K., and Schaak, R.E. (2005) One-pot synthesis of hollow superparamagnetic CoPt nanospheres. *Journal of the American Chemical Society*, **127** (36), 12504–12505.

78 Hong, J.W., Kang, S.W., Choi, B.-S., Kim, D., Lee, S.B., and Han, S.W. (2012) Controlled synthesis of Pd–Pt alloy hollow nanostructures with enhanced catalytic activities for oxygen reduction. *ACS Nano*, **6** (3), 2410–2419.

79 Guo, S., Dong, S., and Wang, E. (2009) Raspberry-like hierarchical Au/Pt nanoparticle assembling hollow spheres with nanochannels: an advanced nanoelectrocatalyst for the oxygen reduction reaction. *The Journal of Physical Chemistry C*, **113** (14), 5485–5492.

80 Liu, Z., Zhao, B., Guo, C., Sun, Y., Xu, F., Yang, H. et al. (2009) Novel hybrid electrocatalyst with enhanced performance in alkaline media: hollow Au/Pd core/shell nanostructures with a raspberry surface. *The Journal of Physical Chemistry C*, **113** (38), 16766–16771.

81 Jin, Y., Shen, Y., and Dong, S. (2004) Electrochemical design of ultrathin platinum-coated gold nanoparticle monolayer films as a novel

nanostructured electrocatalyst for oxygen reduction. *The Journal of Physical Chemistry B*, **108** (24), 8142–8147.

82 Ye, H. and Crooks, R.M. (2005) Electrocatalytic O_2 reduction at glassy carbon electrodes modified with dendrimer-encapsulated Pt nanoparticles. *Journal of the American Chemical Society*, **127** (13), 4930–4934.

83 Lee, D., Jang, H.Y., Hong, S., and Park, S. (2012) Synthesis of hollow and nanoporous gold/platinum alloy nanoparticles and their electrocatalytic activity for formic acid oxidation. *Journal of Colloid and Interface Science*, **388** (1), 74–79.

84 Gao, J., Ren, X., Chen, D., Tang, F., and Ren, J. (2007) Bimetallic Ag–Pt hollow nanoparticles: synthesis and tunable surface plasmon resonance. *Scripta Materialia*, **57** (8), 687–690.

85 Yang, J., Lu, L., Wang, H., and Zhang, H. (2006) Synthesis of Pt/Ag bimetallic nanorattle with Au core. *Scripta Materialia*, **54** (2), 159–162.

86 Hu, Y., Wu, P., Zhang, H., and Cai, C. (2012) Synthesis of graphene-supported hollow Pt–Ni nanocatalysts for highly active electrocatalysis toward the methanol oxidation reaction. *Electrochimica Acta*, **85** (0), 314–321.

87 Zhang, W., Yang, J., and Lu, X. (2012) Tailoring galvanic replacement reaction for the preparation of Pt/Ag bimetallic hollow nanostructures with controlled number of voids. *ACS Nano*, **6** (8), 7397–7405.

88 Mayrhofer, K.J.J., Juhart, V., Hartl, K., Hanzlik, M., and Arenz, M. (2009) Adsorbate-induced surface segregation for core–shell nanocatalysts. *Angewandte Chemie, International Edition*, **48** (19), 3529–3531.

89 Zhou, X.W., Zhang, R.H., Zeng, D.M., and Sun, S.G. (2010) One-dimensional CoPt nanorods and their anomalous IR optical properties. *Journal of Solid State Chemistry*, **183** (6), 1340–1346.

90 Chen, Q.-S., Sun, S.-G., Zhou, Z.-Y., Chen, Y.-X., and Deng, S.-B. (2008) CoPt nanoparticles and their catalytic properties in electrooxidation of CO and CH_3OH studied by *in situ* FTIRS.
Physical Chemistry Chemical Physics, **10** (25), 3645–3654.

91 Zhou, X.-W., Zhang, R.-H., Zhou, Z.-Y., and Sun, S.-G. (2011) Preparation of PtNi hollow nanospheres for the electrocatalytic oxidation of methanol. *Journal of Power Sources*, **196** (14), 5844–5848.

92 Wang, L. and Yamauchi, Y. (2011) Synthesis of mesoporous Pt nanoparticles with uniform particle size from aqueous surfactant solutions toward highly active electrocatalysts. *Chemistry – A European Journal*, **17** (32), 8810–8815.

93 Zhang, H., Yin, Y., Hu, Y., Li, C., Wu, P., Wei, S. *et al.* (2010) Pd@Pt core–shell nanostructures with controllable composition synthesized by a microwave method and their enhanced electrocatalytic activity toward oxygen reduction and methanol oxidation. *The Journal of Physical Chemistry C*, **114** (27), 11861–11867.

94 Hu, Y., Shao, Q., Wu, P., Zhang, H., and Cai, C. (2012) Synthesis of hollow mesoporous Pt–Ni nanosphere for highly active electrocatalysis toward the methanol oxidation reaction. *Electrochemistry Communications*, **18** (0), 96–99.

95 Bae, S.J., Yoo, S.J., Lim, Y., Kim, S., Lim, Y., Choi, J. *et al.* (2012) Facile preparation of carbon-supported PtNi hollow nanoparticles with high electrochemical performance. *Journal of Materials Chemistry*, **22** (18), 8820–8825.

96 Imai, H., Matsumoto, M., Miyazaki, T., Kato, K., Tanida, H., and Uruga, T. (2011) Growth limits in platinum oxides formed on Pt-skin layers on Pt-Co bimetallic nanoparticles. *Chemical Communications*, **47** (12), 3538–3540.

97 Stamenkovic, V.R., Fowler, B., Mun, B.S., Wang, G., Ross, P.N., Lucas, C.A. *et al.* (2007) Improved oxygen reduction activity on $Pt_3Ni(111)$ via increased surface site availability. *Science*, **315** (5811), 493–497.

98 Minch, R. and Es-Souni, M. (2011) On-substrate, self-standing hollow-wall Pt and PtRu-nanotubes and their electrocatalytic behavior. *Chemical Communications*, **47** (22), 6284–6286.

99 Gasteiger, H.A., Marković, N., Ross, P.N., and Cairns, E.J. (1994) Temperature-dependent methanol electro-oxidation on well-characterized Pt–Ru alloys. *Journal of The Electrochemical Society*, **141** (7), 1795–1803.

100 Ye, F., Yang, J., Hu, W., Liu, H., Liao, S., Zeng, J. et al. (2012) Electrostatic interaction based hollow Pt and Ru assemblies toward methanol oxidation. *RSC Advances*, **2** (19), 7479–7486.

101 Zhang, H., Jin, M., Wang, J., Kim, M.J., Yang, D., and Xia, Y. (2011) Nanocrystals composed of alternating shells of Pd and Pt can be obtained by sequentially adding different precursors. *Journal of the American Chemical Society*, **133** (27), 10422–10425.

102 Yuan, Q., Zhou, Z., Zhuang, J., and Wang, X. (2010) Pd-Pt random alloy nanocubes with tunable compositions and their enhanced electrocatalytic activities. *Chemical Communications*, **46** (9), 1491–1493.

103 Liang, Y., Li, Y., Wang, H., Zhou, J., Wang, J., Regier, T. et al. (2011) Co_3O_4 nanocrystals on graphene as a synergistic catalyst for oxygen reduction reaction. *Nature Materials*, **10** (10), 780–786.

104 Wu, G., More, K.L., Johnston, C.M., and Zelenay, P. (2011) High-performance electrocatalysts for oxygen reduction derived from polyaniline, iron, and cobalt. *Science*, **332** (6028), 443–447.

105 Sánchez-Sánchez, C.M., Solla-Gullón, J., Vidal-Iglesias, F.J., Aldaz, A., Montiel, V., and Herrero, E. (2010) Imaging structure sensitive catalysis on different shape-controlled platinum nanoparticles. *Journal of the American Chemical Society*, **132** (16), 5622–5624.

106 Zhang, J., Yang, H., Fang, J., and Zou, S. (2010) Synthesis and oxygen reduction activity of shape-controlled Pt_3Ni nanopolyhedra. *Nano Letters*, **10** (2), 638–644.

107 Chu, Y.-Y., Wang, Z.-B., Jiang, Z.-Z., Gu, D.-M., and Yin, G.-P. (2012) Facile synthesis of hollow spherical sandwich PtPd/C catalyst by electrostatic self-assembly in polyol solution for methanol electrooxidation. *Journal of Power Sources*, **203** (0), 17–25.

108 Yu, X., Wang, D., Peng, Q., and Li, Y. (2011) High performance electrocatalyst: Pt–Cu hollow nanocrystals. *Chemical Communications*, **47** (28), 8094–8096.

109 Xia, B.Y., Wu, H.B., Wang, X., and Lou, X.W. (2012) One-pot synthesis of cubic $PtCu_3$ nanocages with enhanced electrocatalytic activity for the methanol oxidation reaction. *Journal of the American Chemical Society*, **134** (34), 13934–13937.

110 Wang, D. and Li, Y. (2011) Bimetallic nanocrystals: liquid-phase synthesis and catalytic applications. *Advanced Materials*, **23** (9), 1044–1060.

111 Xu, D., Liu, Z., Yang, H., Liu, Q., Zhang, J., Fang, J. et al. (2009) Solution-based evolution and enhanced methanol oxidation activity of monodisperse platinum–copper nanocubes. *Angewandte Chemie, International Edition*, **48** (23), 4217–4221.

112 Ge, J., Xing, W., Xue, X., Liu, C., Lu, T., and Liao, J. (2007) Controllable synthesis of Pd nanocatalysts for direct formic acid fuel cell (DFAFC) application: from Pd hollow nanospheres to Pd nanoparticles. *The Journal of Physical Chemistry C*, **111** (46), 17305–17310.

113 Liu, Z., Zhao, B., Guo, C., Sun, Y., Shi, Y., Yang, H. et al. (2010) Carbon nanotube/raspberry hollow Pd nanosphere hybrids for methanol, ethanol, and formic acid electro-oxidation in alkaline media. *Journal of Colloid and Interface Science*, **351** (1), 233–238.

114 Zhou, W.P., Lewera, A., Larsen, R., Masel, R.I., Bagus, P.S., and Wieckowski, A. (2006) Size effects in electronic and catalytic properties of unsupported palladium nanoparticles in electrooxidation of formic acid. *The Journal of Physical Chemistry B*, **110** (27), 13393–13398.

115 Hammer, B. and Nørskov, J.K. (1995) Electronic factors determining the reactivity of metal surfaces. *Surface Science*, **343** (3), 211–220.

116 Mavrikakis, M., Hammer, B., and Nørskov, J.K. (1998) Effect of strain on the reactivity of metal surfaces. *Physical Review Letters*, **81** (13), 2819–2822.

117 Ruban, A., Hammer, B., Stoltze, P., Skriver, H.L., and Nørskov, J.K. (1997) Surface electronic structure and reactivity of transition and noble metals. *Journal of Molecular Catalysis A: Chemical*, **115** (3), 421–429.

118 Hu, C., Guo, Y., Wang, J., Yang, L., Yang, Z., Bai, Z. *et al.* (2012) Additive-free fabrication of spherical hollow palladium/copper alloyed nanostructures for fuel cell application. *ACS Applied Materials & Interfaces*, **4** (9), 4461–4464.

119 Jiang, Y., Lu, Y., Han, D., Zhang, Q., and Niu, L. (2012) Hollow Ag@Pd core–shell nanotubes as highly active catalysts for the electro-oxidation of formic acid. *Nanotechnology*, **23** (10), 105609.

120 Bai, Z., Yang, L., Zhang, J., Li, L., Lv, J., Hu, C. *et al.* (2010) Solvothermal synthesis and characterization of Pd–Rh alloy hollow nanosphere catalysts for formic acid oxidation. *Catalysis Communications*, **11** (10), 919–922.

121 Lee, C.-L., Tseng, C.-M., Wu, R.-B., and Yang, K.-L. (2008) Hollow Ag/Pd triangular nanoplate: a novel activator for electroless nickel deposition. *Nanotechnology*, **19** (21), 215709.

122 Wang, W., Zhao, B., Li, P., and Tan, X. (2008) Fabrication and characterization of Pd/Ag alloy hollow spheres by the solvothermal method. *Journal of Nanoparticle Research*, **10** (3), 543–548.

123 Yang, L., Hu, C., Wang, J., Yang, Z., Guo, Y., Bai, Z. *et al.* (2011) Facile synthesis of hollow palladium/copper alloyed nanocubes for formic acid oxidation. *Chemical Communications*, **47** (30), 8581–8583.

124 Trogadas, P., Nigra, M.M., and Coppens, M.-O. (2016) Nature-inspired optimization of hierarchical porous media for catalytic and separation processes. *New Journal of Chemistry*, **40** (5), 4016–4026.

4
New Frontiers in Biocatalysis

John M. Woodley and Nicholas J. Turner

4.1
Introduction

There is no doubt that in the last two decades biotechnology has had an increasingly important impact on the chemical industry, both for the production of low-priced bulk chemicals, as well as higher priced pharmaceuticals. Naturally, the drivers for these developments are dependent upon the industry sector. For lower value chemicals, the start of the shift from petrochemical fossil-based feedstocks, as used for the past century, toward renewable bio-based materials is widely acknowledged [1]. Likewise, for higher value pharmaceuticals, there are extraordinary opportunities brought about by the application of enzymatic approaches to simply and precisely catalyze key steps in the synthesis of the most complex molecules (containing many functional groups and stereogenic centers) [2]. In both industry sectors, cheaper processes result and further benefits accrue from a better sustainability profile. Nevertheless, this upbeat message is frequently tempered by long development times since such processes also need to be designed to incorporate a biological catalyst. Using molecular biology tools, discovered around 30 years ago, today such catalysts are characterized by the ability to program their catalytic properties through their genetic code. This opens very many possibilities and it is only in the last decade that the control of this process is starting to yield rewards.

Although Nature has perfected the design of biological catalysts through evolution over millennia, the specific molecules to be converted and the conditions under which the catalysis takes place in Nature are usually far from those of industrial interest. Unfortunately, synthetically designed schemes to achieve the appropriate catalyst in an industrial setting are still far from complete or perfect. Nevertheless, enormous progress is being made. Each of the three formats of biological catalyst can be manipulated and tailored to some extent. First, engineered microbial cells in fermentation can be improved by metabolic engineering, allowing alterations to pathways to improve the achieved yield of product on feedstock, to be as close as possible to the maximum theoretical yield (calculated by redox balance) [3]. This has immediate benefits for fermentation-based chemical production, but also for assisting effective expression of enzymes (which may subsequently be used in an isolated catalyst format). In this way many enzymes are now available at sufficient scale and low enough cost to be useful as the catalyst of first choice for many process chemists. Excellent examples include ketoreductases and lipases. Second, engineered microbial cells in nonviable mode can be used to house novel pathways. In principle, there are few limits to the pathways that can be engineered via the so-called synthetic biology, although substrate still needs to enter, and product to leave, the cells effectively [4]. Ultimately, the rate may be limited by the space available in the reactor. Higher concentrations of catalyst can be achieved in the third class where enzymes are isolated from the cells in which they were originally expressed, giving a higher active site density. Here too, molecular biology can be used to protein engineer enzymes to tailor their properties for a given synthesis [5]. The latter biocatalyst format, frequently used in the synthesis of high-value products (and more generally termed biocatalysis), serves process chemists well since the catalysts can be used "off the shelf." Alongside an increasing range of companies able to supply such biocatalysts is a realization that more and more chemistry can be done using enzymes in artificial pathways (never before seen in Nature) and on nonnatural substrates. From a synthetic perspective, this opens many opportunities to complement conventional approaches. Likewise, protein engineering can be used to alter enzymes to work more effectively under the most favorable process conditions, to complement the work of the process engineer.

4.2
Recent Advances in Biocatalysis

One of the most important areas of growth in biotechnology in recent years has been in the field of biocatalysis. On the one hand, the application of enzymes to useful synthetic problems is perhaps the simplest of all biotechnological approaches, but it is also the one making greatest traction in industry today. There are two reasons for this. The first is the ownership of the field by the synthetic chemistry community, meaning that new and relevant target molecules are forthcoming and that the synthetic problems addressed using enzymes are those that are complementary to conventional organic synthesis [6]. The second is the introduction of protein engineering as a tool to improve the catalytic properties of enzymes to match given synthetic needs. Some excellent examples highlight the power of such approaches in the pharmaceutical sector such as the synthesis of sitagliptin [7] and atorvastatin [8]. Indeed, success has been widespread with some 150 or more processes implemented in the pharmaceutical sector today. The use of enzymes is now the first choice of catalyst in the pharmaceutical sector for the synthesis of chiral alcohols and amines. In short, progress has been spectacular.

However, for the further development of biocatalysis, new approaches will be required. For example, three recent developments highlight the need for completely new thinking across the disciplines. The first concerns the systematic evaluation of new pathways to target molecules. Termed biocatalytic retrosynthesis, this enables entirely new routes and feedstocks to be used to reach a given product [9–11]. Using the vast array of engineered enzymes now available (for a wide range of nonnatural substrates), the possibilities of creating novel multienzymatic or chemoenzymatic pathways are almost unlimited. The creation and evaluation of such pathways will become a major activity in organic synthesis in the future and to support this, process evaluation tools will also be required (assisted by computer-aided technologies). The second area concerns the use of protein engineering to assist process engineering and development. This is a step beyond matching the conditions of the biocatalyst to the processes, as already established [12–14]. The third area we wish to highlight concerns the application of biocatalysis using new process technology. Here, the application of flow chemistry and continuous processing will enable biological catalysts to match developments in the conventional catalysis field for the production of pharmaceuticals and fine chemicals. Today, this is now complemented by the development of new kinetic analysis tools to capitalize upon all that protein engineering can offer. In this chapter, these developments will be discussed to illustrate the latest developments in biocatalysis and the clear impact this will have at the interface with chemical engineering.

4.3
Biocatalytic Retrosynthesis

The concept of retrosynthesis in organic synthesis was first introduced in the early 1960s by E.J. Corey as a tool to aid the planning of synthetic routes to target molecules (see Ref. [15]). Retrosynthesis involves a formal and strategic analysis of how to synthesize a molecule by envisaging a series of disconnections that result in the generation of simpler building blocks, or synthons, which are then amenable to synthesis. Retrosynthesis has not only provided a valuable training tool for students of organic chemistry but it has also been instrumental in the development of new reagents and catalysts in organic synthesis. The application of retrosynthetic analysis often highlights where gaps exist in the repertoire of organic synthesis and hence stimulates the discovery of new transformations to expand the synthetic toolbox. Today, retrosynthesis is an essential component of undergraduate teaching in Chemistry degrees enabling students to develop the required skills for planning synthetic routes to target molecules.

In 2013, it seemed timely to us to expand the concept of retrosynthesis to include biocatalytic transformations, a decision that was prompted by a number of important developments in the field.

- First, as a result of significant advances in enzyme discovery, protein engineering, and directed evolution, the toolbox of biocatalysts available for organic synthesis had expanded significantly. Whereas in the early 1990s, the biocatalytic landscape was dominated by hydrolases (e.g., lipases, esterases, acylases, and proteases) and redox enzymes (principally alcohol dehydrogenases); the next 20 years witnessed the arrival of a much broader range of biocatalysts, including engineered transaminases, nitrile hydrolases, aldolases, lyases, oxynitrilases, oxidases, P450 monooxygenases, Baeyer–Villiger monooxygenases, ammonia lyases, ene reductases and so on, enabling a wider range of C—C bond forming reactions and functional group interconversions (e.g., C=O to C—N; CN to CO_2H; C=C to CH—CH etc.). In the past 5 years, this expansion of the biocatalytic toolbox has increased even more rapidly with the recent availability of amine dehydrogenases, imine reductases, terpene cyclases, carboxylic acid reductases, halogenases, alkyltransferases, and so on.

- Second, the application of directed evolution technologies had resulted in not only a broader range of biocatalysts but importantly new engineered enzymes with substrate scope that more closely matched the requirements of organic chemists wishing to apply biocatalysis to target molecule synthesis. The development of transaminase biocatalysts emphasizes this latter point very clearly. The early transaminase enzymes showed great promise but were somewhat restricted in substrate scope, which limited their application. Through several years of protein engineering and screening, variants were identified that were found to be able to catalyze the transamination of increasingly bulky and structurally demanding ketone substrates. These variants were able to catalyze the formation of both (R)- and (S)-amines and could be applied at late stages in the synthesis of chiral amines as illustrated by the application of transaminase technology for the manufacture of Sitagliptin [7].
- A third and equally important factor was the much greater commercial availability of engineered biocatalysts, particularly in formats that enabled rapid screening (e.g., 96-well plates, crude lysate) to identify potential hits for further enzyme development. In order for biocatalysts to become more widely used, it was important that they be made available in convenient forms (e.g., freeze-dried powders) such that they could be easily incorporated in high-throughput screening platforms alongside chemocatalysts.
- A final point, which is a consequence of a combination of the first three developments outlined above, was the enhanced awareness of biocatalysis within the synthetic organic community coupled with the confidence that once a suitable biocatalyst had been identified then it should be possible to subsequently improve the enzyme through directed evolution and ultimately apply the biocatalyst at scale for preparative synthesis of the target molecule.

At this juncture it is important to emphasize that the real power of biocatalysis lies in the extent to which it can be employed to make step changes in the way in which a synthetic route to a target molecule is developed, hence the need to develop guidelines and rules for biocatalytic retrosynthesis. This point can be illustrated by examining two examples where biocatalysis has been applied in the manufacture of an active pharmaceutical ingredient (API). The first case involves the synthesis of a key intermediate for the manufacture of Montelukast,

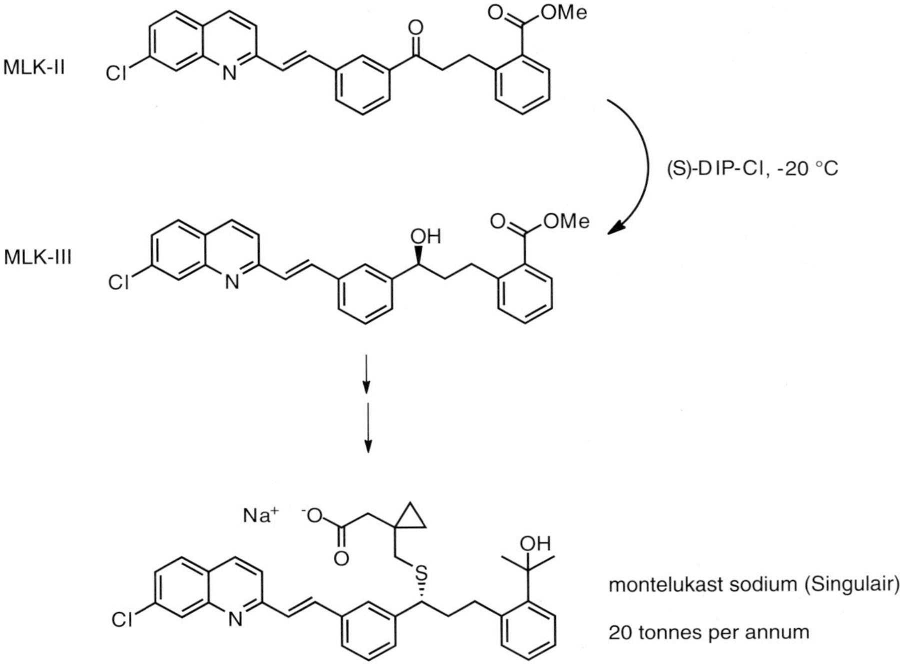

Figure 4.1 Use of chiral boron reagent for the asymmetric reduction of MLK-II to MLK-III.

which is a generic pharmaceutical used for treating asthma [16]. In one of the original synthetic routes employed, the chirality of the key secondary alcohol was set using a chiral borane reagent (Figure 4.1). In the biocatalytic approach, the boron reagent was replaced by a ketoreductase (KRED) with *in situ* cofactor recycling using isopropanol (Figure 4.2). Table 4.1 provides a comparison of the two different routes that reveals how different each process is in many respects, especially in terms of temperature, solvent usage, and waste by-products. From the perspective of biocatalytic retrosynthesis, the replacement of the boron reagent with a KRED biocatalyst is fairly easy to envisage, since the remainder of the synthetic route is unchanged and both processes use the same starting material and give the same product, albeit under vastly different conditions.

The second case involves the synthesis of an intermediate for the antihypertensive drug perindopril. In this example, the use of a biocatalyst leads to a completely different manufacturing route, with a different starting material, different intermediates, less waste, and fewer unit operations. The original process utilizes a Fischer indole synthesis followed by classical resolution to separate the enantiomers resulting in a large amount of waste product (Figure 4.3).

In the biocatalytic process, the enzyme phenylalanine ammonia lyases (PAL) is employed in the initial step for enantioselective amination of *ortho*-bromocinnamic acid for the generation of the corresponding alpha-amino acid in high

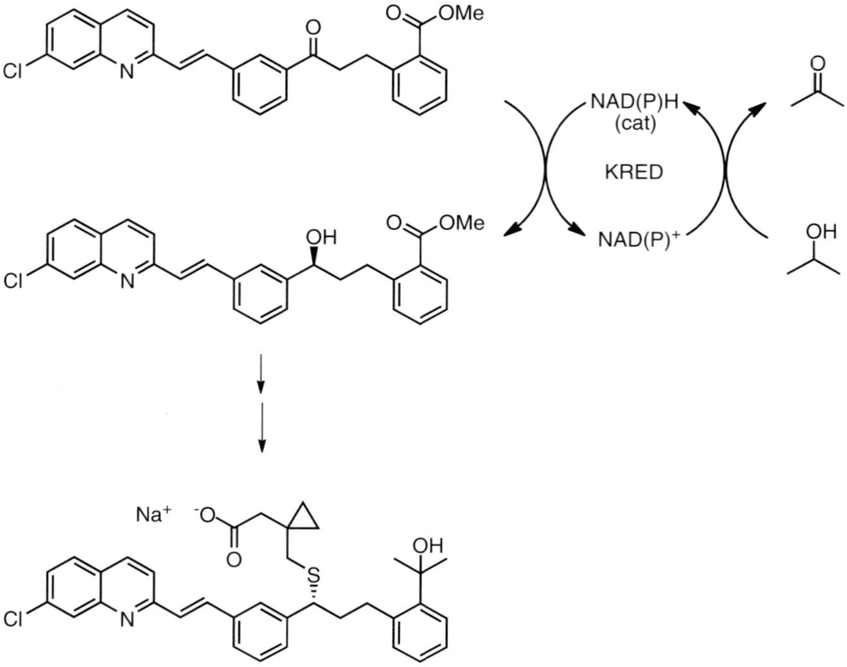

Figure 4.2 Biocatalytic reduction of MLK-II using a KRED.

Table 4.1 Comparison of (S)-DIP-Cl and biocatalytic route to an intermediate for Montelukast.

Parameter	Biocatalytic process	(S)-DIP-Cl Process
Ketone concentration (g l^{-1})	100	100
Catalytic/stoichiometric	Catalytic	1.8 equiv. DIP-Cl
Temperature (°C)	45	−25
Conversion (%)	99.3	Not provided
Product isolation	Direct filtration	Extraction with high dilution
Enantiomeric excess (%)	>99.9	99.2 (after recryst.)
Solvent/MLK-III (l kg^{-1})	6	30–50
Solvents used	Isopropyl alcohol, water, toluene	Dichloromethane, tetrahydrofurane
Other waste generation	Biodegradable enzyme, cofactor	Nonbiodegradable borate salts. Other inorganics, 3.6 equiv. pinene

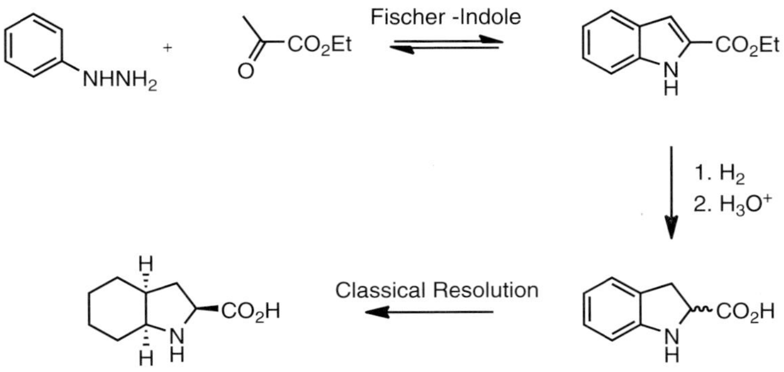

Figure 4.3 Use of Fischer indole synthesis followed by classical resolution to prepare an intermediate for Perindopril.

yield and enantiomeric excess. Thereafter, Cu-catalyzed intramolecular amination leads to the bicyclic compound that can be further reduced [17]. Comparison of Figures 4.3 and 4.4 highlights how the use of a biocatalyst can lead to a major change in the synthetic route, emphasizing the power of biocatalytic retrosynthesis when applied in its widest sense.

Since more engineered biocatalysts have become available with broader substrate scope, the use of more than one enzyme in a synthetic route have become possible, thereby addressing the point made above of developing substantially altered synthetic routes to target molecules. Due to the high-reaction specificity and orthogonality of biocatalysts, they are increasingly being applied in cascade

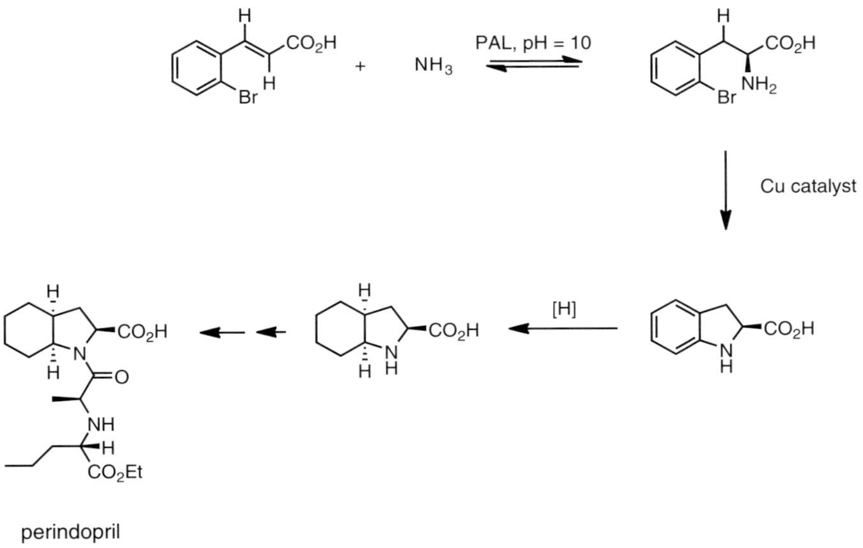

Figure 4.4 Use of phenylalanine ammonia lyase (PAL) to generate an intermediate for perindopril.

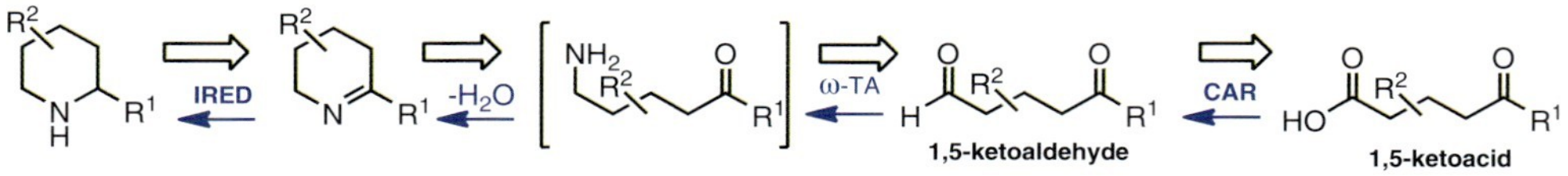

Figure 4.5 Cascade route employing three different biocatalysts for the conversion of substituted 1,5-keto acids to piperidines.

processes in which the product of the first biocatalyst becomes the substrate for the second biocatalyst and so on, mimicking closely the way in which natural products are synthesized in biosynthetic pathways. All biocatalysts are proteins and hence the reaction conditions for their application (i.e., solvent, pH, temperature, pressure) are essentially the same independent of the reaction they are catalyzing. This situation should be contrasted with chemical catalysts in which substantial differences in solvent required or temperature are often experienced. Careful design and implementation of cascade reactions (either *in vitro* or *in vivo*), involving two or more different biocatalysts, leads to the orchestrated, programmable synthesis of target compounds. Crucial to this successful application of multienzyme organic synthesis is the initial retrosynthetic design in order to identify the required biocatalyst for each step and ensure an efficient synthesis of the target molecule. An example of multibiocatalyst synthesis is given by the synthesis of a series of substituted piperidines using a combination of three different enzymes, namely, transaminases, imine reductases, and carboxylic acid reductases (Figure 4.5) [18]. In this cascade, each biocatalyst possesses very high fidelity for the specific functional group that it is transforming, allowing all three biocatalysts to be used in conjunction with no side reactions.

In summary, biocatalytic retrosynthesis is fast becoming a standard tool in the application of enzymes for preparative organic synthesis, particularly in cases where the use of one or more biocatalysts allows the development of a new synthetic route to a target molecule with attendant reduction in the costs of the starting material, reduced wastage, higher process efficiency, and lower environmental impact.

4.4
Process-Driven Protein Engineering

It is clear that introducing the possibility to design catalytic properties desirable for synthetic problems at an industrial scale will bring great opportunities. However, such technologies also require careful implementation. Today, protein engineering is used primarily in the earliest of reaction scoping experiments to extend substrate range and convert molecules that would otherwise be difficult to convert, in large part since the enzymes have never encountered such molecules in Nature. Beyond this, improvements to enzyme tolerance to higher concentrations, temperatures, and extremes of pH ultimately need to be balanced by sufficient stability. Excellent progress has been made in this field such that today the methodology for doing this is to a large extent established, although

Table 4.2 Potential industrially relevant reaction conditions and their effects on enzymatic catalysts.

Enzyme effect	Potential industrial reaction conditions
Interface effects	Presence of organic solvent droplets containing high concentrations of poorly water-soluble substrates and/or products
	Presence of gas bubbles containing gaseous substrates and/or products
Inhibition effects	High concentration of substrates, cosubstrates, products, and by-products
Stability effects	Heterogeneity of conditions at large scale
	Extreme pH, T, and high concentrations of substrates, cosubstrates, products, and by-products

automating all the steps and establishing easy assays are still in need of refinement. Progress is also significant in the introduction of computer-aided techniques to guide algorithms, especially in the area of stability improvement. Some examples of the relevant industrial conditions for which enzymes need to be engineered are given in Table 4.2.

Such technology opens many opportunities but it has become clear that use of an enzyme on a synthetic problem never before seen in Nature (under industrial conditions) requires development work.

The key to success is to establish how to do that development work in the most effective way possible. Against this background it is interesting to observe that it is sometimes quicker to implement process changes rather than biocatalyst changes. Most interesting is that although enzyme limitations and problems can be solved by process engineering or protein engineering, in some cases this may be inadequate and the solution needs to come from a combination. Some examples of process and protein engineering improvement targets are given in Table 4.3.

Table 4.3 Examples of process and protein engineering methods for the improvement of biocatalysts and their processes to enable industrial implementation.

Improvement target	Process engineering method	Protein engineering method
Increased substrate concentration	Controlled feed of highly concentrated substrate	Engineer enzyme to operate at higher substrate concentration
Increased product concentration	Controlled removal product via ISPR	Engineer enzyme to operate at higher product concentration
Increased operational stability	Enzyme immobilization	Engineer enzyme to operate at suitable reactor conditions
Effective operation with oxygen as substrate	Use of pressurized reactor	Engineer kinetic properties with respect to oxygen, K_{MO}

These points are well illustrated by two examples.

The first example concerns the supply of oxygen to oxidases, monooxygenases, and dioxygenases. In each case, the addition of molecular oxygen to these enzymes enables valuable chemistry, but with the added advantage of using air, rather than more hazardous oxidants. In Nature, the reactions do not need to achieve a commercial reaction rate, but it rapidly becomes clear that in order to achieve even modest rates in an industrial context, the supply of oxygen can become limiting. The problem arises because the solubility of oxygen in water (the usual medium for biocatalysis) is very limited (250 µM at 30 °C). The consequence is that the supply rate of oxygen is also limited. For example, in a large-scale reactor in industry, a mass transfer coefficient of 500 h^{-1} is achievable (with around 1.5 W l^{-1} of energy input via agitation). Using such numbers implies a maximum oxygen transfer rate of around 100 mmol l^{-1} h^{-1}, only just sufficient for many industrial processes. Process solutions to this problem involve the enriching of air with oxygen and/or increasing the pressure (increasing the partial pressure by increasing the volume fraction of oxygen and/or the total pressure in the system) [19]. The resulting improvements in oxygen transfer rate are illustrated in Figure 4.6. Clearly, such improvements come at a cost and the necessary economic balance needs to be evaluated.

It is clear that a simple (although potentially costly) process modification can in principle allow sufficient improvement. In this case protein engineering would not appear to be useful. However, it is assumed that oxygen-dependent enzymes work effectively with respect to oxygen, their natural substrate. In reality, this is rarely tested and recent studies reveal that in some cases the K_M for oxygen is high, relative to oxygen solubility in water (in equilibrium with air). The consequence is that the enzyme works far away from saturation and therefore

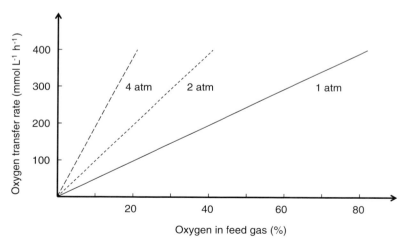

Figure 4.6 Effect of increasing partial pressure (by oxygen in feed gas (%) or total pressure (atm)) on oxygen transfer rate (mmol l^{-1} h^{-1}). Calculation based on assuming room temperature and dissolved oxygen in solution at 20% of equilibrium concentration with the gas.

ineffectively. In this way an excess of enzyme needs to be supplied. Here, it is clear that engineering lower Michaelis constant for oxygen (K_{MO}) on the enzyme will result in more effective use of the enzyme supplied [19]. Interestingly, the extent of the problem is dependent upon the process solution. This illustrates well the need to define the process (and the required productivity) ahead of biocatalyst testing.

A second example concerns the removal of an inhibitory product from enzymes used in the synthesis of nonnatural products at high concentrations. The need for high ($50\,g\,l^{-1}$ or more) concentrations of a nonnatural product in solution is set by the downstream process requirements, but is potentially problematic for all but the most robust of enzymes. Strategies to protein engineer product concentration tolerance into enzymes have been developed, although the identification of suitable assays and the need to improve the enzyme stepwise make the procedure difficult. An alternative pursued by the process engineers for some years has been the strategy of removing the product as it is produced in a hybrid reaction–separation scheme (widely discussed in the process intensification literature). For practical reasons such schemes in bioprocesses frequently contain a recycle loop, although termed in situ product removal (ISPR) [20]. In such a scheme, the rate of product removal in the separation is dependent upon the driving force. Hence, the maximum theoretical rate of removal will be when the product concentration is at its highest. However, this also corresponds to the concentration when the product is at its most damaging to the enzyme. Hence, although the implementation of a process change (ISPR) can reduce the effect of inhibition on the enzymes, it may be at the expense of rate. The trade-off is dependent upon the cost of the downstream processing relative to the enzyme. In cases where further improvement is required, protein engineering to improve tolerance may enable an effective use of the ISPR concept. This illustrates well the case where an integrated approach gives the best solution. As in the previous case, it is essential to consider both the protein engineering and process engineering together at the earliest possible stage in process development.

4.5
Process Developments

4.5.1
Continuous Processes

The recent possibilities to engineer improved stability into enzymes, also opens the opportunity to operate reactions in continuous mode. Established biocatalytic processes such as the production of high-fructose corn syrup (HFCS) using immobilized glucose isomerase already operates at a large scale in this way, and have done so for many years. However, many biocatalytic reactions, especially in the fine chemical and pharmaceutical sector, operate in batch (or more likely fed-batch mode). Since the last decade, the pharmaceutical industry has increasingly adopted flow chemistry and continuous operational approaches with the idea of making smaller processes more intensified and with better control [21, 22]. Here too there is an interface with chemical engineering driven by improvements in enzyme stability as a result of protein engineering. Several more complex biocatalytic processes have now been operated in continuous mode, including the application of amino transferases to the synthesis of optically pure chiral amines [23]. Multiphasic systems have also been used in tube-in-tube reactors, at a small scale to facilitate mass transfer [24]. The further development in this field will also benefit from improvements in stability and kinetics. New challenges will arise not only from the handling of multiphasic mixtures, but also to allow for cofactor recycle and operation of enzymatic cascades.

4.5.2
Kinetic Analysis

Chemical engineers have long been driven by the need for rigorous kinetic modeling of reactors to operate in an optimal way. In recent years attempts at a similar approach for biocatalytic conversions have also been developed [25]. The kinetic models of biocatalytic conversions are particularly interesting since the Michaelis–Menten kinetics, and derivatives thereof, imply a changing reaction order dependent upon the substrate concentration. For example, this makes operation in continuous stirred-tank reactors rather unattractive, compared to plug-flow operation. Protein engineering provides the opportunity to tune the kinetic parameters and this inspired us in recent years to consider the development of an apparatus for the precise measurement of complex multisubstrate, multiphase kinetics. For this, a modified tube-in-tube reactor has been used, linked with suitable analytical and software tools to extract kinetic constants accurately and rapidly [26]. The further development of such tools in the future will be of great importance in allowing chemical engineers to truly capitalize upon the developments in protein engineering, and likewise to allow a more quantitative analysis of the benefits of modifying a given enzyme.

4.6
Future Perspectives

These examples illustrate well that an integrated approach to the design of pathways, biocatalysts, and processes can result in great opportunities, as well as solving immediate problems for industrial implementation of new sustainable and green chemistry. What is more challenging is that such an approach can only come from an interdisciplinary dialog between synthetic chemists, protein engineers, and process engineers. An excellent dialog already exists between the first two disciplines and the challenge now is to integrate the third discipline.

The driver here is a little different, since the objective is effective implementation of an industrial process rather than new catalysis and novel chemistry. This practical necessity means that it would seem that those in industry should drive this development to a large extent. Nevertheless, there are not only scientific principles underlying such an approach but also the need to use reaction engineering and kinetic studies in a far more rigorous way than has been done to date in the bioprocess industries. This means that the industry drive must be supported by significant academic work.

The implementation of biological catalysis into the synthetic organic curriculum has already started. Biocatalysis, by the application of enzymes to synthetic problems, represents a new opportunity with a complementary set of sustainable catalysts. On the other hand, biochemical engineering, as the branch of process engineering with a focus on bioprocesses, has had a particular focus on the needs of supplying proteins (as low-value enzymes or as high-value therapeutic reagents). Today, biochemical engineering needs to develop and grow to also embrace chemical production using biocatalysts and enzymes. This can provide an enormous opportunity for conventional chemical engineering where tools have already been built. What lies ahead is a new process design paradigm where biocatalyst design needs to be added as an extra degree of freedom to the design problem [14, 27]. Ultimately, this will enable the use of more standardized equipment, ideally suited to fast and flexible continuous production.

References

1 Woodley, J.M., Breuer, M., and Mink, D. (2015) *Chemical Engineering Research and Design*, **91**, 2029–2036.
2 Pollard, D.J. and Woodley, J.M. (2007) *Trends in Biotechnology*, **25**, 66–73.
3 Dugar, D. and Stephanopoulos, G. (2011) *Nature Biotechnology*, **29**, 1074–1078.
4 France, S.P., Hepworth, L.J., Turner, N.J., and Flitsch, S.L. (2017) *ACS Catalysis*, **7**, 710–724.
5 Turner, N.J. (2009) *Nature Chemical Biology*, **5**, 567–573.
6 Reetz, M.T. (2016) *Chemical Record*, **16**, 2449–2459.
7 Savile, C.K., Janey, J.M., Mundorff, E.C., Moore, J.C., Tam, S., Jarvis, E.R., Colbeck, J.C., Krebber, A., Fleitz, F.J., Brands, J., Devine, P.N., Huisman, G.W., and Hughes, G.J. (2010) *Science*, **329**, 305–309.
8 Ma, S.K., Gruber, J., Davis, C., Newman, L., Gray, D., Wang, A., Grate, J., Huisman, G.W., and Sheldon, R.A. (2010) *Green Chemistry*, **12**, 18–86.
9 Turner, N.J. and O'Reilly, E. (2013) *Nature Chemical Biology*, **9**, 285–288.
10 Green, A.P. and Turner, N.J. (2016) *Perspectives in Science*, **9**, 42–48.
11 Carreira, E.M., Turner, N.J., Hönig, M., and Sondermann, P. (2017) *Angewandte Chemie, International Edition*, **56**, 8942–8973.
12 Bornscheuer, U.T., Huisman, G.W., Kazlauskas, R.J., Lutz, S., Moore, J.C., and Robins, K. (2012) *Nature*, **485**, 185–194.
13 Woodley, J.M. (2013) *Current Opinion in Chemical Biology*, **17**, 310–316.
14 Burton, S.G., Cowan, D.A., and Woodley, J.M. (2002) *Nature Biotechnology*, **20**, 37–45.
15 Corey, E.J. (1991) *Angewandte Chemie, International Edition*, **30**, 455–465.
16 Rozzell, D. and Liang, J. (2008) *Speciality Chemicals Magazine*, **28**, 36–38.
17 de Lange, B., Hyett, D.J., Maas, P.J.D., Mink, D., van Assema, F.B.J., Sereinig, N.,

de Vries, A.H.M., and de Vries, J.G. (2011) *ChemCatChem*, **3**, 289–292.

18 France, S.P., Hussain, S., Hill, A.M., Hepworth, L.J., Howard, R.M., Mulholland, K.R., Flitsch, S.L., and Turner, N.J. (2016) *ACS Catalysis*, **6**, 3753–3759.

19 Toftgaard Pedersen, A., Birmingham, W.R., Rehn, G., Charnock, S.J., Turner, N.J., and Woodley, J.M. (2015) *Organic Process Research & Development*, **19**, 1580–1589.

20 Woodley, J.M., Bisschops, M., Straathof, A.J.J., and Ottens, M. (2008) *Journal of Chemical Technology and Biotechnology*, **83**, 121–123.

21 McQuade, D.T. and Seeberger, P.H. (2013) *Journal of Organic Chemistry*, **78**, 6384–6389.

22 Adamo, A., Beingessner, R.L., Behnam, M., Chen, J., Jamison, T.F., Jensen, K.F., Monbaliu, J.-C.M., Myerson, A.S., Revalor, E.M., Snead, D.R., Stelzer, T., Weeranoppanant, N., Wong, S.Y., and Zhang, P. (2016) *Science*, **352**, 61–67.

23 Andrade, L.H., Kroutil, W., and Jamison, T.F. (2014) *Organic Letters*, **16**, 6092–6095.

24 Tomaszewski, B., Schmid, A., and Buehler, K. (2014) *Organic Process Research & Development*, **18**, 1516–1526.

25 Ringborg, R.H. and Woodley, J.M. (2016) *Reaction Chemistry & Engineering*, **1**, 10–22.

26 Ringborg, R.H., Toftgaard Pedersen, A., and Woodley, J.M. (2017) *ChemCatChem*, **9**, 3285–3288.

27 Woodley, J.M. (2017) *Computers & Chemical Engineering*, **105**, 297–307.

Part Two
Innovations in Design, Unit Operations, and Manufacturing

5
Conceptual Process Design and Process Optimization

Alexander Mitsos, Ung Lee, Sebastian Recker, and Mirko Skiborowski

5.1
Introduction

The conceptual design of chemical processes represents an important and complex task. Its importance is emphasized by the fact that the choices made in this early state of process development often account for up to 80% of the entire cost of the process [1]. Sustainable process design does not only have to account for economic aspects, but also has to consider environmental and social aspects. Consequently, conceptual process design presents a multilayered problem with subproblems on different scales, which can efficiently be addressed by mathematical programming techniques. The possibilities and challenges have been highlighted in several review articles in recent years [2–4]. This chapter provides the basic principles and an overview of currently available methods for conceptual process design with an emphasis on optimization-based approaches. Section 5.2 introduces the mathematical background, before Section 5.3 presents an overview of the different approaches for synthesis of reactor networks, separation systems, and the overall flowsheet design. Superstructure-based approaches for heat exchanger network design and process flowsheet optimization are subsequently presented in Section 5.4. Finally, the integration of process design and control and the design of water and energy processes are shortly described in Sections 5.5 and 5.6, before the chapter ends with some conclusions and an outlook in Section 5.7.

5.2
Mathematical Background

Model-based process design relies on a number of mathematical problems. This section gives a brief summary of the problems and algorithms for their numerical solution.

5.2.1
System of Nonlinear Equations

Process simulation, that is, the analysis of a given design, after fixing the degrees of freedom, is mathematically formulated as a system of n coupled nonlinear equations with n unknowns,

$$h(x) = 0 \tag{5.1}$$

where $h : \mathfrak{R}^n \rightarrow \mathfrak{R}^n$, has vector x as solution for which all components of h are zero, simultaneously. Nonlinear equation systems are typically solved using iterative methods.

The special case of $n = 1$ (single nonlinear equation) is relatively easy to handle, for example, through the bisection method. Solutions of systems with multiple nonlinear equations, however, are much more difficult to obtain. A basic method to solve nonlinear equations is the Newton–Raphson, or simply Newton method. If h is continuously differentiable on $[x^L, x^U]$, the nonlinear function can be approximated as $h(x_{k+1}) \approx h(x_k) + J(h_k)(x_{k+1} - x_k)$, where J is the Jacobian matrix of h (matrix of partial derivatives). The Newton iteration is obtained by setting $h(x_{k+1}) = 0$ and solving for x_{k+1}.

Thus, to perform one Newton iteration it is necessary to evaluate the Jacobian and invert it. These steps are often challenging in process design problems due to the large size of the problems and the fact that not all functions are known analytically, but rather are often calculated by subroutines, such as physical property packages. Several methods have been proposed as alternatives to Newton's method to avoid the calculation of Jacobian and its inversion [5–7].

Newton's method has superlinear convergence rate in the vicinity of a solution. However, it may not converge to a root without a good initial guess. Modifications of the Newton method have been considered with the aim to improve global convergence, that is, ensure finding a solution for a wide range of initial guesses. A comprehensive review of globalization of Newton method is provided by Deuflhard [8].

5.2.2
Nonlinear Programming (NLP)

In contrast to simulation wherein the degrees of freedom are fixed, optimization allows identifying the optimal values of the variables. Optimization of the design variables is one of the most crucial steps during model-based conceptual design. When an optimization problem is represented in algebraic form with the continuous variables x, it can be expressed as

$$\begin{aligned} \min \quad & f(x) \\ \text{s.t.} \quad & g(x) \leq 0 \\ & h(x) = 0 \end{aligned} \tag{5.2}$$

Note that the variables x contain both degrees of freedom and states of the system. In Eq. (5.2) these are confined to be continuous such as sizes of equipment, temperature, pressure, and flowrates. Thus, this formulation is applicable to fixed process structures. The objective function, $f(x)$, describes the performance metric of the process, for example, energy consumption, operating cost, capital cost, or total cost of the system. The equality constraints ($h(x) = 0$) typically originate in the physical model of the process, in particular balance and system equations. The inequality constraints ($g(x) \leq 0$) originate in design specifications and technical constraints. When all the functions are in linear form, the problems are reduced to linear programming (LP), which can be solved very efficiently [9]. However, typically design problems are inherently nonlinear and moreover involve nonconvex functions. Nonconvexity typically implies the presence of suboptimal local minima. This can be seen, for instance, in energy balances, where, for example, the heat flow is calculated as the product of specific heat capacity, flowrate, and temperature difference, which results in a bilinear term, if both flowrate and temperature are variables. In order to solve nonlinear programming (NLP) problems, several practical methods were presented.

A local optimum of the NLP (Eq. (5.2)) can be obtained using gradient-based optimization algorithms. Successive linear programming (SLP) [10], successive quadratic programming (SQP) [11, 12], and generalized reduced gradient (GRG) [13] are popular methods that share similar characteristics for finding solutions. These solution methods rely on solving a series of simpler subproblems. For instance, in SQP the subproblems involve a quadratic approximation of the objective function and a linear approximation of the constraints. A number of major codes use gradient-based optimization algorithms such as OPT by Vasantharajan et al. [14], MINOS by Murtagh and Saunders [13], CONOPT by Drud [15], SNOPT by Gill et al. [16], and IPOPT by Wächter [17].

A limitation of gradient-based methods is that they cannot distinguish between local and global optima. The methods to obtain global optimal solutions of nonconvex NLP can be categorized as heuristic/stochastic and deterministic approaches. Stochastic global optimization methods include an element of randomness, allowing to escape local optima and thus to identify a global optimum. Tabu search [18], simulated annealing [19], and genetic algorithms [20] are popular heuristic/stochastic global optimization methods. For instance, in genetic algorithms, a population (a set of candidate solution points) is obtained in each step of iteration, and it is modified at the start of operation by selection, crossover, and mutation. Stochastic/heuristic algorithms have only probabilistic guarantees of finding an optimum.

Most deterministic global optimization methods include a convex relaxation of the original problem, that is, constructing an optimization problem with convex feasible set that contains the original feasible set and convex objective function that underestimates the objective function. This typically involves constructing convex relaxations of the functions involved, which can be done in a number of ways such as αBB relaxations [21] or McCormick [22] and auxiliary variable methods [23]. Mitsos et al. [24] demonstrated that the relaxation can also be performed when the function is not known analytically but rather calculated by an algorithm and this is particularly important for design problems, as aforementioned. One of the simplest techniques for deterministic global optimization is branch and bound (B&B). At each iteration two subproblems are solved, both using local solvers: the original optimization is solved locally to obtain an upper bound and the relaxation is solved to obtain a lower bound. The bounds converge by branching, typically by choosing a node and generating two children nodes via subdivision of the host set of one of the variables. The B&B procedure is terminated when the difference between upper and lower bounds is smaller than a tolerance. Deterministic methods are theoretically guaranteed to find a global optimum but are inherently expensive.

5.2.3
Mixed Integer Programming

Mixed integer optimization is a powerful technique for process design since both continuous and discrete variables can be optimized at the same time. Thus, optimal flowsheet connectivity, including selection of unit operations or number of stages, can be performed simultaneously with the optimization of continuous design and operating variables.

The basic form of a mixed-integer nonlinear program (MINLP) is

$$\begin{aligned} \min \quad & Z = f(x,y) \\ \text{s.t.} \quad & h(x) = 0 \\ & g(x,y) \leq 0 \\ & x \in X, y \in \{0,1\}^m \end{aligned} \quad (5.3)$$

where x and y are the continuous and discrete (often binary) variables, respectively. The inclusion of discrete variables in the constraints enables logical relations, for example, if a catalyst is used in the reactor then no catalyst poisoning chemical species are allowed in the reactor inlet. Often the objective and constraints are linear in the binary variables.

MINLPs are very challenging as they involve the combinatorial aspect of integer optimization and the potential nonconvex NLPs. One global solution method is to extend the B&B of NLPs and combine it with B&B of the integer variables. Another technique is outer approximation (OA) [25, 26] that combines the solution of an NLP and a mixed-integer linear programming (MILP) problem. The NLP subproblem is formed by fixing the integer variables and the optimization is performed over the continuous variables. The optimal objective value of this NLP is an upper bound on the MINLP optimal value. At the current iteration, the MILP master problem is formed by linearizing nonlinear functions on the optimal solution of the NLP subproblem. The optimal value of the MILP master problem is a lower bound of the MINLP. The OA procedure is terminated when the difference between upper and lower bound is smaller than a tolerance. A more general formulation than MINLP is the so-called generalized disjunctive programs (GDP) [27]. A comprehensive review of MINLP formulation and solution technique is provided by Trespalacios and Grossmann [28].

5.3 Synthesis

Process synthesis describes the creative task of determining a combination of unit operations capable of providing the desired products from available raw materials, subject to given constraints regarding safety or environmental concerns. In the presence of alternative configurations, the objective of process synthesis is to determine the most promising process variant, that is, the one that maximizes the profit or minimizes the cost of production. The development of process synthesis methods constitutes an active area of research for more than 50 years, with more than 2000 publications on the topic [4]. While modern simulation software allows the definition and simulation of complex flowsheets with a multitude of unit operations and recycle streams, design engineers still resort to heuristic rules involving problem decomposition. Examples for such decomposition are the means-ends analysis [29] or the hierarchical approach to conceptual design [30]. Consequently, most of the approaches to process synthesis deal with specific subproblems, for example, the design of the reactor networks, distillation trains, and heat exchanger networks [4]. The subproblems are addressed either by means of heuristics, mathematical programming, or a combination of both. Approaches based on mathematical programming (numerical optimization) vary strongly in the level of modeling detail of the single unit models. Thermodynamically motivated shortcut methods, or empirical surrogate models, present a valuable tool for the evaluation of a large number of process variants and complex flowsheets with multiple unit operations and recycles. However, in order to obtain accurate results the underlying assumptions for the applied shortcut models need to be taken into account and verified. More rigorous models increase the accuracy of performance estimates and of the economic assessment. However, such models require additional data and pose challenges to numerical algorithms. Daichendt and Grossmann [31] proposed an integration of hierarchical decomposition and mathematical programming for the synthesis of process flowsheets. Such a sequential solution strategy, starting with simple mass balance calculations and increasing modeling depth by including shortcut models and later on equilibrium-based or kinetic process models, has been further elaborated in, for example, the process synthesis framework of Marquardt and coworkers [32, 33] and is a logical consequence for an efficient and systematic conceptual process design. The following three subsections present an overview and an introduction to selected approaches to reactor network design, separation process design, and their complex integration in a simultaneous approach to overall flowsheet synthesis. Heat exchange networks are explicitly addressed in Section 5.4.1.

5.3.1 Reactor Networks

The design methods for reactor networks can be subdivided into four main areas: heuristics, attainable region concepts, elementary process function approaches, and superstructure optimization. Reactor selection heuristics are based on experience [34] and model evaluation [35, 36]. In many cases they are easy to apply and they result in an appropriate reactor network for common reaction classes. However, since they derive design decisions from known results only, heuristics cannot determine innovative reactor concepts, such that their application to complex systems is limited. In order to determine achievable product concentrations of a complete family of reactor networks, Horn [37] introduced the attainable region (AR) concept. This geometrical method relies on a given feed and known reaction kinetics. The AR is defined by the set of all product concentrations that can be obtained by any possible reactor network employing only reaction and mixing. Parallel, serial, or serial–parallel configurations of plug flow reactors (PFRs) [38], differential side stream reactors (DSRs) [39], or continuous stirred-tank reactors (CSTRs) [40] are considered for the construction of the AR. Computational techniques have been developed to apply AR concepts to multiple dimensions [41–43]. Once the AR has been constructed, a reactor network can be identified optimizing for some objective, such as maximal production rate.

A different idea to determine achievable product concentrations of reactor networks has been proposed by Denbigh [44] and has further been extended by several authors [45–47]. It uses the concept of dynamic optimization of an elementary process function. A fluid element is traced on its way along a (spatial or temporal) reaction coordinate. During its travel, the state of the fluid element defined by concentrations, temperature, and pressure is affected by the reaction kinetics and can additionally be manipulated by heat and mass fluxes. Consequently, an optimal state trajectory of the fluid element in thermodynamic state space is determined. An analysis of the optimal heat and mass fluxes provides insight into the choice of promising reactor networks and can help to develop innovative reactor concepts [48, 49]. However, the optimized state trajectory does not provide detailed information about equipment sizing.

Such information can be determined by optimization of a reactor network superstructure, for which the reactor network, the operating conditions, and the equipment sizes are simultaneously optimized. A superstructure for such a reactor network design is generally composed of a set of predefined reactor types, for example, isothermal [50] or nonisothermal CSTRs [51] and PFRs, or CSTRs and cross flow reactors [52]. In contrast to the AR, the possible reactor types and their interconnections have to be specified *a priori*. However, the existence of the specified reactors and their combinations are identified as the result of a superstructure optimization. Superstructure optimization is further described in Section 5.4. A particular challenge for superstructure-based design of reactor networks is that the performance characteristics of many reactor networks are very similar, for example, a cascade of several CSTRs may approximate a PFR quite accurately, which complicates finding a solution.

5.3.2
Separation Systems

Heuristic rules that compile the design knowledge from previous experience into practical guidelines have also been the basis for separation process design [29, 53]. On the basis of thermodynamic insight, potential separation principles and corresponding unit operations can be selected. Therefore, physical properties of the pure components and the mixture to be separated have to be available. Table 5.1 lists a number of separation techniques and the corresponding most relevant pure component properties. Differences in these properties indicate the suitability of a separation technique for the desired separation. A selection of potentially suitable separation techniques provides the basis for the generation of feasible flowsheet variants. Rule-based expert systems [54] and case-based reasoning [55] try to compile the thermodynamic analysis and the heuristic rules into software packages that support the design engineer in the process of flowsheet generation, which is a complex task. There are a tremendous number of potential alternatives, based on different separation principles and potential unit operations, which should be considered in order to determine the most promising flowsheet. While the utilization of simple shortcut models allows for an efficient evaluation of these flowsheet alternatives, an accurate prediction of the separation performance of the different unit operations becomes severely complex for nonideal mixtures.

Table 5.1 List of common separation techniques and relevant pure component properties, indicating their suitability for a desired separation.

Separation technique	Relevant pure component properties
Absorption/stripping	Solubility parameter
Distillation	Vapor pressure, heat of vaporization, boiling point
Crystallization	Melting point, heat of fusion at melting point
Liquid–liquid extraction	Solubility parameter
Molecular sieve adsorption	Kinetic diameter, van der Waals volume, dipole moment, polarizability
Azeotropic distillation	Vapor pressure, solubility parameter, azeotrope formation
Extractive distillation	Vapor pressure, heat of vaporization, boiling point, solubility parameter
Microfiltration/ultrafiltration	Size, molecular weight
Gas separation membranes	Critical temperature, van der Waals volume
Pervaporation/vapor permeation	Molar volume, solubility parameter, dipole moment

Source: Adapted from Ref. [56]. Copyright 1995, Elsevier.

In the context of bulk chemical or fuel production processes, typically distillation is the default unit operation. The potential application is examined based on boiling temperatures, the determination of azeotropes, and miscibility gaps.

Missing physical property data may be estimated using (semi)empirical correlations like group contribution methods [57] and quantitative structure–property relationships (QSPR) or quantum chemical methods like COSMO-RS [58]. If possible, relevant property data should always be validated by means of experimental data. On the basis of equilibrium calculations, feasible distillation sequences can be identified by residue curve maps in combination with balance lines in case of ternary mixtures [59]. Significant progress has been made toward an automatic generation of separation sequences for zeotropic mixtures, for which all basic and thermally coupled configurations can be derived [60]. A transfer to azeotropic mixtures is however complicated due to the inherent distillation boundaries that limit the attainable degree of separation. Figure 5.1a depicts the three potential separation sequences for the separation of a zeotropic ternary mixture, while Figure 5.1b depicts one feasible distillation sequence for an azeotropic mixture with one distillation boundary, which operates without changing the pressure or introducing an additional auxiliary compound. The necessary recycle stream does not only complicate the synthesis of such sequences, but also complicates design calculations.

Only few publications address the generation of process variants for azeotropic mixtures. By introducing (piecewise) linear approximations of the distillation boundaries, Ryll *et al.* [61] determine feasible distillation sequences and minimum recycle flow rates based on an optimization of the mass balance lines. However, such a method does not provide any information on column sizes or the required energy demand. By treating azeotropes as pseudocomponents, classical shortcut methods like the Fenske–Underwood–Gilliland method [62] can be applied to azeotropic mixtures [63]. By integrating this approach into a superstructure it is in principle possible to determine the optimal distillation sequence

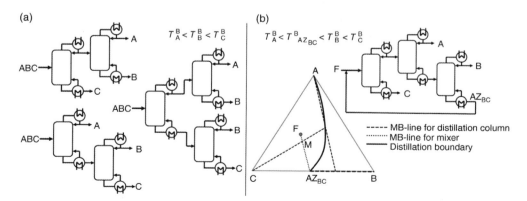

Figure 5.1 The three distillation sequences for the separation of a ternary zeotropic mixture (a) and a possible distillation sequence with recycle for a hypothetical azeotropic mixture (b).

for the separation of azeotropic mixtures by means of an optimization problem [64]. However, in order to accurately describe the nonideality of azeotropic mixtures, more sophisticated distillation column models need to be utilized. These can either be shortcut methods, which rely on simplifying assumptions like an infinite number of equilibrium trays in order to determine the minimum energy demand for a separation, or more rigorous superstructure models based on the *m*ass, *e*quilibrium, *s*ummation, and ent*h*alpy equations (MESH model). Skiborowski *et al.* [65] present an overview of the different methods and their extensions for extractive and heteroazeotropic distillation processes. Both extensions require the selection of an auxiliary compound (the so-called entrainer). A suitable entrainer can be selected from a predefined database or rigorously synthesized from molecular building blocks by means of a computer-aided molecular design (CAMD) method [66]. Solvents with desirable properties, such as distribution coefficients or selectivities, are designed by a generate-and-test procedure [67] or by formulating and solving a MINLP [68]. However, the power of such CAMD approaches is limited by the accuracy of property predictions, necessitating an experimental validation.

Alternative unit operations, as listed in Table 5.1, have to be considered, if a separation by means of distillation alone is not possible or inefficient. In case that a separation is difficult or even impossible using a single type of unit operation, different separation principles can be combined to a hybrid separation process. A hybrid separation process is defined as the combination of at least two different unit operations, contributing to the same separation task [69]. While such processes offer significant potential for improvement, their consideration also extends the design space dramatically and thus makes design more challenging. Skiborowski *et al.* [70] present a review of available computational methods for the conceptual design of distillation-based hybrid processes for the separation of liquid mixtures, that is, hybrid processes that involve distillation. They show that although a lot of progress has been made in the area of conceptual process design, several open problems still prevail, especially concerning variant generation and rate-based modeling, which have to be addressed in order to establish a structured process synthesis framework for such processes. Especially for bio-based production processes, downstream processing can account for a major share of the production cost [71]. There is still a need for efficient and accurate conceptual design methods for separation techniques such as flotation, foam separation, electrokinetic separation processes, and different types of chromatography, which are considered as promising options for the separation of such sensitive and complex product mixtures [71].

5.3.3
Overall Flowsheets

For the design of an overall flowsheet consisting of reaction and separation processes, design methods for reactor networks and separation processes should be used in combination, in order to account for the integration in case of recycles.

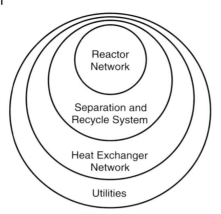

Figure 5.2 The onion model for process design. (Adapted from Ref. [72]. Copyright 2008, Elsevier.)

Common heuristics for overall flowsheets, therefore, decompose the design task into such subproblems [30, 72]. According to the hierarchical approach by Douglas [30], at first a suitable reactor network is determined, while the separation and recycle system as well as the heat exchanger network and utilities are considered as dependent subproblems, which are assessed in subsequent steps. The same approach has been proposed in the onion model [72], which is illustrated in Figure 5.2. Feasibility and economic performance of such heuristically designed flowsheets are usually determined by repetitive simulation studies, whereas first estimates of operating points and unit specifications are often determined by spreadsheet calculations or rules of thumbs [73].

Graphical approaches, for example, GH-plots [74], can help to facilitate the design of efficient process variants. Enthalpy (H) and Gibbs free energy (G) are represented as vectors in the GH-space, depicting the flows of heat and work for all subprocesses. Thus, the design engineer can interpret processes and identify major losses and potential modifications to improve the efficiency. Achievable process efficiencies can also be determined by extending the concept of elementary process functions for reactor networks (Section 5.1) to reaction and separation processes [75] or by means of a steady-state design using only sharp splits and ideal CSTRs [76]. While the determination of such process performance bounds provides an understanding of potential improvements, it does not directly enable the identification of optimal process structures. Instead, it can help identify process bottlenecks.

Process flowsheets can be synthesized based on an analysis and optimization of operating windows, defined as a set of intervals describing the feasible ranges of properties in the material streams linking the process units [77]. The operating windows are gradually narrowed until all process units have been fixed and the flowsheet can be evaluated. However, without accurate performance models for the single unit operations, even the feasibility of the process flowsheet is not guaranteed. Consequently, rigorous models for reaction and separation should be introduced by combining the methods for reactor network design and separation process synthesis for the synthesis of potentially feasible process variants. While the incorporation of such complex models into a superstructure optimization is challenging, it is in principle possible and is further addressed in Section 5.4.2.

In order to narrow down the large number of possible variants and to reduce the effort for rigorous process design, shortcut methods can be used for a fast screening of the process performance. A combination of reactor network and separation process shortcut methods allows for a ranking of possible flowsheet alternatives [78–80], or even an optimization of flowsheet superstructures that embed several process alternatives [81]. However, most of these methods do not consider recycle streams and rely on one or several simplifying assumptions, like constant molar overflow and constant relative volatilities for distillation or equilibrium reactors. In order to determine the best performance of a flowsheet for complex nonideal reactive mixtures, it is important to consider potential recycle streams, rigorous thermodynamic models, as well as reaction and potentially mass transfer kinetics. For such processes, a simultaneous optimization should be performed, because changing operating conditions in any unit will affect the performance of the entire process [33].

The most promising variants from the shortcut screening can finally be selected in order to determine an optimal equipment design, based on the optimization of more rigorous process models. The combination of a heuristic/stochastic optimization method (Section 5.2.2) and a process simulator constitutes one potential and convenient approach, which is often proposed in literature [82, 83]. However, application is still limited to a small number of degrees of freedom, which is already exceeded for small flowsheets [84]. Surrogate models (also known as reduced order models) can be trained by rigorous simulation for later use in optimization [85, 86]. However, a compromise between simplicity and highly accurate surrogate models has to be found, since complex functional forms of the surrogate model impede an efficient optimization [87]. A systematic approach with a stepwise model refinement and initialization strategy, for example, proposed in the process synthesis framework by Recker *et al.* [33], allows addressing the complex optimization problem with rigorous process models in an organized and efficient way.

Besides the combination of reactor network and separation process design, also the use of reactive separations and other forms of intensified processes should be considered in the conceptual design. These advanced concepts provide the potential for energy, mass, and material integration. For instance, reactive distillation (RD) integrates a reactor and a distillation column and can increase conversion and selectivity (even beyond equilibrium limits) and may reduce the required heat duty. For a thorough overview of the fundamentals, modeling approaches, and potential benefits and limitations of RD refer to the article of Keller [88]. Reactive separations like RD can lead to a significant extension of the AR [89, 90], while at the same time result in a significant material integration,

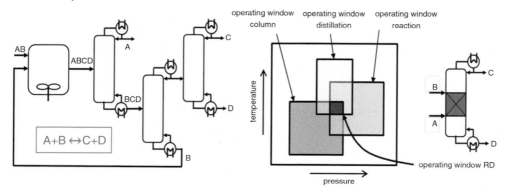

Figure 5.3 Illustration of the potential for material integration by means of reactive distillation and the relationship between the different operating windows. (Adapted from Ref. [91]. Copyright 2003, Elsevier.)

as illustrated in Figure 5.3. However, reactive separations are only feasible if the operating windows of the reaction, the separation, and the materials overlap. Consequently, the feasible region for the degrees of freedom for the design of a reactive separation is substantially constrained in comparison with a sequential arrangement, complicating synthesis and design [91].

There are only a few approaches dealing with the conceptual design of intensified processes. Schembecker and Tlatlik [91] propose a process synthesis procedure for integrated reaction–separation units based on a general analysis of physical properties and the reaction, as well as operating windows of the reaction and separation. Based on a similar analysis of physical properties, Holtbruegge *et al.* [92] introduced an automated tool for the synthesis of (reaction) separation processes, which incorporates several hybrid separation and integrated reaction separation techniques. However, both approaches consider only predefined unit operations. In order to be truly innovative and synthesize novel process configurations or even unit operations, process synthesis has to be performed on a microscopic level, similar to the elementary process functions for reactor–network design. The phenomena-based synthesis method of Lutze *et al.* [93, 94] can be used to derive innovative process units and processes. Phenomena are classified into mixing, phase contact, phase transition, phase change, phase separation, reaction, energy transfer, and stream division and connected to phenomena building blocks, which are then combined to obtain the desired conversion and separation in the most beneficial way. Finally, unit operations have to be identified, which match the combination of phenomena building blocks. While this approach is most general, the resulting problem is of maximum complexity and has so far only been solved in a manual and sequential procedure with several simplifying assumptions concerning modeling. Further developments may facilitate a more general application of such innovative design methods.

5.4
Superstructure-Based Techniques

The methodologies based on superstructure optimization aim to mitigate the problems that may be encountered during hierarchical process design (Section 5.3) and automate the process. A superstructure contains a multitude of potential structural options of the process. The optimal design is decided through mathematical modeling and optimization. The optimal structure, operating conditions, and equipment size are optimized at the same time. However, a number of difficulties arise from the implementation of superstructures. Only those structural designs are considered that are included in the superstructure. However, increasing the number of options included typically results in much more expensive computations. Also, if the unit operations are described by comprehensive first-principle models, the resulting optimization problem is extremely challenging and current optimization algorithms may fail to furnish a global optimum in tractable computational times. A particular challenge is nonconvexity that typically arises in process design formulations and thus local algorithms cannot guarantee global optima or may even fail in finding good feasible points. This can be sometimes overcome by either simplifying the model to convex functions or by using heuristic/deterministic global optimization methods (Section 5.2.2) to find the global optimum.

In this section, superstructure optimization approaches for the synthesis of heat exchanger networks and process flowsheet will be introduced.

5.4.1
Heat Exchange Networks

Heat exchange network (HEN) synthesis is an important subject of process design because optimized HENs may reduce the total cost substantially. While HEN synthesis is challenging, it is much more tractable than overall design because of the comparably easier modeling and the simpler optimization formulation. During the last half century, a number of studies have been published on superstructure-based techniques for HEN synthesis. They are reviewed in the following section.

One of the earliest applications of superstructures to HEN synthesis is the derivation of network configurations during the sequential HEN design. Similar to the sequential design approaches for overall flowsheets (Section 5.3), the sequential design approach for HENs divides the original problem into a series of subproblems, namely, minimum utility target, minimum number of heat exchanger units, and minimum network cost. The utility target and the number of heat exchanger units can be determined by LP and MILP transshipment models, respectively. The superstructure is, then, built with the utility target and number of units and it contains all possible interconnections between hot and cold streams. The stream interconnections are treated as decision variables and determined by solving a NLP. Investment cost is typically used as the objective function. This can be represented as $\min \sum CA^{\alpha}$ where C and α are cost parameters, and A is the heat exchange area. The NLP model can be solved using the methods introduced in Section 5.2.2. The LP and MILP formulation to obtain minimum utility target and number of heat exchangers can be found in Refs [95, 96]. The NLP formulation using superstructures and its extension are provided by Floudas et al. [97]. A comprehensive review for such sequential approaches can be found in Ref. [1].

While sequential approaches of HEN design provide convenient methods to design and optimize HENs, the interaction among the energy targeting, number of units, and heat exchanger area are not rigorously taken into account due to the decomposition. Simultaneous superstructure methodologies have been proposed to optimize heat exchange network without decomposing the problem [98, 99].

The objective function of the optimization problem of these models can be expressed as following:

$$\min \sum \text{utility cost} + \sum \text{fixed charge for exchangers} + \sum \text{area cost}$$
(5.4)

In Eq. (5.4), the utility cost, heat exchanger area, and configuration of the HEN can be simultaneously optimized. Incipient superstructure models proposed by Floudas and Ciric [98] and Yuan et al. [99], however, have drawbacks of not allowing stream splitting and mixing or requiring predetermined temperature intervals.

Yee and Grossmann [100] proposed a stagewise superstructure model that can account for any heat exchange between hot and cold streams. The number of stages in this model can be decided as maximum number of hot or cold streams and it is assumed that each stream has the same temperature at the end of the stage. With these assumptions, the superstructure model can be simplified significantly and all constraints can be expressed with linear relationships. In this model, the equality constraints include energy balances of the complete HEN at each stage, inlet temperature assignment, and utility requirements. The temperature specifications at each stage are expressed as inequality constraints. Similar to the MILP transshipment model, the logical constraints determine stream matches and the minimum approach temperature. In this MINLP model, only the objective function (Eq. (5.4)) contains nonlinear terms and constraints defining feasible space are expressed with linear relationship.

The stagewise superstructure model is easily expanded and applied to several cases involving flexible multiperiod operation [101, 102], detailed heat exchanger design models [103, 104], and isothermal streams with phase change [105]. The flexible HEN model defines the region of feasible operation rather than fixed hot and cold stream temperatures, and a set of optimal conditions provides the flexibility to handle a range of operating conditions. HENs including detailed heat exchanger model consider the contribution of pressure drop along units. By removing the assumption of constant pressure and heat transfer coefficient, the resulting heat exchanger network reflects industrial reality more closely, although the MINLP model is more difficult to solve.

HENs also play an important role during optimization-based process design. Duran and Grossmann [106] addressed a simultaneous optimization method handling heat integration while performing optimization of process flowsheets. In this model, the amount of available heat of the process is identified by using a set of nonlinear inequality constraints. The flowrate and the inlet and outlet temperature of process streams are no longer fixed values but rather variables throughout the optimization, and they may introduce numerical complexity to obtain an optimal solution. Grossmann et al. [107] presented a disjunctive model for the simultaneous heat integration.

5.4.2
Process Flowsheet Optimization

As mentioned in Section 5.3, the design of overall flowsheets consisting of reaction and separation systems can also be performed by superstructure-based techniques. In superstructure-based process design, the aggregation of different flowsheet alternatives to one superstructure is one of the most important steps because of the complexity of the problems. Yeomans and Grossmann [108] propose that superstructures in optimization-based process design can be formulated based on *states, tasks,* and *equipment.* States represent physical and chemical properties of streams. Thus, composition, temperature, pressure, and flowrate can be considered as states. Tasks indicate physical and chemical transformation between the states such as temperature/pressure change or mass transfer. Equipment corresponds to physical devices that execute a given task. In superstructure-based process design, two types of superstructure representations have been introduced: STN and SEN. STN stands for state task network design and was first introduced by Kondili et al. [109]. In this representation, states and tasks are identified first, and equipment is assigned to given tasks. It is also possible to specify states and equipment first and assign different tasks to equipment. This type of superstructure is called state equipment network (SEN). Depending on the problem at hand, STN and SEN formulations differ in size and complexity. Yeomanns and Grossmann [110] illustrate this difference for the separation of a quaternary zeotropic mixture by means of a distillation train. All separation trains for sharp splits can be presented either as a STN with 10 tasks or as a SEN with 3 columns (equipment). However, each task may be evaluated *a priori*, such that the complexity of the STN model is reduced significantly [111].

With the developed superstructure model, the superstructure-based process design problem can be formulated as a MINLP. The objective function is the total cost associated with selecting units and continuous terms according to size/flowrates. The equality constraints represent the modeling equations of the different units, as well as the stream interconnections between units, and the inequality constraints correspond to operation limits, design specifications, physical limits, and logical constraints. In some cases, the optimization-based process design problem can be expressed as a MILP. It is also possible to formulate an equivalent continuous problem (NLP) in some cases, as for instance done by Dahdah and Mitsos [112] for the design of seawater desalination processes.

As described in Section 5.2.3, the solution of MINLPs is very challenging. A particular difficulty is that the model equations ensure that flowrates to non-existing units are zero; however, the corresponding submodels still need to be solvable. In order to reduce the complexity of the problem and handle nonconvexities introduced by binary variables, decomposition methods are widely used. In these methods, splitter and mixer represent interconnection nodes and all other process units in the flowsheet are considered as process unit nodes. The constraints in the MINLP model are partitioned into two sets (i.e., interconnection nodes and process units), and the process units within interconnection nodes can be handled explicitly depending on the value of binary variables. Kocis and Grossmann [113] introduced a modeling and decomposition algorithm, which decomposes a superstructure into an initial flowsheet and subsystems of nonexisting units. Accordingly, NLP subproblems can be solved only for an existing flowsheet.

Several other decomposition methods of superstructures have been proposed to determine the subsystem systematically [31, 114]. This technique combines hierarchical decomposition with mathematical programming. Each level of the decomposition (i.e., input–output level, reaction level, separation level, heat integration level) is formulated as MINLP and optimized. In this way, large MINLP problems that cannot be solved at once can be optimized with a combination of simple and detailed models.

It is also worth to note several examples of process flowsheet optimization that utilize superstructures. For example, several distillation processes have been modeled and optimized using STN superstructures [115, 116], or heat integration processes have been optimized using SEN models [117]. Recently, an optimization framework for superstructures based on surrogate models [118] for process synthesis was introduced. In this model, complex unit models are replaced with surrogate models built from data generated from simulation (Section 5.3). Gassner and Marechal proposed superstructure models that used empirical relations and found an optimal process flowsheet for a synthetic natural gas production process [119]. GDP formulations have also been successfully employed to process flowsheet synthesis. Raman and Grossmann [120] first proposed GDP models for process flowsheet synthesis. Turkay and Grossmann extended the application of GDPs on process flowsheet synthesis with discontinuous cost of equipment [121, 122].

Superstructure-based formulations are one of the major methodologies in the area of conceptual process design. There has been extensive development on heat exchanger networks, distillation systems, reactor networks, and process flowsheet synthesis. These models can contribute to finding an optimal flowsheet structure as well as developing automated design tools. There are several drawbacks such as simplification or expensive computation cost. However, the process design with superstructures can be a promising option to develop optimal processes with advance of optimization algorithm.

5.5
Integrated Process Design, Operation, and Control

While significant progress in model-based process design has been achieved over the last years, the integration of process design, operation, and control is still largely an open challenge as no generally accepted integrated methodology exists [123]. There is however a need to optimize the long-term decisions, midterm decisions, and short-term decisions of a plant as well as their interactions in an integrated approach as these affect design. Long-term decisions are decisions regarding the plant design, since they are the less likely to change. Model-based approaches to support the decision-making on the process design level have been reviewed in the previous sections. Midterm decisions are decisions regarding the operational schedule. They are especially important when uncertainties originating from raw material shortages, fluctuation in pricing, demand changes, or equipment failure have to be taken into account. Most commonly, the operational scheduling and process design problems are decomposed and the operational scheduling is optimized based on a fixed process design. However, there exist different approaches for integrating the decisions on both levels, for example, by formulation as two-stage problem [124] or as MILP [125]. The control strategies constitute the short-term decisions of a process. An integrated approach to process design and control has already been introduced by Lee *et al.* [126]. For an overview on further progress in this area, we refer to Yuan *et al.* [127].

5.6
Water and Energy Processes

Water and energy systems are also important topics of process systems engineering. While the processes are quite different from traditional chemical processes, the methodologies introduced in earlier sections are applicable and have been successfully applied. In this section, few examples are introduced in order to demonstrate the contribution of process design and optimization techniques on water and energy processes.

Water systems can be divided into desalination and water processing networks. Desalination can be accomplished by using thermal or membrane methods. Several technologies are available in thermal desalination such as multistage flash (MSF), multieffect distillation (MED), and thermal vapor compression (TVC). Membrane desalination processes have been rapidly developed over the past 40 years, and are now dominating new plant installation because of their low-energy consumption, higher recovery factor, and lower water cost. An overview of available thermal and membrane desalination technologies and their advantages and disadvantages are provided in Ref. [128].

Modeling and simulation of thermal desalination process is relatively established [112, 129, 130]. However, model-based design methodologies face the problem that the models are nonlinear and nonconvex. Shortcut methods have been developed to simplify the modeling of the thermal desalination process and these techniques are summarized in Ref. [131]. Superstructure optimization also plays an important role in design and optimization of the thermal process. For instance, Dahdah and Mitsos developed a superstructure of MED-MSF [112] and MED–MSF-TVC [132] configurations and structurally optimized combined thermal desalination systems.

In the membrane desalination process modeling, mechanical and process constraints related to membrane characteristics should also be considered. Here, the focus is on reverse osmosis (RO). One of the key design decisions is the number of stages of the membrane desalination system and there are many configurations, often with trade-off between configuration complexity, membrane quality, product quality, and energy requirement. Integration between thermal and membrane desalination techniques [133, 134] and power and water processes [135] have also been actively studied using process systems engineering techniques. The integrated processes may be thermodynamically more efficient and/or economically more attractive, but process design and optimization is much more complicated than conventional desalination processes. Current technologies and available design options are reviewed in Ref. [136].

A comprehensive review of water network design methods can be found in Ref. [137]. The superstructure-based techniques (Section 5.4) can solve the relatively simple water distribution problems (e.g., water using network [28]), but more complex water treatment networks pose substantial challenges as the corresponding MINLPs are large and nonconvex. Grossmann reviewed recent techniques [28].

Process design and optimization techniques can also contribute to energy systems. Organic Rankine cycles (ORC) are a popular method extracting thermal energy from various low-temperature sources such as geothermal, biomass, and solar. Superstructure optimization has been successfully applied to identify optimum structures and operating conditions of ORCs. Gerber and Marechal use a superstructure to propose a systematic methodology for optimal design of a geothermal system [138]. Hipólito-Valencia et al. [139] proposed a stagewise superstructure technique to integrate ORCs with various industrial processes. Mathematical modeling and optimization techniques also contribute to finding working fluid [140] and operating conditions of ORCs [141].

Power generation systems based on fuel cell technology are another interesting field that process design technique has been successfully applied. Mitsos et al. [142] used a superstructure-based formulation to develop optimum micro-power generation configurations and optimized promising process concepts accounting for variable power demand levels [143]. A brief review for design and optimization of energy systems is available in Ref. [144].

5.7 Conclusions and Outlook

Depleting natural resources, environmental concerns, and the desire for sustainability foster the chemical industry to pursue more efficient processes. As reviewed in this chapter, various approaches to support the design process have been developed in the last decades. Such methods have also been utilized for designing processes with biorenewable feedstocks [145, 146]. However, the optimal (e.g., the most sustainable) process for any particular application will depend on many factors and incorporate decisions on optimal reactor networks, separation and recycle systems, heat exchanger networks, and utilities (Section 5.3 and Figure 5.2). Especially, processes that use catalysts involve a particularly large number of additional degrees of freedom. An approach outlined by Hintermair *et al.* [147] breaks the degrees of freedom for catalytic processes into three scales: molecular scale, meso scale, and macro scale. Despite the large progress already made in model-based process design in the last decades, the subproblems on these different scales still have to be decomposed and special attention has to be laid on the interconnection between the scales to identify an optimal process [148]. The ultimate goal of process design is, for sure, the simultaneous optimization of all degrees of freedom in an integrated approach, but this is still an open challenge.

References

1 Biegler, L.T., Grossmann, I.E., and Westerberg, A.W. (1997) *Systematic Methods of Chemical Process Design*, Prentice Hall, Old Tappan.
2 Li, X. and Kraslawski, A. (2004) Conceptual process synthesis: past and current trends. *Chemical Engineering and Processing*, **43**, 589–600.
3 Westerberg, A.W. (2004) A retrospective on design and process synthesis. *Computers & Chemical Engineering*, **28**, 447–458.
4 Cremaschi, S. (2015) A perspective on process synthesis: challenges and prospects. *Computers & Chemical Engineering*, **81**, 130–137.
5 Brent, R.P. (1973) Some efficient algorithms for solving systems of nonlinear equations. *SIAM Journal on Numerical Analysis*, **10** (2), 327–344.
6 Marquardt, D.W. (1963) An algorithm for least-squares estimation of nonlinear parameters. *Journal of the Society for Industrial & Applied Mathematics*, **11** (2), 431–441.
7 Wolfe, P. (1959) The secant method for simultaneous nonlinear equations. *Communications of the ACM*, **2** (12), 12–13.
8 Deuflhard, P. (2011) *Newton Methods for Nonlinear Problems: Affine Invariance and Adaptive Algorithms*, Springer Science & Business Media, Berlin.
9 Bertsimas, D. and Tsitsiklis, J.N. (1997) *Introduction to Linear Optimization*, Athena Scientific, Belmont.
10 Palacios-Gomez, F. and Lasdon, L.E.M. (1982) Nonlinear optimization by successive linear programming. *Managment Science*, **28** (10), 1106–1120.
11 Han, S.P. (1976) Superlinearly convergent variable metric algorithms for general nonlinear programming problems. *Mathematical Programming*, **28** (10), 263–282.

12 Powell, M.J.D. (1978) A fast algorithm for nonlinearly constrained optimization calculations, in *Numerical Analysis*, Springer, Berlin, pp. 144–157.

13 Murtagh, B.A. and Saunders, M.A. (1978) Large-scale linearly constrained optimization. *Mathematical Programming*, **14** (1), 41–72.

14 Vasantharajan, S., Viswanathan, J., and Biegler, L.T. (1990) Reduced successive quadratic programming implementation for large-scale optimization problems with smaller degrees of freedom. *Computers & Chemical Engineering*, **14** (8), 907–915.

15 Drud, A. (1985) CONOPT: A GRG code for large sparse dynamic nonlinear optimization problems. *Mathematical Programming*, **31** (2), 153–191.

16 Gill, P.E., Murray, W., and Saunders, M.A. (2002) SNOPT: An SQP algorithm for large-scale constrained problems. *SIAM Journal on Optimization*, **12** (4), 979–1006.

17 Wächter, A. (2002) An interior point algorithm for large-scale nonlinear optimization with applications in process engineering. Ph.D. thesis, Carnegie Mellon University, Pittsburgh.

18 Glover, F. and Laguna, M. (1997) *Tabu Search*, Springer, New York.

19 Floquet, P., Pibouleau, L., and Domenech, S. (1994) Separation sequence synthesis: how to use simulated annealing procedure? *Computers & Chemical Engineering*, **18** (11), 1141–1148.

20 Reeves, C.R. (1995) A genetic algorithm for flowshop sequencing. *Computers & Operations Research*, **22** (1), 5–13.

21 Adjiman, C.S., Androulakis, I.P., Maranas, C.D., and Floudas, C.A. (1996) A global optimization method, αBB, for process design. *Computers & Chemical Engineering*, **20**, 419–424.

22 McCormick, G.P. (1976) Computability of global solutions to factorable nonconvex programs: part I – convex underestimating problems. *Mathematical Programming*, **10** (1), 147–175.

23 Smith, E.M.B. and Pantelides, C.C. (1999) A symbolic reformulation/spatial branch-and-bound algorithm for the global optimisation of nonconvex MINLPs. *Computers & Chemical Engineering*, **23** (4), 457–478.

24 Mitsos, A., Chachuat, B., and Barton, P. (2009) McCormick-based relaxations of algorithms. *SIAM Journal on Optimization*, **20** (2), 573–601.

25 Duran, M.A. and Grossmann, I.E. (1986) An outer-approximation algorithm for a class of mixed-integer nonlinear programs. *Mathematical Programming*, **36** (3), 307–339.

26 Kesavan, P., Allgor, R.J., Gatzke, E.P., and Barton, P.I. (2004) Outer approximation algorithms for separable nonconvex mixed-integer nonlinear programs. *Mathematical Programming*, **100** (3), 517–535.

27 Raman, R. and Grossmann, I.E. (1994) Modelling and computational techniques for logic based integer programming. *Computers & Chemical Engineering*, **18** (7), 563–578.

28 Trespalacios, F. and Grossmann, I.E. (2014) Review of mixed-integer nonlinear and generalized disjunctive programming methods. *Chemie Ingenieur Technik*, **86** (7), 991–1012.

29 Siirola, J.J. and Rudd, D.F. (1971) Computer-aided synthesis of chemical process designs. From reaction path data to the process task network. *Industrial & Engineering Chemistry Fundamentals*, **10** (3), 353–362.

30 Douglas, J.M. (1985) A hierachical decision procedure for process synthesis. *AIChE Journal*, **31**, 353–362.

31 Daichendt, M.M. and Grossmann, I.E. (1998) Integration of hierarchical decomposition and mathematical programming for the synthesis of process flowsheets. *Computers & Chemical Engineering*, **22** (1), 147–175.

32 Marquardt, W., Kossack, S., and Kraemer, K. (2008) A framework for the systematic design of hybrid separation processes. *Chinese Journal of Chemical Engineering*, **16** (3), 333–342.

33 Recker, S., Skiborowski, M., Redepenning, C., and Marquardt, W. (2015) A unifying framework for optimization-based design of integrated reaction-separation processes.

Computers & Chemical Engineering, **81**, 260–271.

34 Smith, R. (1995) *Chemical Process Design*, McGraw-Hill, New York.

35 Schembecker, G., Dröge, T., Westhaus, U., and Simmrock, K.H. (1995) Readpert – development, selection and design of chemical reactors. *Chemical Engineering and Processing: Process Intensification*, **34**, 317–322.

36 Till, J., Sand, G., Engell, S., Schembecker, G., and von Trotha, T. (2004) READOPT: reaktor-design-optimierung durch heurisitkgestützte MINLP-methoden. *Chemie Ingenieur Technik*, **76** (8), 1105–1110.

37 Horn, F. (1964) Attainable and non-attainable regions in chemical reaction techniques. *Chemical Reaction Engineering. Proceedings of the Third European Symposium*, Pergamon Press, Oxford, pp. 293–303.

38 Feinberg, M. and Hildebrandt, D. (1997) Optimal reactor design from a geometrical viewpoint - I. Universal properties of the attainable region. *Chemical Engineering Science*, **52**, 1637–1665.

39 Feinberg, M. (2000) Optimal reactor design from a geometric viewpoint. Part II. Critical sidestream reactors. *Chemical Engineering Science*, **55**, 2455–2479.

40 Feinberg, M. (2000) Optimal reactor design from a geometrical viewpoint - III. Critical CFSTRs. *Chemical Engineering Science*, **55**, 3553–3565.

41 Rooney, W., Hausberger, B., Biegler, L.T., and Glasser, D. (2000) Convex attainable region projections for reactor network synthesis. *Computers & Chemical Engineering*, **24**, 225–229.

42 Abraham, T. and Feinberg, M. (2004) Kinetic bounds on attainability in the reactor synthesis problem. *Industrial & Engineering Chemistry Research*, **43**, 449–457.

43 Ming, D., Hildebrandt, D., and Glasser, D. (2010) A revised method of attainable region construction utilizing rotated boundary hyperplanes. *Industrial & Engineering Chemistry Research*, **49**, 1803–1810.

44 Denbigh, K.G. (1944) Velocity and yield in continuous reaction systems. *Transactions of the Faraday Society*, **40**, 352–373.

45 Bilous, O. and Amundson, N.R. (1956) Optimum temperature gradients in tubular reactor: I. General theory and methods. *Chemical Engineering Science*, **5**, 81–92.

46 Aris, R. (1961) *The Optimal Design of Chemical Reactors*, Elsevier, Burlington.

47 Horn, F. (1961) C1. Optimale Temperatur- und Konzentrationsverläufe. *Chemical Engineering Science*, **14**, 77–88.

48 Peschel, A., Freund, H., and Sundmacher, K. (2010) Methodology for the design of optimal chemical reactors based on the concept of elementary process functions. *Industrial & Engineering Chemistry Research*, **49**, 10535–10548.

49 Hentschel, B., Kiedorf, G., Gerlach, M., Hamel, C., Seidel-Morgenstern, A., Freund, H., and Sundmacher, K. (2015) Model-based identification and experimental validation of the optimal reaction route for the hydroformulation of 1-dodecene. *Industrial & Engineering Chemistry Research*, **54** (6), 1755–1765.

50 Kokossis, A.C. and Floudas, C.A. (1990) Optimization of complex reactor networks - I. Isothermal operation. *Chemical Engineering Science*, **45**, 595–614.

51 Kokossis, A.C. and Floudas, C.A. (1994) Optimization of complex reactor networks - II. Nonisothermal operation. *Chemical Engineering Science*, **49**, 1037–1051.

52 Schweiger, C. and Floudas, C. (1999) Synthesis of optimal chemical reactor networks. *Computers & Chemical Engineering*, **23**, 47–50.

53 Barnicki, S.D. and Fair, J.R. (1990) Separation system synthesis: a knowledge-based approach. 1. Liquid mixture separations. *Industrial & Engineering Chemistry Research*, **29**, 421–432.

54 Kirkwood, R.L., Locke, M.H., and Douglas, J.M. (1988) A prototype expert system for synthesizing chemical process

flowsheets. *Computers & Chemical Engineering*, **12**, 329–343.

55 Seuranen, T., Hurme, M., and Pajula, E. (2005) Synthesis of separation processes by case-based reasoning. *Computers & Chemical Engineering*, **29**, 1473–1482.

56 Jaksland, C.A., Gani, R., and Lien, K.M. (1995) Separation process design and synthesis based on thermodynamic insights. *Chemical Engineering Science*, **50**, 511–530.

57 Joback, K. and Reid, R. (1987) Estimation of pure-component properties from group-contributions. *Chemical Engineering Communications*, **57**, 233–243.

58 Klamt, A. and Eckert, F. (2000) COSMO-RS: a novel and efficient method for the *a priori* prediction of thermophysical data of liquids. *Fluid Phase Equilibria*, **172**, 43–72.

59 Fien, G.J.A.F. and Liu, Y.A. (1994) Heuristic synthesis and shortcut design of separation processes using residue curve maps: a review. *Industrial & Engineering Chemistry Research*, **33**, 2502–2522.

60 Shah, V.H. and Agrawal, R. (2010) A matrix method for multicomponent distillation sequences. *AIChE Journal*, **56**, 1759–1775.

61 Ryll, O., Blagov, S., and Hasse, H. (2012) ∞/∞-Analysis of homogeneous distillation processes. *Chemical Engineering Science*, **84**, 315–332.

62 Doherty, M.F. and Malone, M.F. (2001) *Conceptual Design of Distillation Systems*, McGraw-Hill, Boston.

63 Liu, G.L., Jobson, M., Smith, R., and Wahnschafft, O.M. (2004) Shortcut design method for columns separating azeotropic mixtures. *Industrial & Engineering Chemistry Research*, **43**, 3908–3923.

64 Yang, X., Dong, H.G., and Grossmann, I.E. (2012) A framework for synthesizing the optimal separation process of azeotropic mixtures. *AIChE Journal*, **58** (5), 1487–1502.

65 Skiborowski, M., Harwardt, A., and Marquardt, M. (2014) Conceptual design of azeotropic distillation processes.

Distillation, Elsevier, Amsterdam, pp. 305–355.

66 Gani, R. and Brignole, E.A. (1983) Molecular design of solvents for liquid extraction based on UNIFAC. *Fluid Phase Equilibria*, **13**, 331–340.

67 Harper, P.M. and Gani, R. (2000) A multi-step and multi-level approach for computer aided molecular design. *Computers & Chemical Engineering*, **24**, 677–683.

68 Karunanithi, A.T., Achenie, L.E.K., and Gani, R. (2005) A new decomposition-based computer-aided molecular/mixture design methodology for the design of optimal solvents and solvent mixtures. *Industrial & Engineering Chemistry Research*, **44**, 4785–4797.

69 Franke, M., Gorak, A., and Strube, J. (2004) Design and optimization of hybrid separation processes. *Chemie Ingenieur Technik*, **76**, 199–210.

70 Skiborowski, M., Harwardt, A., and Marquardt, M. (2013) Conceptual design of distillation-based hybrid separation processes. *Annual Review of Chemical and Biomolecular Engineering*, **4** (1), 45–68.

71 Chmiel, H. (ed.) (2011) *Bioprozesstechnik*, 3rd edn, Spektrum, Heidelberg.

72 Smith, R. and Linnhoff, B. (1988) The design of separators in the context of overall processes. *Computers & Chemical Engineering*, **66**, 195–228.

73 Luyben, W. (2010) Heuristic design of reaction/separation processes. *Industrial & Engineering Chemistry Research*, **49** (22), 11564–11571.

74 Sempuga, B.C., Hausberger, B., Patel, B., Hildebrandt, D., and Glasser, D. (2010) Classification of chemical processes: a graphical approach to process synthesis to improve reactive process work efficiency. *Industrial & Engineering Chemistry Research*, **49**, 8227–8237.

75 Freund, H. and Sundmacher, K. (2008) Towards a methodology for the systematic analysis and design of efficient chemical processes: Part 1. From unit operations to elementary process functions. *Chemical Engineering and Processing: Process Intensification*, **47**, 2051–2060.

76 Feinberg, M. and Ellison, P. (2001) General kinetic bounds on productivity and selectivity in reactor-separator systems of arbitrary design: principles. *Industrial & Engineering Chemistry Research*, **40**, 3181–3194.

77 Steimel, J., Harrmann, M., Schembecker, G., and Engell, S. (2013) A framework for the modelling and optimization of process superstructures under uncertainty. *Chemical Engineering Science*, **115**, 225–237.

78 Ryll, O., Blagov, S., and Hasse, H. (2014) Thermodynamic analysis of multicomponent distillation-reaction processes based on piecewise linear models. *Chemical Engineering Science*, **109**, 284–295.

79 Kossack, S., Refinius, A., Brüggemann, S., and Marquardt, W. (2007) Konzeptioneller Entwurf von Reaktions-Destillations-Prozessen mit Näherungsverfahren. *Chemie Ingenieur Technik*, **79**, 1601–1612.

80 Hentschel, B., Peschel, A., Freund, H., and Sundmacher, K. (2014) Simultaneous design of the optimal reaction and process concept for multiphase systems. *Chemical Engineering Science*, **115**, 69–87.

81 Gassner, M. and Marchéval, F. (2009) Methodology for the optimal thermo-economic, multi-objective design of thermodynamical fuel production from biomass. *Computers & Chemical Engineering*, **33**, 769–781.

82 Gross, B. and Roosen, P. (1998) Total process optimization in chemical engineering with evolutionary algorithms. *Computers & Chemical Engineering*, **22**, 223–236.

83 Vázquez-Ojeda, M., Segovia-Hernández, J.G., Hernández, S., Hernández-Aguirre, A., and Kiss, A.A. (2013) Design and optimization of an ethanol dehydration process using stochastic methods. *Separation and Purification Technology*, **105**, 90–107.

84 Rios, L.M. and Sahinidis, N.V. (2013) Derivative-free optimization: a review of algorithms and comparison of software implementations. *Journal of Global Optimization*, **56**, 1247–1293.

85 Caballero, J.A. and Grossmann, I.E. (2008) An algorithm for the use of surrogate models in modular flowsheet optimization. *AIChE Journal*, **54**, 2633–2650.

86 Fahmi, I. and Cremaschi, S. (2012) Process synthesis of biodiesel production plant using artificial neural networks as the surrogate models. *Computers & Chemical Engineering*, **46**, 105–123.

87 Cozad, A., Sahinidis, N.V., and Miller, D.C. (2014) Learning surrogate models for simulation-based optimization. *AIChE Journal*, **60**, 2211–2227.

88 Keller, T. (2014) Reactive distillation. *Distillation*, Elsevier, Amsterdam, pp. 261–294.

89 Agarwal, V., Thotla, S., and Mahajani, S.M. (2008) Attainable regions for reactive distillation – Part I. Single reactant non-azeotropic systems. *Chemical Engineering Science*, **63**, 2946–2965.

90 Agarwal, V., Thotla, S., Kaur, R., and Mahajani, S.M. (2008) Attainable regions of reactive distillation. Part II: Single reactant azeotropic systems. *Chemical Engineering Science*, **63**, 2928–2945.

91 Schembecker, G. and Tlatlik, S. (2003) Process synthesis for reactive separations. *Chemical Engineering and Processing*, **42**, 179–189.

92 Holtbruegge, J., Kuhlmann, H., and Lutze, P. (2014) Conceptual design of flowsheet options based on thermodynamic insights for (reaction-) separation processes applying process intensification. *Industrial & Engineering Chemistry Research*, **53**, 13412–13429.

93 Lutze, P., Babi, D.K., Woodley, J.M., and Gani, R. (2013) Phenomena based methodology for process synthesis incorporating process intensification. *Industrial & Engineering Chemistry Research*, **52**, 7127–7144.

94 Babi, D.K., Holtbruegge, J., Lutze, P., Górak, A., Woodley, J.M., and Gani, R. (2015) Sustainable process synthesis–intensification. *Computers & Chemical Engineering*, **81**, 218–244.

95 Cerda, J., Westerberg, A.W., Mason, D., and Linnhoff, B. (1983) Minimum utility usage in heat exchanger network

synthesis A transportation problem. *Chemical Engineering Science*, **38** (3), 373–387.

96 Papoulias, S.A. and Grossmann, I.E. (1983) A structural optimization approach in process synthesis—II: heat recovery networks. *Chemical Engineering Science*, **38** (3), 373–387.

97 Floudas, C.A., Ciric, A.R., and Grossmann, I.E. (1986) Automatic synthesis of optimum heat exchanger network configurations. *AIChE Journal*, **32** (2), 276–290.

98 Floudas, C.A. and Ciric, A.R. (1989) Strategies for overcoming uncertainties in heat exchanger network synthesis. *Computers & Chemical Engineering*, **13** (10), 1133–1152.

99 Yuan, X., Pibouleau, L., and Domenech, S. (1989) Experiments in process synthesis via mixed-integer programming. *Chemical Engineering and Processing: Process Intensification*, **25** (2), 99–116.

100 Yee, T.F. and Grossmann, I.E. (1990) Simultaneous optimization models for heat integration—II. Heat exchanger network synthesis. *Computers & Chemical Engineering*, **14** (10), 1165–1184.

101 Konukman, A.E.S., Camurdan, M.C., and Akman, U. (2002) Simultaneous flexibility targeting and synthesis of minimum-utility heat-exchanger networks with superstructure-based MILP formulation. *Chemical Engineering and Processing: Process Intensification*, **41** (6), 501–518.

102 Verheyen, W. and Zhang, N. (2006) Design of flexible heat exchanger network for multi-period operation. *Chemical Engineering Science*, **61** (23), 7730–7753.

103 Frausto-Hernandez, S., Rico-Ramirez, V., Jimenez-Gutierrez, A., and Hernandez-Castro, S. (2003) MINLP synthesis of heat exchanger networks considering pressure drop effects. *Computers & Chemical Engineering*, **27** (8), 1143–1152.

104 Serna-Gonzalez, M., Ponce-Ortega, J.M., and Jimenez-Guiterrez, A. (2004) Two-level optimization algorithm for heat exchanger networks including pressure drop considerations. *Industrial & Engineering Chemistry Research*, **43** (21), 6766–6773.

105 Ponze-Ortega, J.M., Jimenez, G.A., and Grossmann, I.E. (2008) Optimal synthesis of heat exchanger networks involving isothermal process streams. *Computers & Chemical Engineering*, **32** (8), 1918–1942.

106 Duran, M.A. and Grossmann, I.E. (1986) Simultaneous optimization and heat integration of chemical processes. *AIChE Journal*, **32** (1), 123–138.

107 Grossmann, I.E., Yeomans, H., and Kravanja, Z. (1998) A rigorous disjunctive optimization model for simultaneous flowsheet optimization and heat integration. *Computers & Chemical Engineering*, **22**, 157–164.

108 Yeomans, H. and Grossmann, I.E. (1999) A systematic modeling framework of superstructure optimization in process synthesis. *Computers & Chemical Engineering*, **23** (6), 703–731.

109 Kondili, E., Pantelides, C., and Sargent, R. (1993) A general algorithm for short-term scheduling of batch operations—I. MILP formulation. *Computers & Chemical Engineering*, **17** (2), 211–227.

110 Yeomanns, H. and Grossmann, I.E. (1999) Nonlinear disjunctive programming models for the synthesis of heat integrated distillation sequences. *Computers & Chemical Engineering*, **23**, 1135–1151.

111 Harwardt, A., Kossack, S., and Marquardt, W. (2008) Optimal column sequencing for multicomponent mixtures. *Computer Aided Chemical Engineering*, **25**, Elsevier, Amsterdam, pp. 91–96.

112 Dahdad, T.H. and Mitsos, A. (2014) Structural optimization of seawater desalination: I. A flexible superstructure and novel MED–MSF configurations. *Desalination*, **344**, 252–265.

113 Kocis, G.R. and Grossmann, I.E. (1989) A modelling and decomposition strategy for the MINLP optimization of process flowsheets. *Computers & Chemical Engineering*, **13** (7), 797–819.

114 Daichendt, M.M. and Grossmann, I.E. (1994) Preliminary screening procedure for the MINLP synthesis of process

systems – I. Aggregation and decomposition techniques. *Computers & Chemical Engineering*, **18** (8), 663–677.

115 Andrecovic, M.J. and Westerberg, A. (1985) An MILP formulation for heat-integrated distillation sequence synthesis. *AIChE Journal*, **31** (9), 1461–1474.

116 Sargent, R. and Gaminibandara, K. (1976) Optimum design of plate distillation columns. *Optimization in Action*, 267–314.

117 Novak, Z., Kravanja, Z., and Grossmann, I.E. (1996) Simultaneous synthesis of distillation sequences in overall process schemes using an improved MINLP approach. *Computers & Chemical Engineering*, **20** (12), 1425–1440.

118 Henao, C.A. and Maravelias, C.T. (2011) Surrogate-based superstructure optimization framework. *AIChE Journal*, **57** (5), 1216–1232.

119 Gassner, M. and Maréchal, F. (2009) Methodology for the optimal thermo-economic, multi-objective design of thermochemical fuel production from biomass. *Computers & Chemical Engineering*, **33**, 769–781.

120 Raman, R. and Grossmann, I.E. (1993) Symbolic integration of logic in mixed-integer linear programming techniques for process synthesis. *Computers & Chemical Engineering*, **17** (9), 909–927.

121 Turkay, M. and Grossmann, I.E. (1998) Structural flowsheet optimization with complex investment cost functions. *Computers & Chemical Engineering*, **22** (4), 673–686.

122 Gassner, M. and Maréchal, F. (2009) Thermo-economic process model for thermochemical production of synthetic natural gas (SNG) from lignocellulosic biomass. *Biomass and Bioenergy*, **33**, 1587–1604.

123 Pistikopoulos, E.N., Diangelakis, N.A., and Manthanwar, A.M. (2015) Towards the integration of process design, control and scheduling: are we getting closer? *Proceedings of the 12th International Symposium on Process Systems Engineering and 25th European Symposium on Computer Aided Process Engineering*, Elsevier, Copenhagen, pp. 41–48.

124 Yunt, M., Chachuat, B., Mitsos, A., and Barton, P.I. (2008) Designing man-portable power generation systems for varying power demand. *AIChE Journal*, **54** (5), 1254–1269.

125 Hadara, H., Harjunkoski, I., Sand, G., Grossmann, I.E., and Engell, S. (2015) Optimization of steel production scheduling with complex time-sensitive electricity cost. *Computers & Chemical Engineering*, **76**, 117–136.

126 Lee, H.H., Koppel, L.B., and Lim, H.C. (1972) Integrated approach to design and control of a class of countercurrent processes. *Industrial & Engineering Chemistry Process Design and Development*, **11** (3), 376–382.

127 Yuan, Z., Chen, B., Sin, G., and Gani, R. (2012) State-of-the-art and progress in the optimization-based simultaneous design and control for chemical processes. *AIChE Journal*, **58** (6), 1640–1659.

128 Macedonio, F., Drioli, E., Gusev, A., Bardow, A., Semiat, R., and Kurihara, M. (2012) Efficient technologies for worldwide clean water supply. *Chemical Engineering and Processing: Process Intensification*, **51**, 2–17.

129 Abdel-Jabbar, N.M., Qiblawey, H.M., Mjalli, F.S., and Ettouney, H. (2007) Simulation of large capacity MSF brine circulation plants. *Desalination*, **204** (1–3), 501–514.

130 Darwish, M.A., Al-Juwayhel, F., and Abdulraheim, H.K. (2006) Multi-effect boiling systems from an energy viewpoint. *Desalination*, **194** (1–3), 22–39.

131 El-Dessouky, H.T. and Ettouney, H.M. (2002) *Fundamentals of Salt Water Desalination*, Elsevier.

132 Dahdah, T.H. and Mitsos, A. (2014) Structural optimization of seawater desalination: II novel MED–MSF–TVC configurations. *Desalination*, **344**, 219–227.

133 Skiborowski, M., Mhamdi, A., Kraemer, K., and Marquardt, W. (2012) Model-based structural optimization of seawater desalination plants. *Desalination*, **292**, 30–44.

134 Iaquaniello, G., Salladini, A., Mari, A., Mabrouk, A.A., and Fath, H. (2014) Concentrating solar power (CSP) system integrated with MED–RO hybrid desalination. *Desalination*, **336**, 121–128.

135 Zak, G.M., Mancini, N.D., and Mitsos, A. (2013) Integration of thermal desalination methods with membrane-based oxy-combustion power cycles. *Desalination*, **311**, 137–149.

136 Zak, G.M., Ghobeity, A., Sharqawy, M.H., and Mitsos, A. (2013) A review of hybrid desalination systems for co-production of power and water: analyses, methods, and considerations. *Desalination and Water Treatment*, **51** (28–30), 5381–5401.

137 Jezowski, J. (2010) Review of water network design methods with literature annotations. *Industrial & Engineering Chemistry Research*, **49**, 4475–4516.

138 Gerber, L. and Maréchal, F. (2012) Defining optimal configurations of geothermal systems using process design. *Applied Thermal Engineering*, **43**, 29–41.

139 Hipólito-Valencia, B.J., Rubio-Castro, E., Ponce-Ortega, J.M., and Serna-González, M. (2013) Optimal integration of organic Rankine cycles with industrial processes. *Energy Conversion and Management*, **73**, 285–302.

140 Lee, U., Kim, K., and Han, C. (2014) Design and optimization of multi-component organic Rankine cycle using liquefied natural gas cryogenic exergy. *Energy*, **77**, 520–532.

141 Ghasemi, H., Paci, M., Tizzanini, A., and Mitsos, A. (2014) Modeling and optimization of a binary geothermal power plant. *Energy*, **50**, 412–428.

142 Mitsos, A., Palou-Rivera, I., and Barton, P.I. (2004) Alternatives for micropower generation processes. *Industrial and Engineering Chemistry Research*, **43**, 74–84.

143 Yunt, M., Chachuat, B., Mitsos, A., and Barton, P.I. (2008) Designing man-portable power generation systems for varying power demand. *AIChE Journal*, **54**, 1254–1269.

144 Frangopoulos, C.A., von Spakovsky, M.R., and Sciubby, E. (2002) A brief review of methods for the design and synthesis optimization of energy system. *International Centre for Applied Thermodynamics*, **5** (4), 151–160.

145 Upadhye, A., Qi, W., and Huber, G. (2011) Conceptual process design: a systematic method to evluate and develop renewable energy technologies. *AIChE Journal*, **57**, 2292–2301.

146 Kraemer, K., Harwardt, A., Bronneberg, R., and Marquardt, W. (2011) Separation of butanol from acetone-butanol-ethanol fermentation by a hybrid extraction-distillation process. *Computers & Chemical Engineering*, **35** (5), 949–963.

147 Hintermair, U., Franció, G., and Leitner, W. (2011) Continuous flow organometallic catalysis: new wind in old sails. *Chemical Communications*, **47**, 3691–3701.

148 Recker, S., Gordon, C.M., Peace, A., Redepenning, C., and Marquardt, W. (2017) Systematic Design of a Butadiene Telomerization Process: The Catalyst Makes the Difference. *Chemie Ingenieur Technik*, **89** (11), 1479–1489.

6
Development of Novel Multiphase Microreactors: Recent Developments and Future Challenges

Evgeny Rebrov

6.1
Principles and Features

Various types of microreactors were developed for the gas–liquid–solid reactions. The designs are usually distinguished according to the principle of gas–liquid contacting: continuous-phase contacting and dispersed-phase contacting [1]. In the latter case, one fluid phase is dispersed into another fluid phase. In addition, micro-trickle bed operation is reported following the path of classical chemical engineering. These reactors are expected to find their niche applications in fine chemicals and pharmaceutical synthesis.

In the context of a gas–liquid–solid catalytic reaction with reaction occurring on a solid catalyst, two characteristics are critical in evaluating the advantages and disadvantages of a particular flow regime. First, we desire good mass transport between gas and liquid, which depends on the driving force for mass transport and the gas–liquid interfacial area created. Second, we desire a high liquid–solid interfacial area to effectively utilize the catalyst. The highest liquid–solid interfacial area will be achieved when liquid is the continuous phase.

6.1.1
Continuous Phase Multiphase Microreactors

6.1.1.1 Falling Film Microreactor

Falling film microreactor (FFMR) utilizes a multitude of thin liquid films that move by gravity force providing a typical liquid residence time of few seconds to 1 min. The very large specific gas/liquid interface area of $10\,000–20\,000\,m^2/m^3$ provides excellent mass transfer rate between the phases. The FFMR has been widely applied for different chemical reactions such as chlorination [2], sulfonation [3], and hydrogenation [4]. A structured heat exchanger plate is mechanically connected to the reaction plate for heat removal and a near-isothermal behavior is possible in highly exothermic reactions. Yeong *et al.* [5, 6] used a FFMR for the highly exothermic hydrogenation of nitrobenzene to aniline.

Handbook of Green Chemistry Volume 12: Green Chemical Engineering, First Edition. Edited by Alexei A. Lapkin.
© 2018 Wiley-VCH Verlag GmbH & Co. KGaA. Published 2018 by Wiley-VCH Verlag GmbH & Co. KGaA.

A near-isothermal operation was observed at 60 °C and a partial hydrogen pressure of 0.1–0.4 MPa at residence time of 9–17 s.

A major drawback of FFMR is related to rather short liquid residence time, which in most cases is limited to 10–30 s and it depends on the liquid physical properties and operating conditions. The residence time can be increased when the reactor is tilted with respect to the gravity vector [7]. At the same residence time, the conversion of octanal in the normal position was higher due to a lower thickness of the liquid film. However, a higher residence time of 2.9 s was achieved in the tilted position resulting in an increase of conversion from 82 to 89% [8].

6.1.1.2 Mesh Contactor

In a mesh contactor, gas and liquid or two immiscible liquid flows are fed through two adjacent channels that are separated by meshes [9] or a porous plate with well-defined openings between two fluid compartments [10]. In a mesh contactor, the maximum pressure difference across the porous wall should be below the capillary pressure. These devises are typically fabricated by conventional photolithography and deep reactive ion etching (DRIE). The mesh-to-wall distances are usually 50–100 µm. The length of the openings in the porous wall of 10–15 µm and the pore width of 5 µm provide for good stability of the interface. The shape of the meniscus at the interface defines the surface area for mass transfer and it depends on the contact angle and the pressure drop across the porous membrane. The open arc of the mesh contactor is 20–25%, which provides an interfacial area of 2000–2500 m^2/m^3. This value is significantly higher than that observed in a stirred tank reactor. Both high surface area and a thin stagnant fluid film in the pore openings are responsible for high volumetric mass transfer coefficients ($k_L a$). Abdallah *et al.* [11, 12] reported the $k_L a$ of 0.8–1.6 s^{-1} in hydrogenation of α-methylstyrene over a Pd/Al_2O_3 catalyst. The authors performed theoretical estimation of the volumetric mass transfer coefficient using the film model and CFD simulation. They reported that the both theoretical values underestimated the experimental results.

6.1.2
Dispersed Phase Multiphase Microreactors

6.1.2.1 Segmented Flow Microreactors

All of the flow regimes in microchannels can be conventionally divided into three groups: (1) regimes with the predominance of surface tension forces, (2) regimes with the predominance of inertial forces, and (3) transient regimes between the first two cases. These three groups include six main flow regimes: bubble, slug, slug–annular, annular, churn, and trickle–annular regimes [13–16]. Either of the first five regimes can be observed in microchannels depending on the physical properties of the both fluids, the superficial velocities and the channel diameter, whereas the trickle–annular regime is present only in relatively short channels with a large ratio of width to depth [17, 18]. The gas hold-up, the pressure drop, and the volumetric mass transfer coefficient depend on the flow regime in a microreactor. The feature that distinguishes microchannels from channels of large diameters is an axisymmetric flow in which the vectors of the liquid and gas velocities do not depend on the position of the channel with respect to the gravity vector. Hassan *et al.* reported that an axisymmetric flow is observed at a channel diameter of 0.5–1.0 mm, depending on the physical properties of the both fluids, and the orientation of the channel does not affect the positions of the boundaries of the regimes [19]. In vertical channels of larger diameter, bubbly flow is extended to lower U_L than in horizontal ones because buoyancy in vertical channels favors bubble detachment at the inlet and assists the formation of small bubbles. In addition, annular flow in vertical channels is shifted to lower U_G compared to horizontal ones, indicating a larger effect of inertia assisted by buoyancy over surface tension forces in vertical channels.

Fries *et al.* investigated the segmented flow patterns and demonstrated that mass transfer in meandering microchannels can be considerably enhanced as compared with straight microchannels. In meandering microchannels, longer liquid residence time was realized. The authors showed that the recirculating movement in the slugs depends on the channel geometry [20]. They reported the influence of the superficial velocity, the channels diameter, and the curve radius on the recirculation motion.

The position of the boundaries between different flow regimes depends on the mixer geometry. Warnier *et al.* performed a detailed study of the influence of the geometry of the gas–liquid mixer on the position of the flow regime boundaries [21]. Haverkamp *et al.* used two types of cross flow inlet, one with the phases joining perpendicularly and the other with the phases joining smoothly, and observed a profound effect of the inlet type on the air/water flow patterns formed [22]. The transition from Taylor to churn flow was shifted to higher U_G in the smooth inlet. Shao *et al.* studied N_2/water flow patterns and their transitions in circular capillaries using T-mixers of various sizes and configurations [23]. The size of the inlet was found to affect significantly the transition from Taylor to bubbly flow. No bubbly flow was observed with the inlet diameter of 1.25 mm.

6.1.2.2 Microstructured Packed Beds

Many microreactors use micropacked beds for gas–liquid–solid reactions. An advantage of microstructured packed bed is the commercial availability of catalysts. The catalyst particle size is typically below 250 μm and well suited for microchannel applications. Usually a proper loading of the bed into the reactor is required to provide reliable reactor operation without a substantial pressure built up due to bed densification.

Losey *et al.* [24] constructed a composite silicon–glass microreactor using DRIE. The gas and liquid flows were premixed on-chip using high aspect ratio mixing channels. These channels delivered the reactants to ten reactor chambers filled with catalyst particles with a diameter of 50–75 μm. The authors observed that small bubbles were formed at the entrance at relatively low gas and liquid velocities. The overall volumetric mass transfer coefficient ($k_{ov}a$) was investigated in the hydrogenation of cyclohexane in the microstructured packed bed reactor. The $k_{ov}a$ of 5–15 s^{-1} were reported that were an order of magnitude larger than those measured in laboratory scale reactors.

Different flow regimes in MPBR were investigated in details by van Herk et al. [25]. These authors observed pulsations and the formation of segmented flow regime at high gas flow velocities. The segmented gas–liquid flow was also reported by Tadepalli et al. [26] in a MPBR with a Pd/zeolite catalyst with particles of 45–150 µm and the length of catalyst bed of 6–8 cm.

Composite silicon–glass microreactors safely allow increasing the pressure while providing optical access to the reaction channels for investigation of hydrodynamics. Trachsel et al. [27] constructed a silicon–glass microreactor for high-pressure operations at elevated temperatures. The reactor was mechanically stable and showed no failures during continuous operation at 14 MPa and 80 °C. The authors observed an enhancement of mass transfer coefficient in hydrogenation of cyclohexene as the pressure was increased from ambient to 5.1 MPa.

The main drawback of structured packed bed microreactors is uneven flow distribution due to the nonuniform packing of catalyst particles resulting in channeling of the fluids. A large RTD diminishes the conversion, and often, the product selectivity.

6.1.2.3 Prestructured Microreactors

Prestructured microreactors employ a sequential catalyst-trap design, whereby solid catalyst is suspended in the channel by an arrangement of micropillars. Such a device has advantages that commercial catalysts are supported, and that pressure drop across the bed can be reduced by engineering the packing density. McGovern et al. [28] studied flow regimes in three prestructured microreactors with a channel width of 10 and 18 mm and a distance between trap pillars of 75 and 150 µm. They defined regimes of liquid-dominated, gas-dominated, and transitional flows. Liquid- and gas-dominated flows are characterized by stable patterns in which the continuous phase impedes movement of the noncontinuous phase throughout the channel. For example, in liquid-dominated flow, more than 80% of the channel volume is occupied by liquid, with several small pockets of gas. In gas-dominated flow, large gas pockets are observed with liquid channels around them. The gas–liquid interfaces in these flow regimes are largely stagnant. The transitional pattern is characterized by an unstable pattern in which gas and liquid interfacial areas are periodically (30–90 s) refreshing and significantly increasing. The reactor behavior was characterized in the hydrogenation of nitroanisole to anisidine [29]. Krishnamurthy and Peles investigated nitrogen–water two-phase flow across a bank of staggered circular micropillars with a length of 100 µm, a diameter of 100 µm, and a pitch-to-diameter ratio of 1.5 [30, 31] (Figure 6.1). A unique flow pattern called bridge flow was observed at Re_G above 40. In this flow regime, individual gas slugs merged and the liquid traversed along the channel in the form of bridges (Figure 6.2). While this flow pattern was also reported by Xu [32] in conventional scale systems in vertical up

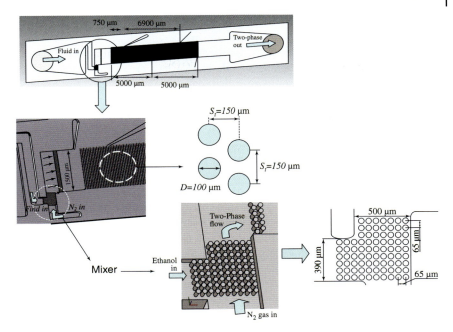

Figure 6.1 The micropillar device overview showing the flow configuration. (Adapted with permission from Ref. [31]. Copyright 2009, Elsevier.)

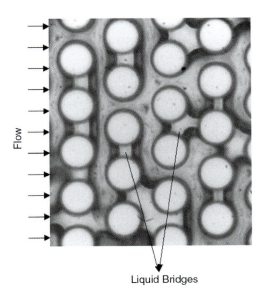

Figure 6.2 Image of liquid bridge nitrogen–water two-phase flow across a bank of staggered circular micropillars. Gas flow rate: 14 m s^{-1}, liquid flow rate: 0.031 m s^{-1}. (Adapted with permission from Ref. [30]. Copyright 2007, the American Institute of Physics.)

and down flow across a tube bundle, in the latter study it was governed by a completely different hydrodynamic mechanism.

The major drawback of the prestructured microreactors with micropillars is their relatively limited surface area [33]. The surface area can be enlarged by deposition of thin films of mesoporous materials.

6.1.2.4 Foam Microreactors

Solid foam packing are available in a variety of pore sizes and materials, including metals, carbon, and ceramics. They have high specific surface area for catalyst deposition. Stemmet *et al.* investigated the gas–liquid mass transfer in solid foam packings in cocurrent upflow and downflow configurations [34–36]. In the countercurrent flow configuration, the trickle bed, bubble, and pulse flow regimes were observed in 5, 20, and 40 ppi solid foam packings, respectively [34]. The flooding points were close to those in Katapak-S packings. The gas–liquid mass transfer coefficient ($k_L a_{GL}$) of $6\,\mathrm{s}^{-1}$ was estimated using penetration theory [35]. The cocurrent flow configuration was studied for 10 and 40 ppi solid foam packings [36]. The bubble and regimes were observed at U_G below and above $0.3\,\mathrm{m\,s}^{-1}$, respectively. The available gas–liquid area for mass transfer increases at larger U_G. The value of k_L decreases with increasing ppi number (smaller pores) of the packing but the value of $k_L a_{GL}$ remains constant due to higher a_{GL} in smaller pores. The gas–liquid mass transfer coefficient was found to increase with increasing liquid and gas velocities, up to a maximum value of $1.3\,\mathrm{s}^{-1}$. The influence of the liquid viscosity and surface tension was studied in the cocurrent upflow and downflow configurations [36]. The correlations for gas–liquid mass transfer were obtained similar to the correlations for packed beds of spherical particles proposed in Fukushima and Kusaka [37].

6.1.2.5 Microreactors with Fibrous Internal Structures

The use of filamentous structured materials (FSMs) made of either composite metal oxides (e.g., ZnO) or carbon nanofibers (CNF) onto reactor internals can enhance the available surface area in microreactors made by the top–down approaches. The diameter of these structured elements assembled by chemical synthesis is of the order of a several hundred nanometers to a few microns and corresponds to the typical diameter of macropores in supported catalysts. Supported catalysts in the form of washcoats and catalytic thin films suffer from internal mass transfer limitations because of the low diffusivity in the meso and micropores in these structures where the active catalyst is present. The FSMs provide a more open and hydrodynamically accessible structure with a large surface area of $100–200\,\mathrm{m}^2\,\mathrm{g}^{-1}$ [38].

An enhancement of the external liquid–solid mass transfer from the liquid bulk to the catalyst support was observed over a CNF layer with a high surface roughness compared to a unsupported flat plate catalyst in the liquid-phase hydrogenation of 3-methyl-1-pentyn-3-ol in a rectangular microchannel [39].

FSMs were tested in the liquid-phase hydrogenation of nitrobenzene to aniline [40] and acetylene alcohols [41, 42]. A reactor based on a bubble column staged by FSMs of Pd/ZnO on sintered metal fibers with integrated cross-flow micro-heat exchangers (HEX) was designed and tested in the solvent free hydrogenation of 2-methyl-3-butyn-2-ol (MBY) aiming on process intensification [43]. The volumetric mass transfer coefficient was $1.2\,\text{s}^{-1}$, which is close to the value reported in mesh contactors. The attained specific reactor performance was several orders of magnitude above the values reported for conventional multiphase reactors.

6.2 Experimental Practice

6.2.1 Flow Regimes

6.2.1.1 Capillary Microreactors

Capillary hydrodynamics has three considerable distinctions from conventional reactors: first, there is a larger contribution from surface forces as compared to volumetric ones; second, a flow is characterized by small Reynolds numbers at which viscous forces dominate over inertial forces; and third, the microroughness and wettability of the channel wall influence flow patterns. Due to these differences, the correlations used in large diameter tubes cannot satisfactorily be used in microchannels. Based on the relative importance of the surface tension over inertial forces, three groups of overall flow regimes are identified, namely, surface tension dominated, inertia dominated, and transitional regimes. These three groups consist of six main flow regimes, namely, bubbly and slug (surface tension dominated), churn and slug–annular (transitional), and dispersed and annular (inertia dominated) [44, 45]. A flow regime map is presented in Figure 6.3 with gas (U_G) and liquid (U_L) superficial velocities as coordinates. In the bubble regime, there is the motion of gas bubbles with the size considerably less than the channel diameter. The bubbles are separated from the channel

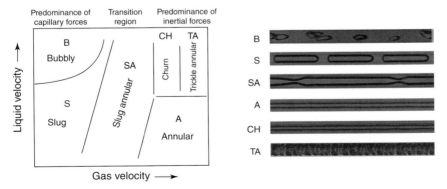

Figure 6.3 Schematic representation of two-phase flow regimes in capillaries.

walls with the liquid film and typically have a wide size distribution. Slug flow (also known as Taylor or plug flow) is characterized by elongated bubbles with an equivalent diameter larger than that of the channel. Liquid slugs separate the gas bubbles, while depending on the channel wettability, a liquid film may form that separates the bubbles from the wall. With increasing U_G at constant U_L, the increased gas void fraction leads to the merging of gas bubbles to form the slug–annular flow. Serizawa et al. observed stable slug–annular flow in microchannels of 25–100 μm in diameter [45]. In this pattern, a gas core is surrounded by a flowing liquid film where large-amplitude solitary waves appear and small bubbles are sometimes present in the liquid. As U_G increases, the elongated Taylor bubbles become unstable near their trailing ends and the transition to the churn flow regime occurs. However, if U_L increases, the long waves disappear and the annular pattern is established. In the churn regime, the liquid film becomes disturbed by the high inertia of the gas core and liquid drops enter the gas core. In the annular regime, the thickness of the liquid film depends on wall wettability. When the liquid film thickness decreases below a certain limit, dry flow occurs in which dry patches develop on the wall [46].

In liquid–liquid microchannel flow, different flow patterns were reported as a function of the volumetric flow ratio of two immiscible liquids, that is, annular flow, parallel flow, drop flow, or slug flow [47, 48]. Annular and parallel flows are observed at Weber numbers above 1 when the inertial forces dominate over the interfacial forces [49]. The transition from slug to parallel flow takes place at increasing flow rate and was observed at We >0.5 and We >0.8 for the water–toluene and ethylene glycol/water–toluene flow, respectively. The latter was composed of 40 wt% of ethylene glycol in demineralized water. Due to the competing nature of the inertial and interfacial forces, annular and parallel flows can be easily destabilized by changing flow rates and volumetric flow ratios [49, 50]. Slug [51, 52] and drop flow [53, 54] are extensively studied due to their easily controllable hydrodynamics and potential applications in fine chemical synthesis and biotechnology. Both Y-mixer [55] and T-mixer [56] provide reproducible segmented slug flow, which allows a high degree of control over the slug size distribution and the liquid–liquid interfacial ratio, which is in the range of 10 000–50 000 m^2/m^3 for a channel diameter in the range from tens to hundreds of micrometers [57].

6.2.1.2 Structured Packed Beds

Visualization studies in prestructured microreactors suggest that different regimes are observed depending on the gas and liquid flow rates [58] (Figure 6.4). At low liquid to gas flow ratios and all gas flow rates, an annular flow was observed in which the liquid was preferential distributed along the wall while the gas passed through the center of the channel. A slug flow with gas slugs passing through a continuous liquid phase exists at high liquid-to-gas flow ratios. At high gas and liquid flow rates, the churn flow regime is observed. The microfabricated pillars in the reactor channels enhance the overall mass transfer coefficient analogous to micropacked bed systems.

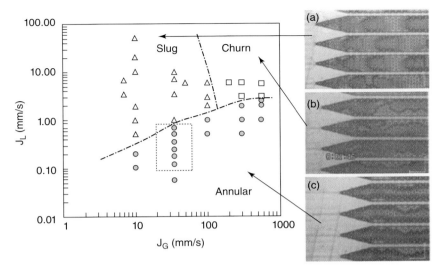

Figure 6.4 Gas–liquid flow regimes observed in the multichannel microreactor with posts. (a) Slug flow (gas as dark area). (b) Churn flow (the interface fluctuates rapidly and liquid periodically spans the entire channel). (c) Annular flow (gas flows at the center of reactor). (Adapted with permission from Ref. [58]. Copyright 2006, American Chemical Society.)

6.2.2
Dispersion and Holdup in Microstructured Packed Bed Reactors

As per the guidelines available in the literature to overcome the wall effects in microstructured packed bed reactors, the maximum catalyst particle diameter that can be used in a reactor with an internal diameter of 10 mm is 0.4 mm [59, 60]. These reactors can be considered as microstructured packed bed reactors, as the flow of gas and liquid is not affected by gravity [61].

6.2.2.1 Liquid Holdup
In the analysis of trickle beds, draining experiments are used to distinguish dynamic holdup from static holdup, where the static fraction remains in the bed after draining has stopped. When the gas and liquid flow are stopped the liquid entirely remains in the microstructured bed. Thus, the bed has zero dynamic holdup and only static holdup. The static holdup (ε_L) is expressed as the fraction of the space between the particles that is, on average, filled with liquid:

$$\varepsilon_L = \frac{F_L \tau}{V_p} \tag{6.1}$$

The liquid holdup does not vary significantly with U_G and remains always above 0.65 (or 0.28–0.37 per unit column volume) in a bed with a mean particle size of 100 μm [62]. High holdup values translate into good wetting characteristics of catalysts and are desired for catalytic reactor applications. These values are considerably higher as compared to those in industrial trickle bed reactors [63]:

$$\varepsilon_L = 3.86 \varepsilon_b \mathrm{Re}_L^{0.565} Ga_L^{-0.42} \left(\frac{a_s d_p}{\varepsilon_b}\right) \quad (6.2)$$

The use of fine particles as a diluent of trickle bed reactor packings is a common method of ensuring good liquid distribution and catalyst wetting in small-scale laboratory test reactors. Kulkarni *et al.* found that liquid holdup (per unit column volume) in a bed with 6 mm glass spheres increased from 0.05 to 0.25 upon mixing with 0.2 mm glass particles [64]. The use of a diluent with a particle size of 0.3 mm increased liquid holdup from 0.05 to 0.20 [65]. In beds diluted with small particles, the liquid is more easily retained between the particles due to the increased surface area, leading to the increased value of the total liquid holdup. These high values are not observed in upflow beds with particles of 3 mm in diameter. Iliuta *et al.* found static holdup in upflow packed beds of 0.03–0.08 and dynamic holdup values in the range 0.10–0.25 [65]. Bej *et al.* found that reactor performance kept increasing for smaller particles, with optimal results obtained for dilution with 0.1 mm particles due to the high holdup values [66].

6.2.2.2 Hydrodynamic Dispersion

Hydrodynamic dispersion, or the spreading of an initially concentrated aliquot of a tracer by the combined action of diffusion and convection, is an important problem in the description of flow in microstructured packed beds. To predict the mixing behavior of the reactor and, hence, the conversion of liquid-phase reactants, knowledge of the residence time distribution (RTD) is required. Dispersion imposes significant restrictions on the assumption of uniform residence times (or plug flow) with important consequences for the conversion, especially as the reaction approaches full conversion. The channel length in microreactors is typically limited to several tens of millimeters. These short lengths seriously limit the conversion levels for which the plug flow assumption holds. The lack of reliable dispersion data in such systems hinders the design of such reactors [67]. It was shown that the axial dispersion–exchange model (ADEM) [68], combining axial dispersion with piston exchange between dynamic and static flow, was suitable only for nonporous particles [69]. Iliuta *et al.* proposed a combination of ADEM with transient diffusion within pores of catalysts [70]. The latter model incorporates all the important factors influencing RTD in porous particles: mass transport between both dynamic and external static liquid zones, mass transfer between liquid flow and particles, diffusion of tracer into pores, and axial dispersion. This model described experimental data relatively well. However, it was also shown [71, 72] that the influence of external transport resistances is important if the catalyst bed is diluted with smaller inert particles due to partial wetting of catalyst particles. The dispersion coefficient decreased by up to 50% for the bed with fine particles, compared to that of an undiluted bed [64].

6.3 Modeling Features

6.3.1 Hydrodynamics

6.3.1.1 Falling Films Microreactors

In a FFMR, chemical reaction is limited by the mass transfer within the liquid phase. Zanfir et al. [73] modeled CO_2 absorption in a NaOH solution with a 2D convection and diffusion model, and compared the results of calculation with experimental data. The authors concluded that carbon dioxide is consumed within a short distance from the gas–liquid interface and neither liquid film thickness nor the thickness of the gas film did significantly influence the CO_2 conversion. As a further approach to intensify mass transfer, pin structures are known that use the re-entrance flow effect caused by constant perturbation of flows through surface structures in the channels [74, 75]. This means a constant renewal of surfaces that was initially used to optimize temperature gradients for highly efficient heat transfer [76–78] and for convective heat transfer intensification in the electronics industry, air-conditioning, and automobile industry [79].

Commenge et al. [80] investigated the gas phase RTD in a FFMR, and observed that the formation of recirculation loops at the gas inlet and a jet effect considerably increased the mixing within the gas phase. Al-Rawashdeh et al. [81] studied the influence of liquid flow distribution on conversion showing that an uneven flow distribution lowers the conversion compared to ideal equal distribution only by 2%. They also investigated the effect of surface wettability of the reaction plate on the actual meniscus shape at the gas–liquid interface during the absorption of CO_2 by an aqueous NaOH solution. The conversion decreased by 20% over a hydrophobic surface as compared to a hydrophilic surface due to a reduction of the interfacial area. Zhang et al. [82] correlated the contact angle with the fluid profile in rectangular microchannels. They reported liquid mass transport coefficients in the range from 5.8 to 13.4×10^{-5} m s^{-1} and demonstrated the influence of surface tension and viscosity.

The flow of the falling liquid film in microchannels differs significantly from that on conventional flat plates. As a result of small channel sizes and the action of capillary forces, the surface of the liquid film in microchannels usually takes the form of a flowing meniscus rather than completely flat film. Meanwhile, the flow of the falling film in microchannels should be treated as a 3D flow and the increased importance of surface tension, evaporation should be addressed.

A 3D convection and diffusion model is formulated to simulate a chemical reaction in a FFMR. The momentum balance in the liquid domain (Figure 6.5a) is described by Eq. (6.3) as the liquid flow is driven by gravity [83].

$$\mu_L \left(\frac{\partial^2 v_{L,z}}{\partial x^2} + \frac{\partial^2 v_{L,z}}{\partial y^2} \right) = -\rho_L g \tag{6.3}$$

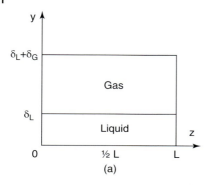 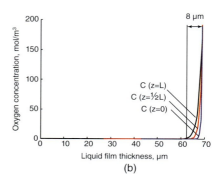

Figure 6.5 (a) Schematic view of the computational domain. (b) Simulated oxygen concentration profiles in the liquid film at 5.0 bar in a channel. Liquid flow rate: 6.0 ml min^{-1}, gas flow rate: 3.5 ml min^{-1} (STP), temperature: 100 °C, pressure: 5.0 bar. (b) Concentration profile octanal at the gas–liquid interface. Liquid residence time: 1.6 s.

The boundary conditions include zero velocity at the wall (no-slip condition) and zero derivative of velocity at the gas–liquid interface (Eq. (6.4)).

$$\left(\frac{\partial v_{L,z}}{\partial y}\right)\bigg|_{y=\delta_l} = 0 \qquad (6.4)$$

The gas flow is driven by a pressure gradient

$$\mu_L \left(\frac{\partial^2 v_{L,z}}{\partial x^2} + \frac{\partial^2 v_{L,z}}{\partial y^2}\right) = -\rho_L g \qquad (6.5)$$

The boundary conditions include no-slip at the top wall and symmetry at the side wall. The internal boundary conditions at the gas–liquid interface imply continuity in the velocity and shear stress. It is usually assumed that the physical properties of the liquid do not change over the reactor length (z) due to low concentration of gaseous component in the solvent and the mass increase due to product formation can be neglected. Then material balance in liquid is described by Eq. (6.6):

$$v_{L,z} \frac{\partial C_{L,i}}{\partial z} = \varphi D_{L,i} \frac{\partial^2 C_{L,i}}{\partial y^2} + jR_i \qquad (6.6)$$

where index i denotes reactants and products and index j denotes their stoichiometric coefficients. The reaction rate is determined by Eq. (6.7)

$$R = E k_L a (C^*_{O_2,L} - C_{O_2,L}) \qquad (6.7)$$

The specific interface area (a) depends on the shape of fluids menisci. In FFMR, surface tension forces dominate over viscous forces and the meniscus shape is determined by minimization of the interface area. If the Fo number is below 1 and $Sh_L > 4$, the mass transfer coefficient is calculated from the penetration theory [84]:

$$k_L = 2\sqrt{\frac{1.5\bar{v}_{L,z}D_{O_2}}{\pi L}} \qquad (6.8)$$

For a fast pseudo nth order reaction, the enhancement factor (E) can be taken equal to the Hatta number if the following inequality is satisfied

$$Ha = \sqrt{\frac{2D_{O_2}k_r C_B^2 (C_{O_2,L}^*)^{n-1}}{(n+1)k_L^2}} > 3 \qquad (6.9)$$

The concentration of the gaseous reactant in the liquid at the gas–liquid interface (C^*) is calculated from the Henry's law:

$$H \cdot P_G = C^*|_{y=\delta_L} \qquad (6.10)$$

The model can be solved numerically using a CFD code [81]. The actual channel profile together with the circular meniscus constitutes the liquid domain. The liquid film is characterized by its central height (δ_L). The gas domain is defined by its height, that is, the distance from the gas–liquid interface to the upper reactor wall (δ_G).

A typical oxygen concentration profile for oxidation of octanal to octanoic acid is shown in Figure 6.5b. It can be seen that the reaction occurs in a narrow liquid film of 8 µm from the gas–liquid interface.

6.3.2
Pressure Drop in Capillary Microreactors

6.3.2.1 Gas–Liquid Microreactors

The pressure drop is an important parameter in the reactor design as it provides crucial information regarding the energy consumption, required pump capacity, and the materials needed for the reactor construction. Accurate prediction of two-phase pressure drop in microstructured reactors is of paramount importance for their design and optimization. If the pressure drop is a significant fraction of the channel inlet pressure, then gas bubbles expand along the channel length. Therefore, the pressure profile needs to be known in order to properly determine the performance of the device. For the annular flow regime, the pressure drop can be described with an accuracy of 15% with the Lockhart–Martinelli–Chisholm correlation [85, 86]. This approach is based on correlating the gas and liquid single-phase pressure drops to obtain the two-phase pressure drop. For the gas phase, the single-phase pressure drop in the laminar flow is given by

$$-\left(\frac{dp}{dz}\right)_g = \frac{32\mu_g U_g}{d_c^2} \qquad (6.11)$$

For the liquid phase, the single-phase pressure drop is given by

$$-\left(\frac{dp}{dz}\right)_l = \frac{32\mu_l U_l}{d_c^2} \tag{6.12}$$

The Lockhart–Martinelli parameter χ is given by Eq. (6.13).

$$-\chi^2 = \left[\left(\frac{dp}{dz}\right)_l\right] / \left[\left(\frac{dp}{dz}\right)_g\right] \tag{6.13}$$

A two phase multiplier for the liquid phase is given by Eq. (6.14)

$$\varphi_l^2 = 1 + \frac{C}{\chi} + \frac{1}{\chi^2} \tag{6.14}$$

The value of C in Eq. (6.14) depends on the regimes of the liquid and gas. When the flow in both the gas and liquid phases are laminar, C is equal to 5. Then the two-phase pressure drop can be calculated by Eq. (6.15).

$$\left(\frac{dp}{dz}\right)_{tp} = \varphi_l^2 \left(\frac{dp}{dz}\right)_l \tag{6.15}$$

The LMH model fails to accurately predict the pressure drop of gas–liquid Taylor flow in small channels, because surface tension effects and/or hydrodynamic details of the flow, such as the spatial gas–liquid distribution and the velocity distributions of both phases, are not taken into account.

In an attempt to better capture the experimental trends and to improve the accuracy and reliability in the prediction of experimental data, Kreutzer *et al.* considered the liquid flow in the slugs to be a fully developed Hagen–Poiseuille flow, which is disturbed by the gas bubbles, causing an excess pressure drop [87]. They consider a gas–liquid Taylor flow moving through a channel with a cross-sectional area A (Figure 6.6a). The gas bubbles have a velocity u_b, a length L_b, and occupy a fraction of the cross-sectional area of the channel A_b/A. The liquid slugs have a length L_s. The flow is divided into unit cells consisting of one gas bubble, its surrounding liquid film, and one liquid slug, and the unit cell length is $L_b + L_s$. They suggested to use a modified friction factor (f_s) for the flow in the liquid slugs:

$$f_s = \frac{16}{Re_{gl}}\left(1 + a\frac{d_c}{L_s}\left(\frac{Re_{gl}}{Ca_{gl}}\right)^{0.33}\right) \tag{6.16}$$

where $0.07 < a < 0.17$ is a fitting parameter. The average velocity of the liquid in the slug is equal to the average cross-sectional velocity in the channel $U_g + U_l$. The equation for the pressure drop over the liquid slugs then becomes

$$\left(\frac{dp}{dz}\right)_s = -\frac{2f_s Re_{gl}\mu_l(U_g + U_l)}{d_h^2} \tag{6.17}$$

Since only a part of the channel length is occupied by liquid slugs, the pressure drop over the liquid slugs must be multiplied with the fraction of channel length

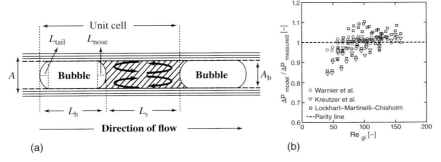

Figure 6.6 (a) Schematic view of Taylor flow in a capillary showing the definitions of the unit cell, bubble length L_b, and the liquid slug length L_s. The lengths of the nose L_{nose} and tail L_{tail} sections of the bubble are also indicated. The diagonally dashed area is the liquid slug. The fluid circulation patterns relative to the bubble movement are also indicated. (b) The pressure difference between experimental and calculated data as a function of Reynolds number. (The calculated data were obtained from Warnier et al. [88], Kreutzer at al. [87], and Lockhart–Martinelli–Chisholm models [85, 86].)

occupied by the slugs to obtain the pressure drop over the channel:

$$\left(\frac{dp}{dz}\right)_c = \left(\frac{dp}{dz}\right)_s \frac{L_s}{L_b + L_s} \tag{6.18}$$

Kreutzer et al. performed both numerical calculations at various flow conditions and experiments at varying gas and liquid velocities using several liquids. The model predictions vary between −15 and +4% of the measured values, while the accuracy increases with increasing Re number (Figure 6.6b).

Warnier et al. proposed a mass balance-based model for predictions of two-phase pressure drop in capillaries [88]. The pressure drop over a unit cell with a length of $L_g + L_l$ is given by

$$-\left(\frac{dp}{dz}\right)_c = \frac{\Delta p_l + \Delta p_g}{L_g + L_l} = \frac{1}{L_g + L_l} \left(\frac{32\mu_l(U_g + U_l)}{d_c^2}(L_l + \delta) + \frac{7.16\sigma(3Ca_g)^{2/3}}{(1 + 3.34Ca_g^{2/3})d_c} \right) \tag{6.19}$$

The length δ is a correction on the bubble length which is equal to $d_b/3$ for gas bubbles with hemispherical caps. Continuity in a capillary requires that the overall average velocity through any cross-section of the channel perpendicular to the direction of flow is equal to the sum of the superficial gas U_g and liquid U_l velocities, which are based on the channel cross-section A. The term $(L_l + \delta)/(L_g + L_l)$ is the volumetric fraction of channel in which liquid is moving at an average velocity of $U_g + U_l$. It can be shown from the mass balance-based model that $(L_l + \delta)/(L_g + L_l)$ can be rewritten as $U_l/(U_g + U_l)$. Then the frictional pressure drop over can be written as follows:

$$-\left(\frac{dp}{dz}\right)_c = \frac{32\mu_l U_l}{d_c^2} \left(1 + \frac{7.16 \cdot 3^{2/3}}{32} \frac{A}{A_b} \frac{1}{(Ca_g^{1/3} + 3.34Ca_g)} \right) \tag{6.20}$$

The cross-sectional area of the bubble A_b depends on the liquid film thickness (d_f). In turn the liquid film thickness is a function of the bubble velocity as given by Eq. (6.19) [89].

$$\frac{d_f}{d_c} = \frac{0.67 Ca_g^{2/3}}{1 + 3.34 Ca_g^{2/3}} \tag{6.21}$$

Therefore, the pressure drop can be found by iterative approach solving system of Eqs (6.18) and (6.19). Expansion of the gas phase along the reactor length makes the superficial gas velocity and the gas bubble velocity to increase over the channel length. Therefore, the pressure is not a linear function of the channel length. The Warnier model describes experimental pressure drop within a range of −4 to +3% of the experimental data, which is a significantly smaller range than those of the Kreutzer model (Figure 6.6b).

6.3.2.2 Liquid–Liquid Microreactors

Salim *et al.* studied two-phase oil–water flows and pressure drop in horizontal microchannels [90]. The pressure drop measurements were interpreted by using the homogeneous and Lockhart–Martinelli models. The two-phase pressure drop was correlated to the single-phase pressure drop of each phase over the whole length of the capillary.

$$\left(\frac{\Delta P}{L}\right)_{TP} = \left(\frac{\Delta P}{L}\right)_c + \eta \varepsilon_d \left(\frac{\Delta P}{L}\right)_d \tag{6.22}$$

where $(\Delta P/L)_{TP}$ is the two-phase pressure drop, $(\Delta P/L)_c$ and $(\Delta P/L)_d$ are the continuous and dispersed single-phase pressure drops, respectively, ε_d is the dispersed phase volume fraction, and η is a fitting factor that depends on the wettability of the capillary wall. The empirical parameter η was determined from the experimental results, with values of 0.67 and 0.80 for the quartz and glass microchannels, respectively. The main drawback of this approach is the absence of the surface tension and slug length influence on the pressure drop. Therefore, this model underestimates the experimental data obtained by other researchers [91].

Jovanovic *et al.* [91] proposed a pressure drop model that accounts for frictional pressure losses in the discrete and continuous phases and the interfacial pressure drop:

$$\Delta P_{slug\,flow} = \Delta P_{Frictional} + \Delta P_{Interfacial} = \Delta P_{Fr,c} + \Delta P_{Fr,d} + \Delta P_I \tag{6.23}$$

The frictional pressure drop in the both phases was calculated from the Hagen–Poiseuille equation for laminar flow in a tube. The interfacial pressure drop is described by the Bretherton's solution for the pressure drop over a single bubble in a capillary (Eq. (6.24)) [92].

$$\Delta P_I = C(3Ca)^{2/3} \frac{\gamma}{d} \tag{6.24}$$

where the constant C, accounting for the influence of the interface curvature, equals 7.16. The theory of Bretherton is in a good agreement with experimental data for $Ca < 5 \cdot 10^{-3}$ and $We \ll 1$. The film velocity was found to be of negligible influence on the pressure drop. The pressure drop model (Eqs. (6.23), (6.24)) was in good agreement with the experimental data, with a mean relative error less than 5%. The value of the curvature parameter of 7.16 provides a good agreement with the experimental results in the 248 μm capillary, while in the 498 μm capillary a value of 3.48 should be used to fit the experimental data due to the asymmetrical distribution of the liquid film [91]. With increasing velocity two effects occur, namely, increase of the film thickness and deformation of the front and back meniscuses. The increase of the film thickness squeezes the slug cap, thus deforming the curvature. As a result the curvature parameter decreases at higher We numbers, which should be accounted for by modifying the curvature parameter values as a function of the slug velocity.

6.3.3
Mass Transfer

6.3.3.1 Capillary Microreactors

The conversion in a gas–liquid reaction catalyzed by a heterogeneous catalyst depends on the mass transfer at the gas–liquid and liquid–solid interfaces. A fast reaction mass transfer may substantially reduce the intrinsic reaction rate and therefore the reactor productivity. In a Taylor flow regime, three mass transfer steps can be identified [93]: (1) the direct transfer from the gas bubble through the thin liquid film to the catalyst surface (GS), (2) the transfer from caps of the gas bubbles to the liquid slug (GL), and (3) the transfer of the dissolved gas from the liquid to the catalyst surface (LS). The latter two steps occur in series and parallel to step 1.

$$k_{OV}a = k_{GS}a_{GS} + \left(\frac{1}{k_{GL}a_{GL}} + \frac{1}{k_{LS}a_{LS}}\right)^{-1} \qquad (6.25)$$

where k is the mass transfer coefficient and a is the interfacial area per unit volume. The expressions for individual values of gas–liquid and liquid–solid mass transfer were also proposed [94]. The gas–liquid mass transfer for Taylor flow through capillaries was obtained using modified relationships by Berčič and Pintar, who studied physical absorption of methane into water [95].

$$k_{GL}a_{GL} = \frac{0.1111(U_{GS} + U_{LS})^{1.19}}{(\varepsilon_L L_S)^{0.57}} \left(\frac{D_i}{D_{BA}}\right)^{0.66} \qquad (6.26)$$

where D_i is the molecular diffusion coefficient of component and D_{BA} is the molecular diffusion coefficient of benzoic acid used as reference [96]. The liquid–solid mass transfer coefficient is calculated by Eq. (6.27).

$$k_{LS}a_{LS} = \frac{0.059(U_{GS} + U_{LS})^{0.63}}{(\varepsilon_L L_S - 9.4 L_S(1 - \varepsilon_L)(\delta/d))^{0.41}} \left(\frac{D_i}{D_{BA}}\right)^{0.66} \quad (6.27)$$

However, the thickness of the liquid film (δ) depends on the capillary and Reynolds numbers [21], which should be taken into account for estimation of mass transfer on overall transformation rates in microchannels. Kreutzer et al. developed a model for the prediction of the slug length (L_s) based on pressure gradient, which is needed in order to predict the slug length [97]. Tsoligalis et al. [98] studied hydrogenation of 4-nitrobenzoic acid to 4-aminobenzoic acid in a circular capillary washcoat with a Pd/Al$_2$O$_3$ catalyst. The observed rate was in good agreement with a three-way hydrogen mass transfer model described by Eq. (6.25).

Baten and Krishna [99] compared the experimental values of the volumetric mass transfer coefficient under Taylor flow from Berčič and Pintar [95] with computational fluid dynamic simulations in 1.5 mm circular capillary diameter and found a remarkably good agreement. Applying Higbie penetration theory [100] to explain gas transfer across the cylindrical bodies and hemispherical caps of Taylor bubbles, they proposed a set of formulas (Eqs (6.28)–(6.32)) for calculating the volumetric mass-transfer coefficient ($k_L a$).

$$k_L a = k_{L,cap} a_{cap} + k_{L,film} a_{film} \quad (6.28)$$

$$k_{L,cap} = 2\left(\frac{\sqrt{2}}{\pi}\right)\left(\frac{DU_b}{d_c}\right)^{0.5} \quad (6.29)$$

$$a_{cap} = \frac{4}{L_{UC}} \quad (6.30)$$

$$k_{L,film} = \left(\frac{\sqrt{2}}{\pi}\right)\left(\frac{DU_b}{L_{film}}\right)^{0.5}, Fo < 0.1 \quad (6.31)$$

$$a_{film} = \frac{4 L_{film}}{d_c L_{UC}} \quad (6.32)$$

where the Fourier number (Fo) was defined as $Fo = Dt_{film}/\delta^2$. $k_L a$ increases with increasing liquid-film length, liquid phase diffusivity, and bubble velocity, whereas it decreases with increasing the liquid-slug length, capillary diameter, and liquid film thickness. The typical values of $k_L a$ were reported in the range of 0.5–2 s^{-1} for a capillary with internal diameter of 1 mm [101]. Vandu et al. confirmed a good agreement between theoretical values of $k_L a$ given by Eqs (6.28)–(6.32) and their experimental results on oxygen absorption dynamics [102]. The corresponding k_L values were estimated using Eq. (6.33)

$$k_L = \frac{k_{L,cap} a_{cap} + k_{L,film} a_{film}}{a_{cap} + a_{film}} \quad (6.33)$$

Equation (6.33) excludes the influence of the gas–liquid specific surface area. The k_L values were reported in the range $4-9 \times 10^{-4}$ m s^{-1} in a 1 mm diameter capillary.

6.3.3.2 Falling Film Microreactors

Commenge *et al.* observed that the formation of recirculation loops at the gas inlet and a jet effect considerably increased the mixing within the gas phase [80]. Al-Rawashdeh *et al.* studied the influence of liquid flow distribution on conversion showing that an uneven flow distribution lowers the conversion compared to ideal equal distribution only by 2% [81]. They also investigated the effect of surface wettability of the reaction plate on the actual meniscus shape at the gas–liquid interface during the absorption of CO_2 by an aqueous NaOH solution. The conversion decreased by 20% over a hydrophobic surface as compared to a hydrophilic surface due to a reduction of the interfacial area. Zhang *et al.* [82] measured liquid mass transfer coefficients in the range from 5.8 to 13.4×10^{-5} m s^{-1} and demonstrated that its value depends on the surface tension and viscosity. Hecht and Kraut [103] studied the dependence of the reaction rate along the reactor length by thermographic imaging method. They reported that enhanced mass transfer rate is a combined result of entrance effects and temperature non-uniformity along the reactor length.

For FFMR, Ziegenbalg *et al.* [83] proposed in-channel mixing structures in new types of reaction plate with herringbone and fins structures to enhance the liquid-sided mass transfer by chaotic mixing. In the former case, grooves with a width of 160 µm and a length of 660–820 µm were machined in the bottom of straight 1200 µm wide and 400 µm deep channels, giving the staggered herringbone structure. In the second approach, fins in the adjacent upstream row are shifted in the direction perpendicular to the flow creating a split and recombine flow pattern. Using plates with herringbone and fins structures, the liquid side mass transfer was increased by 2.2 and 2.1 times, respectively, as compared with the plate with straight 600 µm channels. The practical advantage of the structured plates is that the flow rate could be increased by 60–80% without loss of conversion compared to a standard reaction plate with straight channels.

6.3.4
Two-Phase Flow Distribution

Two-phase flow uniformity depends on the gas–liquid distributing principles and the uniformity of single-phase distributors. Multiphase flow distribution can be realized via branching, internal distribution, or external distribution (Figure 6.7) [104]. In an internal distributor, gas or liquid can be introduced into each microchannel through a buffer reservoir with a volume considerably larger than that of parallel microchannels [105]. In case of slug–annular or annular flow, the dynamic pressure fluctuations are reduced to a large extent, therefore, two-phase flow can be treated as steady. In this case, two single-phase distributors, such as a consecutive manifold structures [106], tree branching [107, 108],

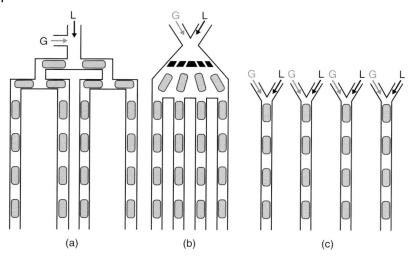

Figure 6.7 Schematic view of multiphase flow distributors. Distributor types are (a) branching, (b) internal distribution, and (c) external distribution.

large manifold volume [109], perforated meshes and grids (thin-wall screens) [110], or thick wall screens [111, 112], can be directly employed without barrier channels.

De Mas *et al.* reported that dynamic pressure fluctuations associated with slug flow in parallel microchannels lead to two-phase flow instability and eventually to channeling when some channels are being filled only with liquid and others with gas [113]. To operate parallel microchannel reactors under uniform slug flow distribution, both phases should be introduced via individual flow distributors with barrier channels that provide a pressure drop significantly higher than that across the reactor channels. The barrier channels prevent gas–liquid channeling and allow operating the reactor in the whole range of gas and liquid flow patterns [94, 113]. The distribution of each phase over the parallel microchannels is realized in a single-phase flow distributors.

Al-Rawashdeh *et al.* proposed a methodology, based on a two-phase resistive network (2PRN) model, for designing the barrier-based flow distributor [114]. A gas and liquid are distributed to parallel channels using a consecutive ladder manifold (Figure 6.8). The gas flow passes through the barrier channels and the liquid flow passes through the reactor channels. Both phases are mixed in a T-mixer and the gas–liquid flow passes out through the reaction channels.

The total flow nonuniformity $\sigma(\tilde{q})$ is split into a cumulative contribution of three flow nonuniformity factors: manifold $\sigma(\tilde{q}_M)$, barrier channels $\sigma(\tilde{q}_B)$, and mixers and reactor channel, $\sigma(\tilde{q}_M)$ nonuniformity factors is given by Eq. (6.34) [114].

$$\sigma(\tilde{q}) = \sqrt{\sigma^2(\tilde{q}_M) + \sigma^2(\tilde{q}_B) + \sigma^2(\tilde{q}_C)} \qquad (6.34)$$

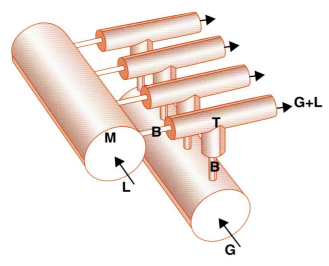

Figure 6.8 Schematic representation of barrier-based gas–liquid flow distributor for parallel microchannels. Symbols used are (G) gas, (L) liquid, (M) manifold, (B) barrier channel, and (T) T-mixer.

where $\tilde{q}$ is the flow rate per channel normalized by average flow rate through all channels, which is computed by Eq. (6.35).

$$\tilde{q}_i = \tilde{q}_{M,i} \times \tilde{q}_{B,i} \times \tilde{q}_{C,i} \tag{6.35}$$

The first term under the square root in Eq. (6.34) accounts for the flow nonuniformity due to the influence of manifold type and dimensions. The second term quantifies the flow nonuniformity due to the variation in the inner diameter of the barrier channels. The third term includes the influence of variations in the inner diameter of the mixers and reaction channels, and the influence of the second phase. It linearly increases as the variation in pressure drop over reactor channels, $\sigma(\Delta P_C)$, increases, while it inversely decreases as normalized pressure drop over the barrier channel, $\Delta \tilde{P}_B$, increases:

$$\sigma(\tilde{q}_C) = \frac{\sigma(\Delta P_C)}{\Delta \tilde{P}_B} \tag{6.36}$$

The influence of the hydraulic flow resistance ratio R_C/R_M and the number of parallel channels N on the manifold flow nonuniformity factor $\sigma(\tilde{q}_M)$ is given by Eq. (6.37).

$$\sigma(\tilde{q}_M) = 14.71 N^{1.90} \left(\frac{R_C}{R_M}\right)^{-0.96} \tag{6.37}$$

In turn, the hydraulic flow resistance is calculated from

$$R = \frac{32\mu L \lambda_{NC}}{d^2 A} \tag{6.38}$$

where λ_{NC} is noncircularity factor that depends on channel geometry [115]. The barrier flow nonuniformity factor $\sigma(\tilde{q}_B)$ is given as a function of the variation in the inner diameter of the barrier channels, $\sigma(d_B)$ for circular (Eq. (6.39)) and slit (Eq. (6.40)) geometry, respectively.

$$\sigma(\tilde{q}_B)_{circular} = 4.0 \frac{\Delta \tilde{P}_B}{1 + \Delta \tilde{P}_B} \sigma(d_B) \tag{6.39}$$

$$\sigma(\tilde{q}_B)_{slit} = 2.65 \frac{\Delta \tilde{P}_B}{1.56 + \Delta \tilde{P}_B} \sigma(d_B) \tag{6.40}$$

It can be seen from Eqs (6.39) and (6.40) that for the same fabrication tolerance the flow uniformity was 20% better with barrier channels with slit geometry as compared to those with a circular geometry.

6.4 Applications

6.4.1 Falling Film Microreactors

The transformation of aldehydes to carboxylic acids is an important reaction in organic synthesis as the carboxylic acids are versatile intermediates in a variety of synthetic transformations. Stoichiometric routes employ explosive oxidation reagents or those containing poisonous metals and they struggle with increasingly stringent environmental laws. Therefore, environmentally benign catalytic oxidation by molecular oxygen is of great importance. Several catalysts for aldehydes oxidation by molecular oxygen have been proposed, including homogeneous catalyst systems, such as $Ni(acac)_2$ [116] and Keggin-type heteropolyanions $[PW_9O_{37}(Fe_{3-x}Ni_x(OAc)_3)]^{(9+x)-}$ [117]. We have studied catalytic octanal oxidation with oxygen at 100 °C and the total pressure of 5 and 10 bar in a falling film microreactor with varying reaction plates bearing different in-channel mixing structures (Scheme 6.1). The liquid flow rate was changed in the range 3.3–17.5 ml min^{-1}. The octanal conversion at a pressure of 5 bar increases from 35 to 70% as the residence time increases from 0.7 to 2.4 s over the standard reactor plate (Figure 6.9). Similar trends were observed over the grooved and finned plates while the absolute values were higher. The full conversion of octanal was not achieved in these experiments due to a relatively short residence time.

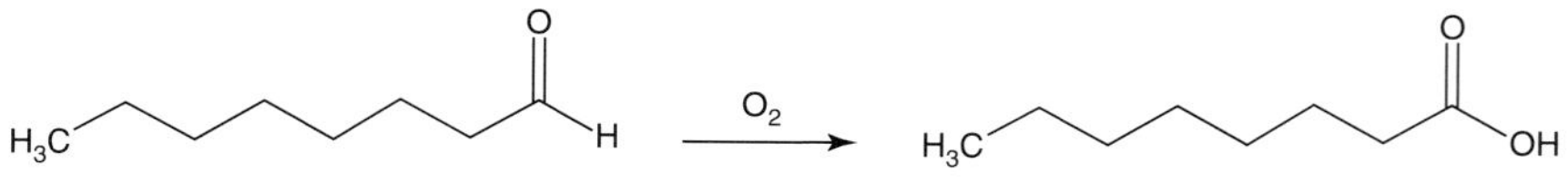

Scheme 6.1 Octanal oxidation with oxygen.

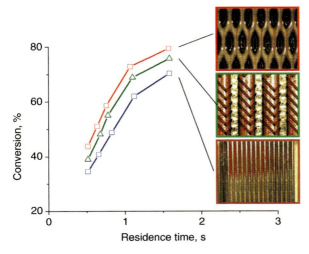

Figure 6.9 Octanal conversion in the FFMR at 100 °C and 5 bar oxygen pressure over standard (bottom image) and structured reactor plates. The grooves in grooved plates (middle image) have a width of 160 μm and a length of 660 from one side and 820 μm from the other side of the channel. In the plates with fins (top image), "streamlined fins" are arranged horizontally in rows over the complete width of the reaction plate. The rows are shifted in a way such that one fin is positioned in the center of the space between two neighbored fins of the up-and downstream row.

The conversion further increased to 89% when the reactor was tilted to an angle of 60° with respect to the gravity vector as a result of a longer residence time.

6.4.2
Capillary Microreactors

6.4.2.1 Wall Coated Catalytic Microreactors

Hydrogenations are typically fast exothermic reaction and the heat should be removed from the reactors to provide near isothermal operation. Hessel and coworkers [118] proposed a multichannel microstructured reactor combined with a heat exchanger. In this assembly, cooling channels were made in a top plate parallel to the reaction channels (Figure 6.10). The reactor was operated under Taylor flow regime. The authors reported that it was difficult to achieve a uniform phase distribution in every channel. Hornung *et al.* demonstrated a palladium-coated microcapillary flow capillary microreactor with a diameter of 146 μm in the transfer hydrogenation of several substrates [119]. They immobilized a Pd catalyst on the inner channel wall of polymeric multicapillary extrudates made from ethylene-vinyl alcohol copolymer.

A different design for multiphase selective hydrogenation reaction was proposed by Rebrov *et al.* In this design, thin mesoporous films were dip-coated onto the inner wall of fused silica capillaries [120, 121] following the original synthetic protocols documented by Grosso *et al.* [122, 123]. This route yields

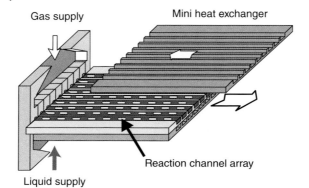

Figure 6.10 Multichannel microstructured reactor combined with a heat exchanger. The liquid and gaseous reactants are contacted through a static micromixer, and, subsequently, fed into the reaction channels. (Adapted with permission from Ref. [118]. Copyright 2000, Elsevier.)

homogeneous films with controlled amounts of catalytic species trapped within their framework [124]. The protocol involves four major steps: (i) synthesis of colloidal (metal or polymetallic) nanoparticles stabilized by a surfactant; (ii) preparation of an inorganic metal precursor sol containing colloidal nanoparticles and structure directing agent; (iii) destabilization of the sol by solvent evaporation to yield a gel; and (iv) surfactant removal and *in situ* reduction of metal nanoparticles [121].

Selective hydrogenation of the carbonyl bond in α,β-unsaturated aldehydes represents an important step in many fine chemical processes, since the desired products, unsaturated alcohols, are widely used in the fragrance and drug industries. A high selectivity of 96% toward unsaturated aldehydes was observed over Pt–Sn bimetallic catalysts supported onto a 90 nm titania film in hydrogenation of citral (3,7-dimethyl-2,6-octadienal) in wall-coated capillary microreactors. In these experiments the liquid flow was varied between 3 and 30 µl min^{-1} and the gaseous flow was varied between 0.2 and 1.0 ml min^{-1} (STP), such that the flow was in the annular regime, whereby the gas core flows through the center of the capillary and the liquid flows as a film near the walls. The authors have shown that capillary microreactors represent a promising way to reduce both the time and costs associated with chemical processes while also achieving increased selectivity and safety.

An alternative approach was proposed by Önal *et al.* [125] where a PTFE capillary microreactor was employed for citral hydrogenation. A Ru(II) homogeneous catalyst was dissolved in aqueous phase while the reactants and product were dissolved in the organic phase. The reaction was performed in a liquid–liquid slug flow regime with hydrogen bubbles dispersed in the organic phase. The productivity of the reactor depends on the liquid–liquid mass transfer rate, which in turn depends on the interfacial surface area and recirculation in the organic phase slugs. A threefold increase in the overall reaction rate was observed when the channel diameter was reduced from 1000 to 500 µm.

Cherkasov *et al.* demonstrated a novel coating method to produce rather thick uniform coatings onto the microchannel walls for selective hydrogenation reactions [126]. A capillary microreactor (10 m long, 0.53 mm internal diameter) wall coated with titania-supported Bi-poisoned Pd catalyst was used for the semihydrogenation of 2-methyl-3-butyn-2-ol providing 93% alkene yield for 100 h on stream without deactivation [127]. The yield was increased to 95% in the presence of 10 mol% pyridine in the reaction mixture. The results demonstrated that capillary reactors provided alkene the same selectivity as an ideal stirred tank reactor. A throughput in the range of 10–50 kg day^{-1} of liquid product was achieved in a single reactor with a diameter of 1.6 mm at a reaction pressure of 50 bar [128]. A further increase in production scale can be realized by numbering up and process intensification using higher temperature and pressure.

6.4.2.2 Phase Transfer Catalysis in Microreactors

Phase transfer catalysis (PTC) is a chemical method employing catalysts with the ability to penetrate the interface between two immiscible phases, and transfer the immiscible reactants into the phase where the reaction takes place. The rate of a PTC reaction is most often limited by the rate of catalyst phase transfer, usually from an aqueous to an organic phase, which leads to requirement of long duration to complete the reaction [129]. Hisamoto *et al.* performed a PTC diazo coupling reaction in a microchannel (Scheme 6.2) [130]. The reaction was performed by introducing ethyl acetate containing 5-methylresorcinol and an aqueous phase containing 4-nitrobenzene diazonium tetrafluoroborate through the two inlets of the microchip under continuous flow conditions. The flow rates of organic and aqueous phases were fixed at 10 µl min^{-1}. The specific interfacial surface area in the microchannel during parallel flow of the phases was twice as high as compared to a stirred batch reactor. This relatively small increase in the interfacial surface area already enabled efficient transfer of 5-methyl resorcinol from the organic phase to the aqueous phase where the diazo coupling reaction was completed with almost 100% conversion.

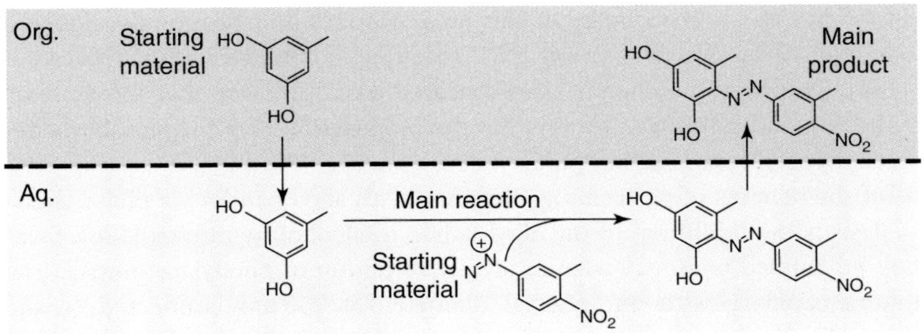

Scheme 6.2 Phase-transfer-catalyzed diazo coupling reaction.

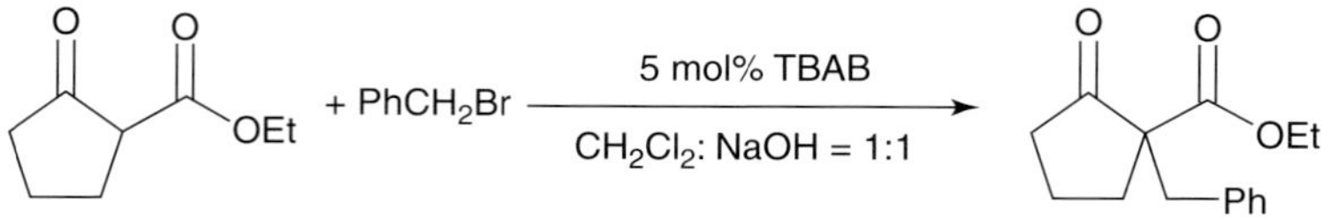

Scheme 6.3 Benzylation reaction of ethyl 2-oxocyclopentanecarboxylate with benzyl bromide.

Alkylation reactions of β-keto esters are among the most important carbon–carbon bond-forming reactions in organic synthesis [131]. Operation in the slug flow regime has been shown to be a useful tool for enchasing mass- and heat-transfer-limited phase-transfer-catalyzed alkylations [132]. Ueno et al. carried out the benzylation reaction of ethyl 2-oxocyclopentanecarboxylate with benzyl bromide in the presence of 5 mol% of tetrabutylammoniun bromide as a phase transfer catalyst in 200 μm internal diameter microchannels in segmented flow (Scheme 6.3) [132]. The flow rate was kept constant at 5.9 and 1.2 μl min^{-1}, for the mean residence time of 2 or 10 min, respectively. The conversion was 54% higher than that obtained in a batch reactor. Okamoto investigated alkylation of malonic ester with iodoethane using a phase transfer catalyst, tetrabutylammonium hydrogen sulfate, in slug flow. The conversion was 21% higher in a 500 μm internal diameter microchannel as compared to that in a batch reactor [133]. Jovanovic et al. applied an interdigital mixer–redispersion capillary assembly to prevent the liquid–liquid bubbly flow coalescence (Figure 6.11) [57]. By controlling the total flow rate and the aqueous-to-organic ratio, the bubbly flow surface-to-volume ratio was increased up to 230 700 m^2 m^{-3}. The capillary assembly was tested on the phase-transfer-catalyzed esterification of sodium benzoate and benzyl bromide into benzyl benzoate in the presence of tetra-n-butylammonium bromide (TBAB) catalyst dissolved in potassium hydroxide (Scheme 6.4). The fluidic control of the surface-to-volume ratio via the aqueous-to-organic ratio allowed for a 2.5-fold increase in conversion from 38 to 95%, at a residence time of 20 s. However, the increase of the surface-to-volume ratio promoted the increase of the hydrolysis of benzyl bromide into benzyl alcohol, thus reducing the selectivity to below 70%. By removing the hydroxide ions from the system, the side reaction of hydrolysis was suppressed, resulting in a constant selectivity of 91%.

Industry has also seen benefits in this area. Monzyk and Brophy developed a multichannel device for performing PTC reactions combined with a separation step [134]. The Merck company demonstrated a microdevice that integrates a mixer, heater, and a sonification unit for the o-alkylation of 5-bromosalicylaldehyde, achieving 60% conversion [135].

One of the benefits of performing alkylation with alkyl halides via phase transfer catalysis is the inhibition of the alkyl halide to alcohol hydrolysis [136]. Jovanovic et al. studied phase-transfer-catalyzed alkylation of phenylacetonitrile in a 10 m long capillary with an internal diameter of 250 μm connected to the Y-mixer [137]. The reaction system is composed of two phases: an organic phase containing the alkylating agent (n-butyl bromide) and phenylacetonitrile, and an

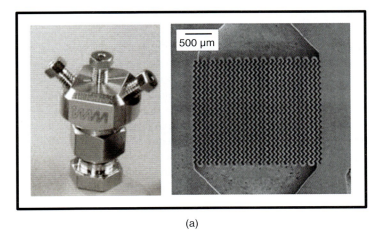

Figure 6.11 Interdigital mixer–redispersion capillary assembly. (a) Stainless steel high-pressure interdigital mixer (HPIMM) coupled with a 0.50 or 0.75 mm internal-diameter redispersion capillary. (b) The redispersion capillary consists of 1 mm long 0.25 mm inner-diameter constrictions spaced 500 mm apart.

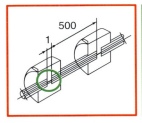
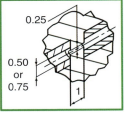

Scheme 6.4 (a) Phase-transfer-catalyzed esterification of sodium benzoate (1) and benzyl bromide (2) into benzyl benzoate (3). (b) Side reaction of hydrolysis of benzyl bromide into benzyl alcohol (4).

aqueous phase containing potassium hydroxide and the phase transfer catalyst (triethylbenzylammonium chloride, TEBA). The reaction rate increased with increasing the organic slug length due to higher mass transfer. However, the rate of the by-product formation in a consecutive reaction increased. An optimum aqueous-to-organic volumetric ratio of 2.3 was found that gave a conversion of 74% with a product selectivity of 99%. The application of the fluidic control methodology by varying the aqueous-to-organic flow ratio from 1 to 6.1 yielded an increase of 1.5–3.8 times in conversion compared to that in a batch reactor.

When kilogram quantities of material are required, large channel diameters are usually preferred (>1 mm). For biphasic reactions, however, the reaction rate can be significantly affected. Fiber reactors and microstructured packed-bed reactors are a good alternative in these cases.

6.4.2.3 Microstructured Packed Bed Reactors

Micropacked bed reactors with a typical pellet size of about 100 μm seem to offer a simple way of catalyst introduction and they are suitable for any catalyst. However, catalytic beds have a number of problems such as bed densification with time resulting in considerable pressure build-up that requires catalyst dilution with a hard inert material. In a highly exothermic reaction, the removal of heat from the reaction zone to the external walls creates considerable radial temperature gradients in the reactor. Finally, liquid channeling may result in very wide residence time distribution and therefore poor product selectivity in consecutive reactions [138]. However, for a limited class of reactions that have no side products, packed bed reactors can successfully be applied [139].

Primary amines are often seen in bioactive compounds and are used as synthetic intermediates of natural products and active pharmaceutical ingredients (APIs). Hydrogenation of nitriles to primary amines was studied by Saito *et al.* over a polysilane/SiO_2-supported Pd catalyst in a continuous micropacked bed reactor with a diameter of 4.8 mm and a length of 10 cm. The process provided almost quantitative product yield under mild reaction conditions [140]. Aromatic nitriles were reduced to primary amines, at mild reaction temperature of 60 °C using a flow rate of 0.2 ml min^{-1}. This provided a product yield of 40 mmol within 18 h continuous-flow reaction.

A process for the safe production of diazoketone, the drug intermediate of (S)-1-benzyl- 3-diazo-2-oxopropylcarbamic acid *tert*-butyl ester, was developed by Pollet *et al.* using trimethylsilyldiazomethane as a replacement for diazomethane (Scheme 6.5) [141]. They used a 4.6 mm tubular reactor with a length of 25 cm packed with 0.5 mm glass beads. By modifying the chemistry and maximizing the mixing and heat transfer, the batch process was successfully converted to a continuous flow process. A 100% yield of the diazoketone product was reported in the packed tubular reactor. Naber and Buchwald demonstrated that the combination of a toluene/water biphasic system and a microstructured packed bed reactor (MSPB) resulted in an effective protocol for palladium-catalyzed amination reactions of aryl halides in the presence of tetrabutylammonium bromide catalyst dissolved in an

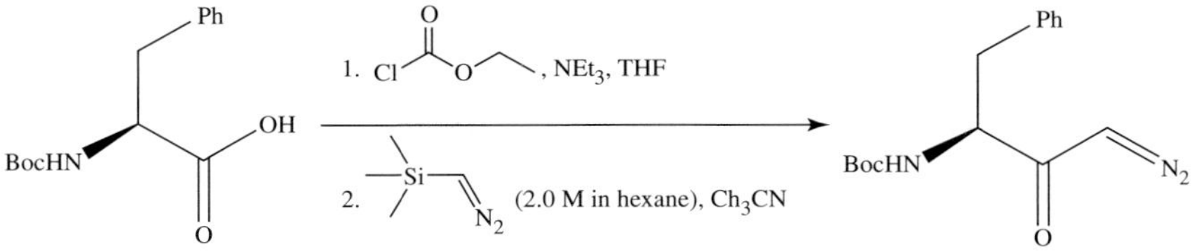

Scheme 6.5 Synthesis of the (S)-1-benzyl-3-diazo-2-oxopropylcarbamicacid *tert*-butyl ester.

aqueous potassium hydroxide solution (Scheme 6.6) [142]. They used a tubular reactor with an internal diameter of 3.8 mm packed with 60–125 μm stainless steel spheres operated in a slug flow regime. A sixfold increase in the product yield was reported in the MSPB as compared with that in a tubular reactor.

We have demonstrated a microreactor with a core–shell composite titania catalyst for direct synthesis of amide from an amine and a carboxylic acid [143, 144]. A higher reaction rate was observed in the packed bed reactor due to efficient heat transfer from the magnetic core to the titania shell. No deactivation was observed in the first 6 h on stream. The reaction rate gradually decreased by 43% after 55 h on stream. The spent catalyst was flushed with a 35 wt% H_2O_2 solution at 90 °C. The catalyst recovered showed the same initial activity as the fresh catalyst.

Irfan *et al.* used a flow reactor for the reduction of pyridines. Picolinic acid was reduced with hydrogen over a Pd/C catalyst at 80 °C. The residence time was 8.4 s. They investigated the effect of the reactor temperature and pressure on the selectivity (Scheme 6.7) [145].

Operation in the explosive regime benefits from higher space–time yield because elevated temperatures, pressures, and reactant concentrations can be used [146]. A mini-trickle bed operation was reported where a premixed H_2/O_2 gas mixture was dispersed with an aqueous solution flowing through a bed made of a supported Pd catalyst [147]. Pd/C was the most selective catalyst for hydrogen peroxide synthesis, partially because of better wetting characteristics than

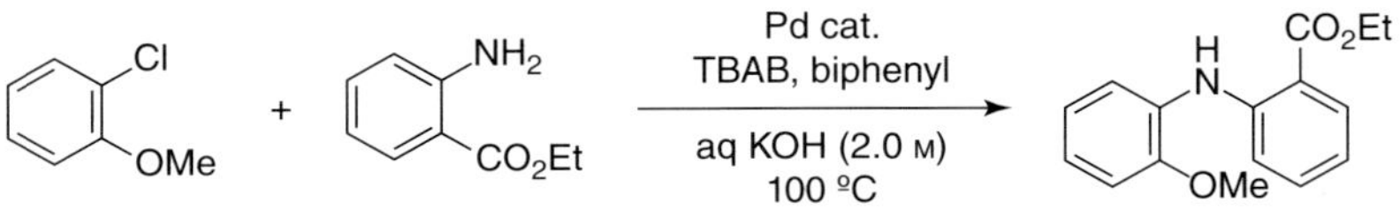

Scheme 6.6 Phase transfer catalyzed C–N cross-coupling reaction.

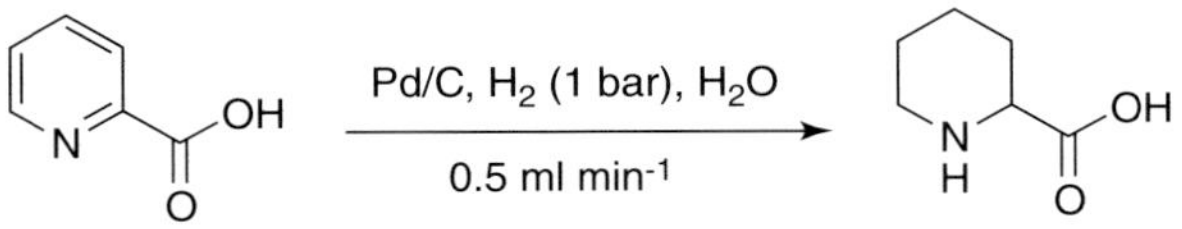

Scheme 6.7 Hydrogenation of picolinic acid.

Pd/Al$_2$O$_3$ and Pd/SiO$_2$ catalysts. Longer residence times in the reactor reduced unused hydrogen and increased production of hydrogen peroxide.

6.5
Conclusions and Outlook

The design and operation of gas–liquid microreactors require comprehensive reactor models that can well represent the interplay between transport phenomena and chemical kinetics therein. Such models are not available as yet, primarily due to complex multiphase flow characteristics in microreactors, although some preliminary efforts have already been seen [148, 149]. The formulation of reactor models that can provide a predictive description of heat and mass transfer effects, and chemical reactions under different gas–liquid flow patterns in microreactors is therefore a must in the near term [150].

One of the latest evolutions in pharmaceutical and fine chemicals manufacturing is continuous processing, and in particular, the conversion of traditional batch processes into continuous operation. The key change driver is the need to achieve cost savings, improve product quality, and process sustainability. Incremental improvements in the current (batch) technology are almost completely exhausted requiring a new disruptive technology.

Industry has been investing in microstructured reactors, ordering equipment and setting up collaborations with academic labs to build up knowledge transfer. Among the achievements in this area, a number of continuous processes, which were successfully designed and carried out in the pilot plants, are described. Although some overstatements may have been made on the advantages of continuous flow, as pointed out by Valera *et al.* [151] some of the advantages are also very clear. The flow chemistry is not going to replace batch completely, but it is strongly believed it can end up being utilized for a significant proportion of synthesis in pharmaceutical industry. The continuous processing strategy allows a just-in-time (JIT) production approach and has a reduced footprint [152]. A JIT approach produces less waste and reduces costs because, among others, the product accumulation and storage requirements are significantly less. The cost benefits for the continuous processing versus batch vary from 15 to 40% depending on the production scale, application of online instruments allowing real-time monitoring of the reaction conditions, and advanced process control [153]. The environmental benefits and safety costs are yet to be fully estimated.

References

1 Hessel, V., Angeli, P., Gavriilidis, A., and Lowe, H. (2005) Gas–liquid and gas–liquid–solid microstructured reactors: contacting principles and applications. *Industrial and Engineering Chemistry Research*, **44**, 9750.

2 Ehrich, H., Linke, D., Morgenschweis, K., Baerns, M., and Jähnisch, K. (2002)

Application of microstructured reactor technology for the photochemical chlorination of alkylaromatics. *CHIMIA International Journal for Chemistry*, **56**, 647.

3 Müller, A., Cominos, V., Hessel, V., Horn, B., Schürer, J., Ziogas, A., Jähnisch, K., Hillmann, V., Großer, V., Jam, K.A., Bazzanella, A., Rinke, G., and Kraute, M. (2005) Fluidic bus system for chemical process engineering in the laboratory and for small-scale production. *Chemical Engineering Journal*, **107**, 205.

4 Dietrich, T. (2009) *Microchemical Engineering in Practice*, John Wiley & Sons, Inc., New Jersey, 487 p.

5 Yeong, K.K., Gavriilidis, A., Zapf, R., and Hessel, V. (2003) Catalyst preparation and deactivation issues for nitrobenzene hydrogenation in a microstructured falling film reactor. *Catalysis Today*, **81**, 641.

6 Yeong, K.K., Gavriilidis, A., Zapf, R., and Hessel, V. (2004) Experimental studies of nitrobenzene hydrogenation in a microstructured falling film reactor. *Chemical Engineering Science*, **59**, 3491.

7 de Bellefon, C., Lamouille, T., Pestre, N., Bornette, F., Pennemann, H., Neuman, F., and Hessel, V. (2005) Asymmetric catalytic hydrogenations at micro-litre scale in a helicoidal single channel falling film micro-reactor. *Catalysis Today*, **110**, 179.

8 Rebrov, E.V., Duisters, T., Löb, P., Meuldijk, J., and Hessel, V. (2012) Enhancement of the liquid-sided mass transfer in a falling film catalytic microreactor by in-channel mixing structures. *Industrial and Engineering Chemistry Research*, **51**, 8719.

9 Wenn, D.A., Shaw, J.E.A., and Mackenzie, B. (2003) A mesh microreactor for 2-phase reactions. *Lab on a Chip*, **3**, 180.

10 TeGrotenhuis, W.E., Cameron, R.J., Viswanathan, V.V., and Wegeng, R.S. (2000) Solvent extraction and gas absorption using microchannel contactors. *Proceeding of the 3rd International Conference on Microreaction Technology*, Springer, Berlin, 541 p.

11 Abdallah, R., Meille, V., Shaw, J., Wenn, D., and de Bellefon, C. (2004) Gas–liquid and gas–liquid–solid catalysis in a mesh microreactor. *Chemical Communications*, **10**, 372.

12 Abdallah, R., Magnico, P., Fumey, B., and de Bellefon, C. (2006) CFD and kinetic methods for mass transfer determination in a mesh microreactor. *AIChE Journal*, **52**, 2230.

13 Mishima, K. and Hibiki, T. (1996) Some characteristics of air–water two-phase flow in small diameter vertical tubes. *International Journal of Multiphase Flow*, **22**, 703.

14 Triplett, K.A., Ghiaasiaan, S.M., Abdel-Khalik, S.I., and Sadowski, D.L. (1999) Gas–liquid two-phase flow in microchannels part I: two-phase flow patterns. *International Journal of Multiphase Flow*, **25**, 377.

15 Yang, C.-Y. and Shieh, C.-C. (2001) Flow pattern of air–water and two-phase R-134a in small circular tubes. *International Journal of Multiphase Flow*, **27**, 1163.

16 Pohorecki, R., Sobieszuk, P., Kula, K., Moniuk, W., Zielinski, M., Cyganski, P., and Gawinski, P. (2008) Hydrodynamic regimes of gas–liquid flow in a microreactor channel. *Chemical Engineering Journal*, **135**, 185.

17 Chinnov, E.A. and Kabov, O.A. (2008) Two-phase flow in a short mini-channel. *Technical Physics Letters*, **34**, 41.

18 Chinnov, E.A., Guzanov, V.V., Cheverda, V., Markovich, D.M., and Kabov, O.A. (2009) Regimes of two-phase flow in short rectangular channel. *Microgravity Science and Technology*, **21**, S199.

19 Hassan, I., Vaillancourt, M., and Pehlivan, K. (2005) Two-phase flow regime transitions in microchannels: a comparative experimental study. *Microscale Thermophysical Engineering*, **9**, 165.

20 Fries, D.M., Waelchli, S., and von Rohr, P.R. (2008) Gas–liquid two-phase flow in meandering microchannels. *Chemical Engineering Journal*, **135**, S37.

21 Warnier, M.J.F., Rebrov, E.V., de Croon, M.H.J.M., Hessel, V., and Schouten, J.C. (2008) Gas hold-up and liquid film thickness in Taylor flow in rectangular micro channels. *Chemical Engineering Journal*, **135**, S153.

22 Haverkamp, V., Hessel, V., Löwe, H., Menges, G., Warnier, M.J.F., Rebrov, E.V., de Croon, M.H.J.M., Schouten, J.C., and Liauw, M. (2006) Hydrodynamics and mixer-induced bubble formation in microbubble columns with single and multiple channels. *Chemical Engineering and Technology*, **29**, 1015.

23 Shao, N., Salman, W., Gavriilidis, A., and Angeli, P. (2008) CFD simulations of the effect of inlet conditions on Taylor flow formation. *International Journal of Heat and Fluid Flow*, **29**, 1603.

24 Losey, M.W., Schmidt, M.A., and Jensen, K.F. (2001) Microfabricated multiphase packed-bed reactors: characterization of mass transfer and reactions. *Industrial and Engineering Chemistry Research*, **40**, 2555.

25 van Herk, D., Kreutzer, M.T., Makkee, M., and Moulijn, J.A. (2005) Scaling down trickle bed reactors. *Catalysis Today*, **106**, 227.

26 Tadepalli, S., Halder, R., and Lawal, A. (2007) Catalytic hydrogenation of o-nitroanilose in a microreactor: reactor performance and kinetic studies. *Chemical Engineering Science*, **62**, 2663.

27 Trachsel, F., Hutter, C., and von Rohr, P.R. (2008) Transparent silicon/glass microreactor for high pressure and high-temperature reactions. *Chemical Engineering Journal*, **135**, S309.

28 McGovern, S., Harish, G., Pai, C.S., Mansfield, W., Taylor, J.A., Pau, S., and Besser, R.S. (2008) Multiphase flow regimes for hydrogenation in a catalyst-trap microreactor. *Chemical Engineering Journal*, **135**, S229.

29 McGovern, S., Harish, G., Pai, C.S., Mansfield, W., Taylor, J.A., Pau, S., and Besser, R.S. (2009) Investigation of multiphase hydrogenation in a catalyst-trap microreactor. *Journal of Chemical Technology and Biotechnology*, **84**, 382.

30 Krishnamurthy, S. and Peles, Y. (2007) Gas–liquid two-phase flow across a bank of micropillars. *Physics of Fluids*, **19**, 043302.

31 Krishnamurthy, S. and Peles, Y. (2009) Surface tension effects on adiabatic gas–liquid flow across micro pillars. *International Journal of Multiphase Flow*, **35**, 55.

32 Xu, J. (1999) Experimental study on gas–liquid two-phase flow regimes in rectangular channels with mini gaps. *International Journal of Heat and Fluid Flow*, **20**, 422.

33 de Loos, S.R.A., van der Schaaf, J., Tiggelaar, R.M., Nijhuis, T.A., de Croon, M.H.J.M., and Schouten, J.C. (2010) Gas–liquid dynamics at low Reynolds numbers in pillared rectangular micro channels. *Microfluidics Nanofluidics*, **9**, 131.

34 Stemmet, C.P., Jongmans, J.N., van der Schaaf, J., Kuster, B.F.M., and Schouten, J.C. (2005) Hydrodynamics of gas–liquid counter-current flow in solid foam packings. *Chemical Engineering Science*, **60**, 6422.

35 Stemmet, C.P., van der Schaaf, J., Kuster, B.F.M., and Schouten, J.C. (2006) Solid foam packings for multiphase reactors: modelling of liquid holdup and mass transfer. *Chemical Engineering Research and Design*, **84**, 1134.

36 Stemmet, C.P., Meeuwse, M., van der Schaaf, J., Kuster, B.F.M., and Schouten, J.C. (2007) Gas–liquid mass transfer and axial dispersion in solid foam packings. *Chemical Engineering Science*, **62**, 5444.

37 Fukushima, S. and Kusaka., K. (1979) Gas–liquid mass transfer and hydrodynamic flow region in packed columns with cocurrent upward flow. *Journal of Chemical Engineering of Japan*, **12**, 296.

38 Li, P., Li, T., Zhou, J.H., Sui, Z.J., Dai, Y.C., Yuan, W.K., and Chen, D. (2006) Synthesis of carbon nanofiber/graphite-felt composite as a catalyst. *Microporous Mesoporous Materials*, **95**, 1.

39 de Loos, S.R.A., van der Schaaf, J., de Croon, M.H.J.M., Nijhuis, T.A., and Schouten, J.C. (2011) Enhanced liquid–solid mass transfer in microchannels by a layer of carbon nanofibers. *Chemical Engineering Journal*, **167**, 671.

40 Höller, V., Radevik, K., Yuranov, I., Kiwi-Minsker, L., and Renken, A. (2001) Reduction of nitrite-ions in water over Pd-supported on structured fibrous materials. *Applied Catalysis B: Environmental*, **32**, 143.

41 Crespo-Quesada, M., Grasemann, M., Semagina, N., Renken, A., and Kiwi-Minsker, L. (2009) Kinetics of the solvent-free hydrogenation of 2-methyl-3-butyn-2-ol over a structured Pd-based catalyst. *Catalysis Today*, **147**, 247.

42 Semagina, N., Grasemann, M., Xanthopoulos, N., Renken, A., and Kiwi-Minsker, L. (2007) Structured catalyst of Pd/ZnO on sintered metal fibers for 2-methyl-3-butyn-2-ol selective hydrogenation. *Journal of Catalysis*, **251**, 213.

43 Grasemann, M., Renken, A., Kashid, M., and Kiwi-Minsker, L. (2010) A novel compact reactor for three-phase hydrogenations. *Chemical Engineering Science*, **65**, 364.

44 Akbar, M.K., Plummer, D.A., and Ghiaasiaan, S.M. (2003) On gas–liquid two-phase flow regimes in microchannels. *International Journal of Multiphase Flow*, **29**, 855.

45 Serizawa, A., Feng, Z., and Kawara, Z. (2002) Two-phase flow in microchannels. *Experimental Thermal and Fluid Science*, **26**, 703.

46 Cubaud, T. and Ho, C.M. (2004) Transport of bubbles in square microchannels. *Physics of Fluids*, **16**, 4575.

47 Chen, L., Tian, Y.S., and Karayiannis, T.G. (2006) The effect of tube diameter on vertical two-phase regimes in small tubes. *International Journal of Heat and Mass Transfer*, **49**, 4220.

48 Baroud, C.N. and Willaime, H. (2004) Multiphase flows in microfluidics. *Comptes Rendus Physique*, **5**, 547.

49 Zhao, Y., Chen, G., and Yuan, Q. (2006) Liquid–Liquid two-phase flow patterns in a rectangular microchannel. *AIChE Journal*, **52**, 4052.

50 Dessimoz, A.L., Cavin, L., Renken, A., and Kiwi-Minsker, L. (2008) Liquid–Liquid two-phase flow patterns and mass transfer characteristics in rectangular glass microreactors. *Chemical Engineering Science*, **63**, 4035.

51 Kashid, M.N., Harshe, Y.M., and Agar, D.W. (2007) Liquid–liquid slug flow in a capillary: an alternative to suspended drop or film contactors. *Industrial and Engineering Chemistry Research*, **46**, 8420.

52 Cherlo, S.K. and Pushpavanam, S. (2010) Experimental and numerical investigations of two-phase (liquid–liquid) flow behavior in rectangular microchannels. *Industrial and Engineering Chemistry Research*, **49**, 893.

53 Nisisako, T., Torii, T., and Higuchi, T. (2002) Droplet formation in a microchannel network. *Lab on a Chip*, **2**, 24.

54 Uada, A.S., Lorenceau, E., Link, D.R., Kaplan, P.D., Stone, H.A., and Weitz, D.A. (2005) Monodisperse double emulsions generated from a microcapillary device. *Science*, **308**, 537.

55 Kashid, M.N. and Agar, D.W. (2007) Hydrodynamics of liquid–liquid slug flow capillary microreactor: flow regimes, slug size and pressure drop. *Chemical Engineering Journal*, **131**, 1.

56 Garstecki, P., Fuerstman, M.J., Stonec, H.A., and Whitesides, G.M. (2006) Formation of droplets and bubbles in a microfluidic T-junction-scaling and mechanism of break-up. *Lab on a Chip*, **6**, 437.

57 Jovanovic, J., Hengeveld, W., Rebrov, E.V., Nijhuis, T.A., Hessel, V., and Schouten, J.C. (2011) Redispersion microreactor system for phase transfer catalyzed esterification. *Chemical Engineering & Technology*, **34**, 1691.

58 Wada, Y., Schmidt, M.A., and Jensen, K.F. (2006) Flow distribution and ozonolysis in gas–liquid multichannel microreactors. *Industrial and Engineering Chemistry Research*, **45**, 8036.

59 Zimmerman, S.P. and Ng, K.M. (1986) Liquid distribution in trickling flow trickle-bed reactors. *Chemical Engineering Science*, **41**, 861.

60 Chu, C.F. and Ng, K.M. (1989) Flow in packed tubes with a small tube to particle diameter ratio. *AIChE Journal*, **35**, 148.

61 Chander, A., Kundu, A., Bej, S.K., Dalai, A.K., and Vohra, D.K. (2001) Hydrodynamic characteristics of cocurrent upflow and downflow of gas and liquid in a fixed bed reactor. *Fuel*, **80**, 1043.

62 Márquez, N., Castaño, P., Makkee, M., Moulijn, J.A., and Kreutzer, M.T. (2008)

Dispersion and holdup in multiphase packed bed microreactors. *Chemical Engineering and Technology*, **31**, 1130.

63 Colombo, A.J., Baldi, G., and Sicardi, S. (1976) Solid–liquid contacting effectiveness in trickle bed reactors. *Chemical Engineering Science*, **31**, 1101.

64 Kulkarni, R.R., Wood, J., Winterbottom, J.M., and Stitt, E.H. (2005) Effect of fines and porous catalyst on hydrodynamics of trickle bed reactors. *Industrial and Engineering Chemistry Research*, **44**, 9497.

65 Iliuta, I., Thyrion, F.C., Muntean, O., and Giot, M. (1996) Residence time distribution of the liquid in gas–liquid cocurrent upflow fixed-bed reactors. *Chemical Engineering Science*, **51**, 4579.

66 Bej, S.K., Dabral, R.P., Gupta, P.C., Mittal, K.K., Sen, G.S., Kapoor, V.K., and Dalai, A.K. (2000) Studies on the performance of a microscale trickle bed reactor using different sizes of diluent. *Energy and Fuels*, **14**, 701.

67 Gunjal, P.R., Ranade, V.V., and Chaudhari, R.V. (2003) Liquid distribution and RTD in trickle bed reactors: experiments and CFD simulations. *Canadian Journal of Chemical Engineering*, **81**, 821.

68 Van Swaaij, W.P.M., Charpentier, J.C., and Villermaux, J. (1969) Residence time distribution in the liquid phase of trickle flow in packed columns. *Chemical Engineering Science*, **24**, 1083.

69 Iliuta, I., Thyrion, F.C., and Muntean, O. (1998) Axial dispersion of liquid in gas–liquid co-current downflow and upflow fixed bed reactors with porous particles. *Chemical Engineering Research and Design*, **76**, 64.

70 Iliuta, I., Thyrion, F.C., Muntean, O., and Giot, M. (1996) Residence time distribution of liquid in the gas–liquid co-current upflow fixed bed reactors. *Chemical Engineering Science*, **51**, 4579.

71 Tsamatsoulis, D. (2001) The effect of particle dilution on wetting efficiency and liquid film thickness in small trickle beds. *Chemical Engineering Communications*, **185**, 67.

72 Hickman, D.A., Weidenbach, M., and Daniel, P. (2004) A comparison of a batch recycle reactor and an integral reactor with fines for scale-up of an industrial trickle bed reactor from laboratory data. *Chemical Engineering Science*, **59**, 5425.

73 Zanfir, M., Gavriilidis, A., Wille, Ch., and Hessel, V. (2005) Carbon dioxide absorption in a falling film microstructured reactor: experiments and modelling. *Industrial and Engineering Chemistry Research*, **44**, 1742.

74 Haji-Sheikh, A. and Beck, J.V. (2007) Entrance effect on heat transfer to laminar flow through passages. *International Journal of Heat and Mass Transfer*, **50**, 3340.

75 Mishan, Y., Mosyak, A., and Pogrebnyak, E. (2007) Effect of developing flow and thermal regime on momentum and heat transfer in micro-scale heat sink. *International Journal of Heat and Mass Transfer*, **50**, 3100.

76 Li, J., Peterson, G.P., and Cheng, P. (2004) Three-dimensional analysis of heat transfer in a micro-heat sink with single phase flow. *International Journal of Heat and Mass Transfer*, **47**, 4215.

77 Yang, Y.T. and Peng, H.S. (2008) Numerical study of pin-fin heat sink with un-uniform fin height design. *International Journal of Heat and Mass Transfer*, **51**, 4788.

78 Hardt, S., Ehrfeld, W., Hessel, V., and van den Bussche, K.M. (2003) Strategies for size reduction of microreactors by heat transfer enhancement effects. *Chemical Engineering Communications*, **190**, 540.

79 Sahiti, N. (2008) *Pin Fin Heat Transfer Surfaces*, VDM Verlag Dr. Müller, Saarbrücken.

80 Commenge, J.M., Obein, T., Genin, G., Framboisier, X., Rode, S., Schanen, V., Pitiot, P., and Matlosz, M. (2006) Gas-phase residence time distribution in a falling-film microreactor. *Chemical Engineering Science*, **61**, 597.

81 Al-Rawashdeh, M., Hessel, V., Lob, P., Mevissen, K., and Schönfeld, F. (2008) Pseudo 3-D simulation of a falling film microreactor based on realistic channel and film profiles. *Chemical Engineering Science*, **63**, 5149.

82 Zhang, H., Chen, G., Yue, J., and Yuan, Q. (2009) Hydrodynamics and mass transfer

of gas–liquid flow in a falling film microreactor. *AIChE Journal*, **55**, 1110.
83 Ziegenbalg, D., Löb, P., Al-Rawashdeh, M., Kralisch, D., Hessel, V., and Schönfeld, F. (2010) Use of 'smart interfaces' to improve the liquid-sided mass transport in a falling film microreactor. *Chemical Engineering Science*, **65**, 3557.
84 Van Elk, E.P., Knaap, M.C., and Versteeg, G.F. (2007) Application of the penetration theory for gas–liquid mass transfer without liquid bulk. *Chemical Engineering Research and Design*, **85**, 516.
85 Lockhart, R.W. and Martinelli, R.C. (1949) Proposed correction of data for isothermal two-phase component flow in pipes. *Chemical Engineering Progress*, **45**, 39.
86 Chisholm, D. (1967) A theoretical basis for the Lockhart–Martinelli correlation for two-phase flow. *International Journal of Heat and Mass Transfer*, **10**, 1767.
87 Kreutzer, M.T., Kapteijn., F., Moulijn, J.A., Kleijn, C.R., and Heiszwolf, J.J. (2005) Inertial and interfacial effects on pressure drop of Taylor flow in capillaries. *AIChE Journal*, **51**, 2428.
88 Warnier, M.J.F., de Croon, M.H.J.M., Rebrov, E.V., and Schouten, J.C. (2010) Pressure drop of gas–liquid Taylor flow in round micro capillaries for low to intermediate Reynolds numbers. *Microfluidics Nanofluidics*, **8**, 33.
89 Aussillous, P. and Quere, D. (2000) Quick deposition of a fluid on the wall of a tube. *Physics of Fluids*, **12**, 2367.
90 Salim, A., Fourar, M., Pironon, J., and Sausse, J. (2008) Oil–water two-phase flow in microchannels: flow patterns and pressure drop measurements. *Canadian Journal of Chemical Engineering*, **86**, 978–988.
91 Jovanovic, J., Rebrov, E.V., Nijhuis, T.A., Hessel, V., and Schouten, J.C. (2011) Liquid–liquid slug flow: hydrodynamics and pressure drop. *Chemical Engineering Science*, **66**, 42.
92 Bretherton, F.P. (1961) The motion of long bubbles in tubes. *Journal of Fluid Mechanics*, **10**, 166.
93 Kreutzer, M.T., Du, P., Heiszwolf, J.J., Kapteijn, F., and Moulijn, J.A. (2001) Mass transfer characteristics of three-phase monolith reactors. *Chemical Engineering Science*, **56**, 6015.
94 Kashid, M., Gupta, A., Renken, A., and Kiwi-Minsker, L. (2010) Numbering-up and mass transfer studies of liquid–liquid two-phase microstructured reactors. *Chemical Engineering Journal*, **158**, 233.
95 Berčič, G. and Pintar, A. (1997) The role of the gas bubbles and liquid length on mass transport in the Taylor flow through capillaries. *Engineering Science*, **52**, 3709.
96 Wilke, C.R. and Chang, P. (1955) Correlation of diffusion coefficients in dilute solutions. *AIChE Journal*, **1**, 264.
97 Kreutzer, M.T., van der Eijnden, M.G., Menno, G., Kapteijn, F., Moulijn, J.A., and Heiszwolf, J.J. (2005) The pressure drop experiment to determine slug lengths in multiphase monoliths. *Catalysis Today*, **105**, 667.
98 Tsoligkas, A.N., Simmons, M.J.H., and Wood, J. (2007) Two phase gas–liquid reaction studies in a circular capillary. *Chemical Engineering Science*, **62**, 5397.
99 Baten, J.M. and Krishna, R. (2004) CFD simulations of mass transfer from Taylor bubbles rising in circular capillaries. *Chemical Engineering Science*, **59**, 2535.
100 Higbie, R. (1935) The rate of absorption of a pure gas into a still liquid during short periods of exposure. *Transactions of the American Institute of Chemical Engineers*, **31**, 365.
101 Liu, D. and Wang, S. (2011) Gas–liquid mass transfer in Taylor flow through circular capillaries. *Industrial and Engineering Chemistry Research*, **50**, 2323.
102 Vandu, C.O., Liu, H., and Krishna, R. (2005) Mass transfer from Taylor bubbles rising in single capillaries. *Chemical Engineering Science*, **60**, 6430.
103 Hecht, K. and Kraut, M. (2010) Thermographic investigations of a microstructured thin film reactor for gas/liquid contacting. *Industrial and Engineering Chemistry Research*, **49**, 10889.
104 Wada, Y., Schmidt, M., and Jensen, K. (2006) Flow distribution and ozonolysis in gas–liquid multichannel

microreactors. *Industrial and Engineering Chemistry Research*, **45**, 8036.
105 Chambers, R.D., Fox, M.A., Holling, D., Nakano, T., Okazoeb, T., and Sandford, G. (2005) Elemental fluorine. Part 16. Versatile thin-film gas–liquid multi-channel microreactors for effective scale-out. *Lab on a Chip*, **5**, 191.
106 Commenge, J.M., Falk, L., Corriou, J.P., and Matlosz, M. (2002) Optimal design for flow uniformity in microchannel reactors. *AIChE Journal*, **48**, 345.
107 Tondeur, D. and Luo, L. (2004) Design and scaling laws of ramified fluid distributors by the constructal approach. *Chemical Engineering Science*, **59**, 1799.
108 Bejan, A. and Lorente, S. (2006) Constructal theory of generation of configuration in nature and engineering. *Journal of Applied Physics*, **100**, 041301.
109 Schenk, R., Hessel, V., Hofmann, C., Löwe, H., and Schönfeld, F. (2003) Novel liquid-flow splitting unit specifically made for numbering-up of liquid/liquid chemical microprocessing. *Chemical Engineering and Technology*, **26**, 1271.
110 Idelchik, I.E. (1991) *Fluid Dynamics of Industrial Equipment: Flow Distribution Design Methods*, Taylor & Francis.
111 Rebrov, E.V., Ekatpure, R.P., de Croon, M.H.J.M., and Schouten, J.C. (2007) Design of a thick walled screen for flow equalization in microstructured reactors. *Journal of Micromechanics and Microengineering*, **17**, 633.
112 Rebrov, E.V., Ismagilov, I.Z., Ekatpure, R.P., de Croon, M.H., and Schouten, J.C. (2007) Header design for flow equalization in microstructured reactors. *AIChE Journal*, **53**, 28.
113 de Mas, N., Gunther, A., Kraus, T., Schmidt, M., and Jensen, K. (2005) Scaled-out multilayer gas–liquid microreactor with integrated velocimetry sensors. *Industrial and Engineering Chemistry Research*, **44**, 8997.
114 Al-Rawashdeh, M., Nijhuis, T.A., Rebrov, E.V., Hessel, V., and Schouten, J.C. (2012) Design methodology for barrier-based two phase flow distributor. *AIChE Journal*, **58**, 3482.
115 Amador, C., Gavriilidis, A., and Angeli, P. (2004) Flow distribution in different microreactor scale-out geometries and the effect of manufacturing tolerances and channel blockage. *Chemical Engineering Journal*, **101**, 379.
116 Howarth, J. (2000) Oxidation of aromatic aldehydes in the ionic liquid [bmim]PF_6. *Tetrahedron Letters*, **41**, 6627.
117 Mizuno, N., Nozaki, C., Hirose, T.O., Tateishi, M., and Iwamoto, M. (1997) Liquid-phase oxygenation of hydrocarbons with molecular oxygen catalyzed by Fe_2Ni-substituted Keggin-type heteropolyanion. *Journal of Molecular Catalysis A: Chemical*, **117**, 159.
118 Jähnisch, K., Baerns, M., Hessel, V., Ehrfeld, W., Haverkamp, V., Löwe, H., Wille, Ch., and Guber, A. (2000) Direct fluorination of toluene using elemental fluorine in gas/liquid microreactors. *Journal of Fluorine Chemistry*, **105**, 117.
119 Hornung, C.H., Hallmark, B., Mackely, M.R., Baxendale, I.R., and Ley, S.V. (2010) A palladium wall coated microcapillary reactor for use in continuous flow transfer hydrogenation. *Advanced Synthesis and Catalysis*, **352**, 1736.
120 Rebrov, E.V., Berenguer-Murcia, A., Skelton, H.E., Johnson, B.F.G., Wheatley, A.E.H., and Schouten, J.C. (2009) Capillary microreactors wall-coated with mesoporous titania thin film catalyst supports. *Lab on a Chip*, **9**, 503.
121 Rebrov, E.V., Berenguer-Murcia, A., Wheatley, A.E.H., Johnson, B.F.G., and Schouten, J.C. (2009) Thin catalytic coatings on microreactor walls: a way to make industrial processes more efficient. *Chimica oggi – Chemistry Today*, **27**, 45.
122 Grosso, D., Balkenende, A.R., Albouy, P.A., Ayral, A., Amenitsch, H., and Babonneau, F. (2001) Two-dimensional hexagonal mesoporous silica films prepared from block copolymers: detailed characterization and formation mechanism. *Chemistry of Materials*, **13**, 1848.
123 Grosso, D., Cagnol, F., Soler-Illia, G.J.A.A., Crepaldi, E.L., Amenitsch, H., Brunet-Bruneau, A., Bourgeois, A., and Sanchez, C. (2004) Fundamental of mesostructuring through evaporation-induced self-assembly. *Advanced Functional Materials*, **14**, 309.

124 Protasova, L.N., Rebrov, E.V., Skelton, H.E., Wheatley, A.E.H., and Schouten, J.C. (2011) Kinetic study of liquid-phase hydrogenation of citral on Au/TiO$_2$ and Pt-Sn/TiO$_2$ thin films in capillary microreactors. *Applied Catalysis A: General*, **399**, 12.

125 Önal, Y., Lucas, M., and Claus, P. (2005) Einsatz eines kapillar-mikroreaktors für die selektive hydrierung von α,β-ungesättigten aldehyden in der wässrigen mehrphasenkatalyse. *Chemie Ingenieur Technik*, **77**, 101.

126 Cherkasov, N., Ibhadon, A.O., and Rebrov, E.V. (2015) Novel synthesis of thick wall coatings of titania supported Bi poisoned Pd catalysts and application in selective hydrogenation of acetylene alcohols in capillary microreactors. *Lab on a Chip*, **15**, 1952.

127 Cherkasov, N., Ibhadon, A.O., and Rebrov, E.V. (2016) Solvent-free semihydrogenation of acetylene alcohols in a capillary reactor coated with a Pd-Bi/TiO$_2$ catalyst. *Applied Catalysis A: General*, **515**, 108.

128 Cherkasov, N., Al-Rawashdeh, M., Ibhadon, A.O., and Rebrov, E.V. (2016) Scale up study of capillary microreactors in solvent-free semihydrogenation of 2-methyl-3-butyn-2-ol. *Catalysis Today*, **273**, 205.

129 Starks, C., Liotta, C., and Halpern, M. (1994) *Phase-Transfer Catalysis: Fundamentals, Applications and Industrial Perspectives*, Chapman & Hall, London.

130 Hisamoto, H., Saito, T., Tokeshi, M., Hibara, A., and Kitamori, T. (2001) Fast and high conversion phase-transfer synthesis exploiting the liquid–liquid interface formed in a microchannel chip. *Chemical Communications*, **24**, 2662.

131 Dehmlow, E.V. and Dehmlow, S.S. (1993) *Phase Transfer Catalysis*, 3rd edn, Wiley-VCH Verlag GmbH, Weinheim, Germany.

132 Ueno, M., Hisamoto, H., Kitamori, T., and Kobayashi, S. (2003) Phase-transfer alkylation reactions using microreactors. *Chemical Communications*, **8**, 936.

133 Okamoto, H. (2006) Effect of alternating pumping of two reactants into a microchannel on a phase transfer reaction. *Chemical Engineering and Technology*, **29**, 504.

134 Monzyk, B. and Brophy, J.H. (2004) Multiphasic microchannel reactions. Patent WO2004037399A2.

135 Wurziger, H., Pieper, G., Schmelz, M., and Schwesinger, N. (2002) Use of a microreaction channel with a piezo element. Patent WO0249737A1.

136 Durst, H.D. and Liebeskind, L. (1974) Phase transfer catalysis. The acetoacetic ester condensation. *Journal of Organic Chemistry*, **39**, 3271.

137 Jovanovic, J., Rebrov, E.V., Nijhuis, T.A., Hessel, V., and Schouten, J.C. (2010) Phase transfer catalysis in segmented flow in a microchannel: fluidic control of selectivity and productivity. *Industrial and Engineering Chemistry Research*, **49**, 2681.

138 Alsolami, B.H., Berger, R.J., Makkee, M., and Moulijn, J.A. (2013) Catalyst performance testing in multiphase systems: implications of using small catalyst particles in hydrodesulfurization. *Industrial and Engineering Chemistry Research*, **52**, 9069.

139 Irfan, M., Glasnov, T.N., and Kappe, C.O. (2011) Heterogeneous catalytic hydrogenation reactions in continuous-flow reactors. *ChemSusChem*, **4**, 300.

140 Saito, Y., Ishitani, H., Ueno, M., and Kobayashi, S. (2017) Selective hydrogenation of nitriles to primary amines catalyzed by a polysilane/SiO$_2$-supported palladium catalyst under continuous-flow conditions. *Chemistry Open*, **2**, 211.

141 Pollet, P., Cope, E.D., Kassner, M.K., Charney, R., Terett, S.H., Richman, K.W., Dubay, W., Stringer, J., Eckert, C.A., and Liotta, C.L. (2009) Production of (S)-1-benzyl-3-diazo-2-oxopropylcarbamic acid *tert*-butyl ester, a diazoketone pharmaceutical intermediate, employing a small scale continuous reactor. *Industrial and Engineering Chemistry Research*, **48**, 7032.

142 Naber, J.R. and Buchwald, S.L. (2010) Packed-bed reactors for continuous-flow C-N cross-coupling. *Angewandte Chemie, International Edition*, **49**, 9469.

143 Liu, Y., Gao, P., Cherkasov, N., and Rebrov, E.V. (2016) Direct amide synthesis over core–shell TiO_2@$NiFe_2O_4$ catalysts in a continuous flow radiofrequency-heated reactor. *RSC Advances*, **103**, 100997.

144 Liu, Y., Cherkasov, N., Gao, P., Fernández, J., Lees, M.R., and Rebrov, E.V. (2017) The enhancement of direct amide synthesis reaction rate over TiO_2@SiO_2@$NiFe_2O_4$ magnetic catalysts in the continuous flow under radiofrequency heating. *Journal of Catalysis*, **355**, 120.

145 Irfan, M., Petricci, E., Glasnov, T.N., Taddei, M., and Kappe, C.O. (2009) Continuous flow hydrogenation of functionalized pyridines. *European Journal of Organic Chemistry*, **2009**, 1327.

146 Pennemann, H., Hessel, V., and Löwe, H. (2004) Chemical micro process technology – from laboratory scale to production. *Chemical Engineering Science*, **59**, 4789.

147 Inoue, T., Schmidt, M.A., and Jensen, K.F. (2007) Microfabricated multiphase reactors for the direct synthesis of hydrogen peroxide from hydrogen and oxygen. *Industrial and Engineering Chemistry Research*, **46**, 1153.

148 Zanfir, M., Gavriilidis, A., Wille, Ch., and Hessel, V. (2005) Carbon dioxide absorption in a falling film microstructured reactors: experiments and modeling. *Industrial and Engineering Chemistry Research*, **44**, 1742.

149 Haverkamp, V., Hessel, V., Liauw, M.A., Löwe, H., and Menges, M.G. (2007) Reactor model for fast reactions in the micro-bubble column and validation. *Industrial and Engineering Chemistry Research*, **46**, 8558.

150 Chen, G., Yue, J., and Yuan, Q. (2008) Gas–liquid microreaction technology: recent developments and future challenges. *Chinese Journal of Chemical Engineering*, **16**, 663.

151 Valera, F.E., Quaranta, M., Moran, A., Blacker, J., Armstrong, A., Cabral, J.T., and Blackmond, D.G. (2010) The flow's the thing. or is it? Assessing the merits of homogeneous reactions in flask and flow. *Angewandte Chemie, International Edition*, **49**, 2478.

152 Warman, M. (2011) Continuous processing in secondary production. *Chemical Engineering in the Pharmaceutical Industry: R&D to Manufacturing* (ed. D.J. am Ende), John Wiley & Sons, Inc., New Jersey.

153 Calabrese, G.S. and Pissavini, S. (2011) From batch to continuous flow processing in chemicals manufacturing. *AIChE Journal*, **57**, 828.

7
Process Intensification through Continuous Manufacturing: Implications for Unit Operation and Process Design

Sebastian Falß, Nicolai Kloye, Manuel Holtkamp, Angelina Prokofyeva, Thomas Bieringer, and Norbert Kockmann

Continuous manufacturing for small- to medium-scale chemical production is a topic currently receiving great attention from industry and academia. The present chapter attempts to give an overview over the status quo of this technology. First, we will consider the general potential of continuous processes and discuss why the technology has experienced an increased interest over the last years (Section 7.1). The implementation of continuous processes heavily depends on the availability of suitable equipment. In a next step, we will therefore address the question what kind of equipment can be used, which advantages it offers, and where we still see white spots (Section 7.2). While equipment availability is a prerequisite for the successful application of continuous processing, efficient process development and implementation taking into account the approach's specific characteristics are required to leverage the full potential of the technology. This topic will be discussed in Section 7.3. Specific examples where continuous processes have been successfully implemented at industrial scale to improve process sustainability are presented in Section 7.4. Finally, Section 7.5 summarizes our perspective on the current status of continuous manufacturing.

7.1
Continuous Processes as a Means of Process Intensification

Stankiewicz and Moulijn aptly define process intensification as "the development of novel apparatuses and techniques that, compared to those commonly used today, are expected to bring dramatic improvements in manufacturing and processing, substantially decreasing equipment size/production capacity ratio, energy consumption, or waste production, and ultimately resulting in cheaper, sustainable technologies." [1] The distinguishing aspect from process optimization is thus the substantial improvement achieved through a paradigm shift of whatever kind compared to the incremental optimization of existing processes.

Table 7.1 Classification of commonly found advantages of continuous.

Logistics/quality	Chemistry/process	Safety
• Increased throughput • Accelerated implementation (shorter lead times, facilitated scale-up) • Reduction of lot-to-lot variations • Minimization of inventory and establishment/validation stocks	• More efficient heating and cooling • Novel process windows • Direct processing of unstable intermediates/products • Implementation of demanding separations (e.g., multistage extraction)	• Improved reaction control for exothermic reactions • Smaller equipment/reduced hold-ups • Operation without vapor space • Chance to use disposable equipment for highly potent or cytotoxic reagents/products

Source: Based on Ref. [5], American Chemical Society.

One particular branch of process intensification is the application of continuous process technologies. From a technological point of view, continuous equipment frequently enables much better control of heat and mass transfer inside the equipment. Therefore, existing processes can be run at higher temperature, pressure, and/or concentration, if advantageous for the reaction, or completely new synthetic pathways become possible that could not be safely carried out in conventional equipment (often referred to as "novel process windows") [2–4].

Drivers for continuous processing have been extensively evaluated and discussed in several publications [5, 6]. According to a study by Poechlauer and coworkers, the commonly stated advantages can be categorized into three groups as illustrated in Table 7.1: logistics/quality, chemistry/process, and safety [5]. Which of these drivers are most pronounced has to be evaluated on a case-by-case basis. However, looking at the pharmaceutical industry for instance, certain drivers are more likely to have an impact depending upon the stage within a product's life cycle as well as a molecule's location along the production chain from starting material over intermediates to the very final steps [7, 8]. In the early stages of pharmaceutical development, the speed of scale-up and implementation is often necessary while process robustness and constant delivery of larger amounts of in-spec product become more important in the late clinical phases and for market entry. Getting closer to patent expiration, the COGS (costs of goods sold) are of increasing importance, possibly even leading to the development of more efficient second-generation manufacturing processes. Current manufacturing in multipurpose batch plants often results in significant amounts of inventory, and thus working capital expenditures, in form of intermediates or the final product to ensure a reliable market supply [9]. End-to-end or at least highly integrated continuous processes offer the chance to reduce lead times and inventories, therefore, reducing this often significant cost driver. Considering the product's position along the production chain, increases in yield and selectivity become more important, the closer to the final product a molecule is located. Such improvements are often achieved through tighter temperature control or improved mixing [10]. Close to the raw materials an increase of selectivity in the range of a small percentage is often not important when a recycle can be used or side products can easily be separated. Often these steps are run at higher concentrations. Here, safety aspects for hazardous reactions (metal organic or exothermic reactions, nitrations, unstable intermediates or decomposition, closed handling of small amounts, toxic intermediates, etc.) or the large or flexible production range and scale-up are of primary interest.

Continuous processes can enable more efficient and sustainable manufacturing, and can therefore be considered as means of process intensification. Additionally, they also enable the implementation of further process intensification such as microwave, sonochemistry, or photochemistry. Drastic improvements of reaction performance for certain reactions have been reported using these technologies [11]. Nonetheless, an important drawback is the limited penetration depth of the electromagnetic or acoustic radiation [12–14]. This renders scale-up difficult in conventional batch vessels. Continuous equipment with its higher surface to volume ratio offers a potential solution to overcome this challenge. This aspect is discussed in Section 7.2.4 in more detail.

Considering the potential advantages of continuous processes, it is not surprising that the American Chemical Society Green Chemistry Institute Pharmaceutical Roundtable has voted continuous manufacturing the most important area of research in green chemical engineering in 2011 [15]. Estimates on the number of chemical reactions that could benefit from continuous processing are in the range of 40% [16] to 50% [8] underlining the significant potential of this technology. Looking at the status quo within the chemical industry, however, the picture is quite different. While the top 30 petrochemicals and most of the top 300 organic chemicals are produced continuously, 90% of those ranked 301–3000 and even 97% for numbers 3001–30 000 are made batch wise [17]. What is the reason for this discrepancy between technology potential and actual implementation? According to another study by Poechlauer and coworkers, the biggest hurdle is the existence of conventional batch plants with sufficient capacities readily available followed by a general uncertainty whether the technology's benefits are sufficient to justify its implementation. Further reasons include a lack of or insufficient technical maturity of available technical systems at different scales and the development processes that starts with a batch synthesis that is often kept and scaled-up even though the process might not be optimal [18]. It should also be mentioned that the reaction time (or process time for unit operations other than reaction) is an additional degree of freedom in batch operation. In continuous equipment changes to the residence time will also affect the hydrodynamic process conditions (e.g., poorer mixing at lower flow rates). Therefore, batch vessels possess a great flexibility and can accommodate a wide range of different processes that is particularly important in multipurpose plants not dedicated to the production of a particular product.

It is interesting to note that many of the recent initiatives in the field of continuous manufacturing come from the pharmaceutical industry that accounts for around 70% of the $100 billion/year fine chemical market [19]. The traditionally conservative nature of this industry when it comes to manufacturing has been underlined by the *Wall Street Journal* stating that "even as it invents futuristic new drugs, its manufacturing techniques lag far behind those of potato-chip and laundry-soap makers." [20] One reason there has been little innovation in the way drugs are produced is the regulatory system that has put high hurdles and little incentives for companies to move past the traditional tried-and-true manufacturing. A striking example of the systems inefficiency is the estimated 5–16% of products that have to be reworked or discarded because they do not meet specification, which is obviously far from a lean manufacturing process [20, 21]. Such quality issues have been one reason for regulatory bodies to start reconsidering the traditional approach. In 2004, the FDA (US Food and Drug Administration) stated "Pharmaceutical manufacturing operations are inefficient and costly. Compared to other industrial sectors, the rate of introduction of modern engineering process design principles, new measurement and control technologies, and knowledge management systems is low. Opportunities for improving efficiency and quality assurance through an improved focus on design and control, from an engineering perspective, are not generally well recognized," [22] which has led them to now encourage the implementation of modern manufacturing approaches and being "fully supportive of the industry moving in the direction of continuous processing." [23]

While regulation has long been one factor stifling innovation, there has been little need for pharmaceutical companies to rethink their manufacturing processes due to healthy margins and profits. Over the last decades, however, the market environment has become more challenging. In the US, for instance, the generic share of prescriptions filled has drastically increased from 19% in 1984 to 86% in 2013 [24] and even modest selling drugs now face an increased competition by generics [25]. At the same time, the average period of exclusivity for blockbuster drugs has decreased from 13.8 years (period 1995–2001) to 11.2 years (2002–2005) [25] and a striking $290 billion of sales are at risk from patent expirations from 2012 to 2018, a situation commonly referred to as patent cliff [26]. In addition to the growing competition, there is an increasing cost pressure from regulatory bodies. 60% of global prescription sales now come from countries with formal cost-effectiveness assessments in drug-funding decisions in place [27]. Finally, in search of new markets, cost-effective provision of drugs is also required to serve the next billion customers in emerging markets that will represent 45% of the world's gross domestic product by 2018 [28].

At the same time the average costs to develop a new drug (including the cost of failures) have drastically increased over the last decades. While the methods and resulting numbers are controversially discussed as well as company and drug dependence [29–32], the overall trend is generally acknowledged and the number of drugs approved per billion dollars spent on R&D (research and development) has approximately halved every 9 years since 1950 [33]. The major reasons are an increased failure rate in clinical trials, a focus shift toward drugs for chronic illnesses requiring larger and longer clinical trials, and greater technological complexity in development and specificity in disease targets [34]. An analysis by Vernon *et al.* estimates that only around 20% of new molecular entities create revenue higher than the average capitalized R&D expenses [35].

Due to the often small marginal costs of drug manufacturing [36, 37], drug production is sometimes considered inexpensive and the focus for improving economic performance is rather on more efficient R&D processes and improving success rates in clinical trials. Nonetheless, considering the entire COGS the picture is different. On average, in research-oriented brand-name pharmaceutical companies, COGS are estimated to sum up to 26% of revenues while they even account for 52% for generics [38]. Obviously, it would be desirable to address the COGS at an early stage of the development process – preferably even before the nomination of a drug candidate [39]. However, as already stated, at this stage time-to-market is of outstanding importance and an optimization of manufacturing conditions often not justified taking into account the uncertain fate of an API (active pharmaceutical ingredient) as commercial product. Therefore, if mere reduction of manufacturing costs is the only benefit of a new production technology, this is unlikely to be sufficient to be widely accepted by the pharmaceutical industry. If, on the other hand, the time-to-market can be reduced through accelerated scale-up or more reproducible quality can be achieved through avoidance of lot-to-lot variations, these levers can be much more significant. As already pointed out, both of these are potential advantages of continuous processing compared to batch. At the same time, however, it is important to develop processes that are sufficiently flexible given the often volatile and difficult to predict markets. For instance, a study by Ahlawat and coworkers has shown that for nearly two-thirds of drug launches, the actual sales during the first year of launch deviated by more than 60% from the forecasted sales one year prior to launch [40]. Flexibility is therefore a crucial point that needs careful consideration when developing continuous equipment and processes in order to be competitive with established multipurpose batch plants.

Many of the major players in the pharmaceutical market have identified continuous manufacturing as a promising technology to address some of the challenges they are currently facing. Globally, companies have spent well over 1 billion € on developing continuous manufacturing processes [41].

Genzyme (now Sanofi) in collaboration with NiTech has built a multihundred tonnes per year API manufacturing plant at their Haverhill site. Previously to installing the NiTech reactor (Section 7.2.1), Genzyme was considering $2 \times 150\,m^3$ reactors that would have had a significantly larger footprint [42]. The reaction rate could be improved by a factor of 40 compared to the batch process and the FDA approved process was commissioned in 2007 [43].

Also in 2007, Novartis and the MIT (Massachusetts Institute of Technology) have launched the Novartis–MIT Center for Continuous Manufacturing. Over a period of 10 years Novartis invested US$ 65 million in research activities at the MIT in order to develop new technologies [44]. A core part was the development of an end-to-end process for the synthesis of the API aliskiren hemifumarate (Section 7.4).

GSK (GlaxoSmithKline) has demonstrated the potential of continuous processing as an alternative to conventional batch processing in a number of projects at pilot scale [45]. The initiative includes a modular pilot plant that has been installed at Stavenage, UK in cooperation with Foster Wheeler [46]. The plant was fully commissioned in 2008 and was designed to produce on an equivalent 3–5 tonnes per annum scale [47].

Apart from these company-specific initiatives, more recently, there is an increasing tendency to develop continuous manufacturing technologies collaboratively. Within Europe, a number of large research projects, such as F^3 Factory, Synflow, Copiride, Polycat, or Impulse, have addressed various challenges of continuous processes. Additionally, initiatives like the EPSRC Center for Innovative Manufacturing in Continuous Manufacturing and Crystallization (CMAC) or the Bayer AG/TU Dortmund joint venture INVITE pursue the approach to jointly develop solutions to still existing challenges on the way to a widespread implementation of continuous manufacturing.

From this noncomprehensive list, it is clear that a lot of companies, including many of the major players in the pharmaceutical market, have a strong interest in continuous manufacturing as new production technology. While conventional batch manufacturing currently remains by far the prevailing mode of operation, companies are willing to invest significant amounts of resources in developing continuous manufacturing as an alternative. Especially in the pharmaceutical sector, the manufacturing technology itself is not generally considered as core competency and companies are increasingly recognizing the potential of collaborative development efforts.

7.2
Equipment for Continuous Processes

From a phenomenological point of view, an ideal batch reactor is the same as an ideal plug flow reactor, the only difference being that the time coordinate is transformed into a local coordinate. Both models assume perfect mixing. In a batch reactor this means a perfectly homogeneous mixture at all times. For a plug flow reactor, perfect mixing signifies a homogeneous distribution in radial direction and negligible mixing along the length of the reactor. The difference arises during scale-up, because it is easier to combine high surface-to-volume ratios required for heat transfer with good radial mixing in continuous equipment than with perfect homogenization in large batch vessels. On the downside, the fluid dynamics are often coupled to the flow rate and thus residence time requiring the adaptation of equipment size for a particular process. A number of equipment suppliers now offer technological solutions for different upstream and downstream unit operations and at different scales that will be briefly evaluated in this chapter. Apart from process performance, scalability, flexibility, and robustness are important factors to fully capitalize on the benefits of continuous processes. Scalability and a sound knowledge of the devices capabilities (e.g., heat transfer coefficient, residence time distribution, power input, mixing times) is required to achieve a quick transfer from laboratory to production scale. A quick implementation on production scale is further supported by flexible equipment that can be taken off the shelf and can be adjusted to different processes with only minor changes as compared to the development of new equipment for a given process. Last, robustness is always crucial for industrial implementation, but even more so when several devices are connected in series as the failure of one piece will directly affect the entire process.

7.2.1
Upstream Equipment

For small-scale production units a broad range of equipment is commercially available. This includes reactors with channel diameters from several ten millimeters down to microstructured devices. Two main groups can be distinguished as equipment where mixing solely relies on the geometrical design and reactors with active energy input for enhanced mixing independent of the flow rate.

7.2.1.1 Reactors without Active Mixing
The broad range of commercially available equipment without active mixing can be clustered according to its basic design.

Empty Tube
The simplest way to provide residence time to reaction systems in flow chemistry is an empty pipe or tube that can be designed as shell and tube heat exchanger. Under laminar flow conditions these systems suffer from poor residence time distribution and limited heat transfer, which can be strongly enhanced by applying secondary flow patterns. Possible workshop solutions are coiled pipes introducing dean vortices to the fluid [48], an effect that can be further enhanced in the coiled flow inverter configuration [49, 50]. However, only increasing the heat transfer coefficient on the reactor inside might be insufficient if the heat transfer becomes limited on the outside, for example, when the coiled tube is simply placed inside a thermostat bath.

Tube with Static Mixers
The introduction of static mixing elements enhances the residence time distribution and heat transfer capacity of tubular reactors under laminar flow conditions. The repeated recombination and twisting of fluid layers allows for good radial mixing and therefore a plug flow-like behavior. Also, the heat transfer is enhanced by a factor of 4–10 compared to empty tube reactors. Heterogeneous flow systems that require a constant delivery of mixing energy throughout the reactor volume can be run in static mixing devices. A broad variety of mixing inlays is available. For low to medium viscosity liquid-phase systems, X-type mixers are most commonly employed. Exemplary reactors in this field are the SMX™ (Sulzer Ltd.) and CSE-X® mixers (Fluitec mixing + reaction solutions

AG) or the Miprowa® reactor (Ehrfeld Mikrotechnik GmbH). For high-throughput systems and highly exothermic reactions, a further intensification of heat transfer can be achieved by the introduction of additional cooling pipes into the reactor. These pipes are used as geometrical structuring while providing additional heat transfer area. This allows for high scale-up factors for the reactor diameter at a constant specific heat transfer area. Examples for these devices are the SMR™ (Sulzer Ltd.) or CSE-XR® mixer (Fluitec mixing + reaction solutions AG). A comprehensive overview on static mixers and their potential application is given by Ghanem et al. [51].

Intensified Mixing
For fast and very fast reactions in homogeneous phase systems, an instantaneous and uniform mixing at the reactor inlet is desired to minimize local concentration deviations, for example, to increase selectivity in case of undesired parallel or subsequent reactions. This can be achieved by intensified static mixing units. Mixing speed is strongly correlated to the specific energy dissipation in the mixing zone. Therefore, it is always a compromise between mixing effectiveness and pressure drop. A comprehensive comparison of various mixing systems is given by Falk et al. [52].

Intensified mixing devices can be distinguished by the applied fluid contacting pattern. SAR (split and recombine) mixers perform multiple sequential splitting, folding, and recombining steps in order to create fluid films with a thickness in the micrometer range, enabling fast mixing by molecular diffusion. At higher flow rates, secondary flow patterns additionally improve the mixing quality. The inner volume is low leading to residence times in the range of 1–100 µs. Typical examples of this mixer type are the Caterpiller mixers (Fraunhofer IMM, Mainz) and the Cascade mixers (Ehrfeld Mikrotechnik GmbH) and for large volumetric flow rates the Starlam mixers (Fraunhofer IMM, Mainz). While SAR mixers provide several serial mixing elements, ML (multilamination) mixers create a multitude of thin fluid layers in parallel. Within these thin layers the diffusive mixing is fast enough to enable homogenization. Typical examples for scalable ML mixers are the Interdigital mixer (Fraunhofer IMM, Mainz) and Slit-Plate mixer (Ehrfeld Mikrotechnik GmbH). An approach to achieve a further thinning of the fluid layers and therefore shorter diffusive mixing times is applied in flow-focusing mixers, for example, the SuperFocus Interdigital Micromixer (Fraunhofer IMM, Mainz). The introduction of secondary flow patterns at higher Reynolds numbers allows for enhanced mixing in SAR as well as ML mixers. For low to very low Reynolds numbers, ML mixers can be more efficient than SAR, as the mixing time mainly depends on the fluid lamella thickness. In ML mixers, however, the pressure drop increases significantly at higher flow rates while the improvements in mixing performance are comparably small because of the mixers' working principle.

A third type of mixing elements fully depends on secondary flow patterns and therefore needs a certain minimum flow rate to achieve and ensure good mixing.

Examples for these mixing devices are T- or Y-junction mixers, SZ-mixing structures, meandering channels, or impinging jet mixers. For special applications there are further devices available. For instance, Ehrfeld Mikrotechnik GmbH provides a valve mixer that is designed to avoid any backflow from the mixing chamber if there is a chance of clogging otherwise. Mixing elements are available as single devices or may be integrated directly into the structure of plate reactors.

Plate Reactors
Enhanced heat exchange can be provided in plate reactors with milli- or micro structured channel geometries. These reactors are designed for excellent heat transfer using high surface-to-volume ratios enabling reactions under harsh conditions in terms of concentration, temperature, and/or pressure. Plate reactors can be configured due to variable stacking of different plate types within one frame. These plates can differ in channel geometry, contacting patterns, and mixing intensity. Exemplarily, a process consisting of the following steps can be realized within one device: Precooling of the starting material, multiple sequential injections of starting material and mixing points, residence time with heat removal, and reaction quench.

Corning Inc. provides the Advanced-Flow™ reactors in different plate sizes with flow rates from <1 ml min^{-1} up to 5000 ml min^{-1}. High-performance materials like ceramics (SiC) can further enhance heat transfer and corrosion resistance for demanding applications. Ehrfeld Mikrotechnik GmbH provides two different single-channel plate reactors: The Lonza FlowPlate® reactor and the ART® reactor, both available with different plate sizes and at different scales. The employed single channel approach is well suited to avoid any maldistribution or undefined flows, which might occur in devices with parallel channels, but also complicates numbering-up.

Plate-type reactors are also available with parallel channel design. These reactors often provide excellent volumetric space usage and a compact design. Examples for these devices are the Compact Heat Exchange Reactors (Chart E&C) or the vacuum brazed stacked plate microreactor (Fraunhofer IMM, Mainz). Novel manufacturing techniques allowing a cost-efficient mass production of large-scale microstructured equipment are currently developed for these devices [53].

7.2.1.2 Reactors with Dynamic Mixing
A common feature of the previously described reactors is that the energy for mixing is supplied in form of pressure drop across the apparatus. The aim for these devices is to develop geometric designs that achieve a high level of mixing and a low-pressure drop at the same time. Particularly, mass transfer-limited multiphase processes sometimes require higher amounts of energy input than achievable through passive mixing devices. Another application for dynamically mixed devices is the handling slurries in flow that is one of the major challenges to continuous processes due to the increased likelihood of fouling and

clogging [8, 16]. Some innovative designs for dynamically mixed, continuous reactors have been developed and are now commercially available. So far, these devices have found less widespread implementation compared to their static counterparts, most likely due to their higher technical complexity.

Spinning Disc Reactor

In spinning disc reactors, the reagents are fed to the center of a fast rotating disc. Driven by centrifugal forces, the fluid starts moving outward forming a thin film in the range of 50 μm for water-like liquids. Such reactors are suited for fast, highly exothermic gas/liquid reactions with low to medium viscosities [54]. Although this reactor concept has not yet been implemented more widely, first results indicate the technology's potential. Oxley et al. have been able to reduce the required reaction time and plant volume by more than 99% while reducing the impurity level by 93% for a phase-transfer-catalyzed Darzen's reaction [55]. The technology is commercialized by the Dutch company Flowid BV.

Spinning Tube in Tube/Taylor–Couette Reactor

Depending on the source, the terms spinning tube in tube and Taylor–Couette reactor are differentiated or used interchangeably to describe a reactor where an inner rotating cylinder is placed inside a static outer cylinder. The reaction takes place in the small annular channel between rotor and stator. Channel widths can be as narrow as 250 μm [56]. The term Taylor–Couette reactor implies that the reactor is run in an operating window such that so-called Taylor vortices are formed inside the annular gap that leads to a narrow residence time distribution as well as improved heat and mass transfer [54]. For other processes, however, the formation of these vortices is undesirable [57]. Hence, one of the challenges is the careful selection of operating conditions to achieve the desired hydrodynamic conditions [54]. A reported application is the Biodiesel production where the short achievable residence time permits to suppress a slower secondary undesired reaction [57]. In fermentation, the evenly distributed shear stress is advantageous to avoid damaging the cells [58].

Continuously Oscillatory Baffle Reactor

The COBR (continuously oscillatory baffle reactor) consists of a pipe that is segmented into several compartments by means of annular baffles. Mixing independent of the net flow rate is achieved through the application of oscillatory motion (range 0.5–10 Hz) to the fluid that superimposes the net flow through the pipe. This concept is mostly suited for processes with residence times above 10 min [54]. The reactor is commercialized by NiTech® Solutions Ltd, a 2003 spin-out company from Heriot-Watt University, Edinburgh [59]. The technology has been used for a variety of processes one of which has already been pointed out in Section 7.1. Due to the narrow residence time distribution that can be achieved independent of the net flow rate and even for long residence times as well as the uniform mixing conditions the technology is also well suited for continuous crystallization [60].

Agitated Cell and Agitated Tube Reactor

Another mixing principle is employed in the Coflore® flow reactors, which are available at different scales from AM Technology. The agitated cell reactor (ACR) is available from 20 to 126 ml while the agitated tube reactor (ATR) covers the range from 0.1 to 10 l. The reactor is separated into a number of chambers each of which contains free moving agitators. These agitators promote mixing when the entire reactor is shaken laterally [61]. The mixing principle prevents phase separation through centrifugal forces and avoids the use of dynamic seals. Different surface-to-volume ratios along the reactor path length can be realized through the use of different mixers [62]. The technology has been applied in different processes, including the functionalization of carbon nanotubes with an aryldiazonium salt where the reaction time could be decreased from 15 h in batch to 30 min in the ACR [63].

7.2.2
Downstream Equipment

Continuous downstream operation can offer significant advantages. This concerns a higher possible degree of automation and reduced downtimes for charging and removing substances as well as more efficient processes such as in multistage countercurrent extraction. Furthermore, continuous downstream processing is also required to telescope continuous reactions and realize end-to-end continuous processes. On the other hand, just like for reaction equipment, batch technology often offers a higher flexibility to cope with different conditions and frequent process changeovers.

For an industrial application, robustness and reliability of equipment is of major importance. On large scale (i.e., petro and bulk chemistry) – where continuous operation is the standard production mode – suitable equipment for the different unit operations is readily available. In principle, most of these unit operations can be scaled down. Equipment down to a scale of $1-2 \, l \, h^{-1}$ is established for most unit operations and constitutes a size from which scale-up is relatively straightforward [64]. This kind of equipment is state of the art on mini- and pilot plant scale since many years and is commercially available from companies such as Normag, QVF®, HiTec Zang, or AP Miniplant. Examples, including distillation, extraction, and crystallization are numerous in the research literature. These units are preferably made of glass due to chemical resistance and the advantage of visual inspection, but require particular care during transport and installation. Also, this kind of equipment is basically a down-scaled version of conventional equipment without much further process intensification. Such equipment is not necessarily cheaper than its larger counterparts due to the specialized nature and manual work involved in the production.

A research area aiming at intensified processes is the exploitation of centrifugal fields [65]. One of the oldest examples of this technology is the Podbielniak liquid–liquid extractor already used for penicillin recovery in 1945 [66]. While centrifugal extractors are commercially available at different scales (e.g., from

CINC Industries, Rousselet Robatel, or GEA Westfalia Separator Group) the application of centrifugal fields is still in the early stages for other unit operations such as distillation [67] or adsorption [68]. First developments in this direction have been made. However, such equipment has to demonstrate its reliability before it will find more widespread use.

The implementation of microstructures in reaction, mixing, and heat transfer applications with the aim to capitalize on significantly shorter diffusion and heat conduction path lengths has received great attention over the last decade. Even though separation processes could equally benefit from enhanced heat and especially mass transfer processes, reported applications in the field of microseparation technology exist, but are still relatively limited [69, 70] for a number of reasons. First, due to the changing nature of relevant forces, the widely used scale-up strategy based on similarity cannot be generally applied to microstructured equipment [70]. Numbering-up instead of scale-up will introduce new challenges, most notably avoidance of flow maldistribution among different channels. Second, phase separation, especially in liquid–liquid extraction can become more difficult [70]. Current research projects address this issue, for example, through the development of membrane-based separation methods [71] or microsettlers [72]. Finally, small-scale continuous equipment is inherently susceptible to fouling and therefore in many cases not applicable for robust long-term operation. Potential solutions include segmented flow regimes [73] which are, however, limited to relatively small operating windows, surface modifications [74], or the use of ultrasound [75].

Even though the handling of solids is particularly difficult in microstructured equipment, the challenge is not limited to this scale. The transport and dosing of solids (mostly as slurries) is one of the most important challenges for continuous processing. Pumps are often a weak point especially at low-flow rates (e.g., a few milliliters per minute). To remedy this, the pumps are often used intermittently [76, 77]. Alternatively, Eli Lilly has proposed the use of pressure swing chambers where the material is transported without using pumps, but through application of pressure and vacuum [76, 78, 79]. However, such solutions imply that oscillating filling levels in the concerned vessels need to be acceptable.

Likewise, continuous solid–liquid separation is a veritable challenge especially at low to medium flow rates. First, surface interaction of solids and walls leads to undesired effects like channeling, adhesion on walls, or inhomogeneity in small filter cakes. As a rule of thumb, the diameter of the filtration area should be larger than 0.1 m to avoid wall effects. Second, solid–liquid separation spans a huge order of magnitude regarding the solid concentration that can range from 0.01 to 40 vol% or the specific filter cake resistance that can vary across multiple orders of magnitude depending on the application. Therefore, each centrifuge, decanter, separator, or filtration device has its own specific operating window. While the aforementioned variability in process conditions makes it difficult to provide general solutions, the development of specialized solutions is often not economically justified for small throughputs and a limited number of applications. The UK-based company, Alconbury Weston Ltd. offers a small-scale continuous filter, but often solid–liquid separation is performed in a semibatch manner, also when the other process steps are working continuously [76] or relies on the development of in-house solutions [80]. Another development worthwhile considering is the use of single-use devices especially regarding cleaning and associated regulatory aspects [81].

Looking at the modern chemical industry, about half of the output is in the form of particulate crystal products [82] and over 90% of APIs are crystals of small organic molecules [83]. Such processes pose additional challenges on the equipment design. Since the solid is the desired product the separation unit often comprises an additional washing step further increasing the unit complexity. In the pharmaceutical sector, cross contamination and the success of cleaning procedures have to be considered, avoidance of dead zones and incrustation is crucial. Eventually, biological efficacy is often a function of the particle size that is adjusted during crystallization. This means, solid–liquid separation should not lead to agglomeration or other changes in particle structure.

The solid–liquid separation should be considered, together with its prior step of crystallization or precipitation. Hence, adjustment of crystal habits for better filtration behavior is mandatory. One of the major research topics at CMAC or the Synthesis and Solid State Pharmaceutical Center is to develop crystal sizes, shapes, and habits that fulfill the requirements for a better application in the patient as well as a better filtration behavior to reduce losses and increase the efficiency of the solid–liquid separation.

Finally, continuous solutions for the last process steps have also been developed. GEA, for instance, offers a continuous tabletting line (ConsiGma^{TM}) to transfer powder into coated tablets at different scales. The system is designed in a modular way so that the customer can choose which process steps are required. Apart from the basic modules for granulation, drying, and evaluation (i.e., measure critical quality attributes inline), additional modules can be included for feeding and blending of additional components, tablet compression, and tablet relaxation and coating [84].

Obviously, it is beyond the scope of this chapter to cover the vast number of downstream unit operations. Continuous manufacturing is particularly beneficial for those unit operations where higher separation efficiencies can be achieved based on the principle of countercurrent exchange. In general, however, it can be said that the development of process-intensified downstream equipment for small- to medium-scale chemical processes is not as far advanced as for reaction equipment. Therefore, at this stage the development of end-to-end processes will often rely on the construction of in-house solutions for particular tasks, especially when solids are involved.

7.2.3
Process Integration

Chemical processes normally consist of many steps. As already discussed, the connection of various steps in a continuous end-to-end process can offer significant advantages. However, continuous processing sometimes also allows for the integration of different steps into a single device. Depending on the process this can further improve the process economics, for example, by achieving conversions beyond the thermodynamic equilibrium or by decreasing investment and operating costs.

An example for such a technology is the combination of reaction and *in situ* separation in membrane reactors. In methane reforming, for instance, lower reaction temperatures at comparable methane conversions can be realized through the continuous removal of the formed hydrogen. This way, cheaper materials can be used for reactor construction and heat losses can be decreased. Lin and coworkes used a palladium-membrane reactor in which they reduced the CO yield to about 2% for methane steam reforming at a methane conversion of 80% [85].

Similarly, thermodynamic limitations can be overcome through sorption-enhanced reforming. For instance, this can be realized through the continuous adsorption of carbon dioxide that shifts the equilibrium of the water–gas shift reaction toward the further production of hydrogen. Obviously, this process requires an additional sorbent regeneration step and much of the current research focuses on testing and improving the sorbents multicycle durability. Fluidized bed reactors or multiple fixed bed reactors that are operated in intervals have been proposed for this process [86].

Another approach is to combine reaction and separation in a piece of equipment conventionally used for separation operations. This is done in reactive separations carried out in distillation or extraction columns [87]. Again, the drivers are shifting the chemical equilibrium, improving selectivity, reducing energy consumption, or decreasing the capital investment, and hence improving the overall process economics. However, the operating window for reactive distillations can be relatively small and multiple steady states may occur requiring proper automation and control to implement such a process [88]. Reactive extraction is widely investigated and of particular interest for thermally sensitive products. An example is the reactive extraction of lactic acid that is used as feedstock for biodegradable polymers [89].

For more detailed information on the integration of reaction and separation, the reader is referred to Ref. [90].

7.2.4
Continuous Equipment as Enabling Technology

It has already been said that some interesting process intensification technologies are difficult to scale-up. These include microwave, sonochemistry, and photochemistry due to the fast attenuation of the respective waves with distance. Here, continuous equipment with its higher surface-to-volume ratio can help to overcome this challenge.

Microwave Chemistry
The first reported applications of microwave irradiation in organic syntheses date back to 1986 [91, 92]. Microwaves offer the potential to rapidly heat a

reaction mixture without the necessity to heat the reactor walls through convection and conduction as well. The technology becomes particularly interesting when reaction rates can be increased through local heating, that is, energy supply exactly where it is needed for the chemical reaction, such as in heterogeneously catalyzed reactions by intense heating of the catalyst particles [54]. However, such examples are relatively rare or of little practical relevance. The existence of truly nonthermal microwave effects claimed in some publications and resulting from a direct, stabilizing interaction between certain molecules and the electromagnetic field is highly questionable. Nonetheless, microwave heating can enable very fast heating and operation in novel process windows and thus enhance the performance of chemical reactions [93]. One challenge with the application of microwaves to chemical reactions is the limited penetration depth in the range of a few centimeters. Therefore, it is unlikely that the technology will be capable to produce industrially relevant quantities when used with conventional batch technology [12, 94]. In the mid-1990s, Cablewski and coworkers reported on the development of a continuous lab-scale microwave reactor that they consider to be more suitable for scale-up as well as for accommodating the high temperature, high pressure regimes often encountered in microwave chemistry [95]. Although larger batch applications have been reported [96], continuous reactors can be considered the preferable option in microwave chemistry for processing volumes larger than 1 l under sealed vessel conditions [93]. In 2012, Morschhäuser et al. presented a high pressure, high temperature (60 bar, 310 °C) with a commercially relevant throughput of $20\,l\,h^{-1}$ that they consider to be the first "general purpose industrial scale continuous flow microwave reactor for organic synthesis." [12]

Sonochemistry
Similar to microwaves, ultrasound has also been reported to enhance the rate of chemical and biological processes. After first reports in the 1920, the field of sonochemistry has received increased attention since the early 1980s. The effect can be explained by acoustic cavitation, that is, the formation, growth, and collapse of microbubbles, in liquid-phase systems, which leads to local temperatures and pressures as high as 5000 K and 50 000 bar [11]. Compared to microwave irradiation (and many other enhanced field techniques) acoustic irradiation has the advantage that it does not require specific fluid properties. It only requires a liquid that propagates the acoustic waves [54]. An additional advantage of ultrasound is its capability to effectively prevent clogging in continuous reactions that has been reported in a number of applications [97–99]. Similar to microwaves, however, the penetration depth is limited, especially at higher frequencies, so that batch reactors are not suitable anymore. Modeling and scale-up of acoustic fields remain challenging due to their complexity in the presence of formed cavitation bubbles or suspended solids in the reaction system. Material erosion is another issue that needs to be considered during equipment design [11, 14, 100]. Despite these drawbacks a number of larger scale sonochemical reactors have been developed and are commercially available [100].

Photochemistry

Photochemical reactions allow running reactions at lower temperatures associated with energy and cost savings and thus greener processes. However, industrial application particularly in the pharmaceutical and fine chemical industry is very limited. As with microwaves and ultrasound, the limited penetration depth is one of the challenges when scaling up batch processes [13]. Coupling of photochemistry with continuous micro or spinning disc reactors therefore constitutes an interesting approach to overcome this challenge [101]. For instance, Loponov *et al.* looked at the oxygenation of α-pinene to pinocarvone with photochemically generated singlet oxygen. Different reactor setup/light source configurations were investigated. A model-based optimization was used and the optimal setup was found to be a microreactor with a channel depth of around 200 µm for this particular case. However, the authors state that the curve of the cost per kilogram of pinocarvone is relatively flat around this optimum such that reactors in the lower millimeter range could as well be employed without affecting the process economics [102]. Another successful example is the synthesis of the antimalaria drug artemisinin from artemisinic acid reported by Lévesque and Seeberger. The authors claim that this process offers significant benefits over the direct extraction from sweet wormwood, but has proven a formidable challenge. A central step in their continuous synthesis is an ene reaction initiated by *in situ*, photochemically generated singlet oxygen. The use of singlet oxygen in batch would have been difficult because of mass transfer limitations and the need for specialized equipment to produce the reagent since the use of photochemistry is restricted through the attenuation of light with distance [103].

The review article by van Gerven *et al.* is a good starting point for further information on the intensification of photocatalytic processes [101].

7.3
Process Development and Implementation for Continuous Processes

In order to capitalize on the benefits of continuous processes and minimize the time-to-market, an efficient process development is required (Section 7.3.1). The developed process often needs to be implemented in a way sufficiently flexible to cope with a volatile market environment. Here, modularization and platform technologies are promising approaches (Section 7.3.2).

7.3.1
Process Development and Scale-Up

Lab-Scale Development

It is unlikely to benefit from some of the advantages of continuous manufacturing, particularly the novel process windows, when early development is done in batch with subsequent transformation to continuous during scale-up. In this case, the continuous process is likely to be relatively close to the batch variant.

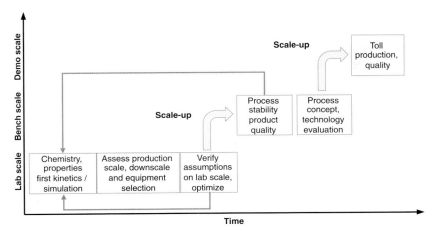

Figure 7.1 Typical steps for process development and scale-up.

Over the years it has been increasingly recognized that the full benefits of continuous processing can only be exploited if employed already at the early stages of process development where understanding of the current process, its chemistry, and kinetics play a crucial role.

Figure 7.1 illustrates typical steps from lab-scale development to demo-scale production. Initial investigations are almost always carried out in batch experiments. The same holds true for the determination of kinetic data that are more easily evaluated in batch as well. However, even at this early stage continuous processes might offer benefits, for example, to test the catalyst stability over an extended period of time or if very quick heating, cooling or mixing is required. Some aspects that need to be considered during continuous kinetic measurements will be discussed further. In general, the development of continuous processes will require more upfront understanding, which, however, will pay out in the later project phases.

Based on these initial investigations, possibly combined with additional simulations, a suitable process setup is designed at lab scale. The targeted production scale should already be kept in mind at this early stage in order to ensure the scalability of the developed process. If continuous processing is identified as promising technology, a rapid transfer of the batch process on laboratory scale can be carried out applying modular equipment such as micromixers, residence time units, heat exchangers, and so on. Such setups can be assembled directly in the laboratory or purchased as modular or stand-alone units from commercial device manufacturers. This increased interest in continuous development on lab scale has led many companies to develop and commercialize such solutions.

Generally, the user can choose between different degrees of automation. A comprehensive product portfolio in the area of modular microreaction systems that allows for a fast evaluation of the potential of a continuous processes for a wide range of reaction classes as well as for product-specific developments is offered by Ehrfeld Mikrotechnik GmbH. Key players in the area of stand-alone units are Chemtrix BV, FutureChemistry Holding BV, Uniqsis Ltd., ThalesNano Inc., Vapourtec Ltd., Syrris Ltd., and Microinnova. The choice of stand-alone system depends on the reaction network and process parameters. An overview over different systems can be found in Ref. [104].

The longest established provider for lab-scale flow systems is Syrris [105]. Their latest product in the field of flow chemistry is the Asia system that is available with different modules and degrees of automation. The system is suitable for reactions with residence from seconds to multihour, temperatures from −15 to 250 °C, and pressures up to 20 bar [106].

In case of gas–solid reactions, such as hydrogenation, the H-Cube® continuous-flow hydrogenation reactor from ThalesNano Inc. is widely applied in industrial laboratories as well as in academia. H-Cube, as a stand-alone unit, ensures a safer and greener continuous process compared to batch via *in situ* generation of hydrogen from water. A broad range of hydrogenation reactions were successfully carried out in the H-Cube [107].

Vapourtec offers two different stand-alone units for continuous chemistry applications. The R-series is often chosen by both academia and industrial researches due to the flexibility in combining modules (pumps, reactors) allowing to set up tailored reaction systems. In addition, the automatic pump monitoring ensures high precision for the flow rates. Separate temperature control of the reactors allows performing multistep syntheses in an easy-to-use system [108].

Scale-Up
Once, the lab-scale development is completed, the first scale-up is performed. At this stage, equipment miniaturization and process intensification should be applied with a proper proportion of sense as "overintensification" can also lead to a lack of technical robustness. For instance, geometrical structuring and miniaturization can incorporate significant disadvantages compared to plain structures, including increased pressure drops, potential for fouling and blocking, difficulty of cleaning, and greater cost per volume. Therefore, the detailed understanding of the occurring limitations and phenomena is required to decide for an appropriate geometrical structuring and scale-up strategy. Depending on the particular process the following information are of particular importance for a successful scale-up:

1) **Reaction Kinetics**
   In order to optimize the process conditions and local stoichiometry, a detailed kinetic understanding is required. Especially, for fast reactions it is challenging to measure the real kinetics of a system, rather than a pseudokinetic that might be influenced by local hotspot formation or mixing limitations. For kinetic measurements under flow conditions, the following considerations should be taken into account:
   – A step response analysis of the kinetic reactor setup at different flow rates can give information on the back mixing behavior, which should be near plug flow conditions. For multiphase systems, it should be considered that

the different phases can have different residence times and residence time distributions inside the apparatus.
- Mixing timescale should be an order of magnitude below the estimated characteristic reaction timescale to avoid mixing limitations.
- Cooling timescale of the reactor should be in a region well below reaction timescale in order to avoid local hotspot formation and enable isothermal operation. The observation of reactor outlet temperature is not a sufficient criterion for isothermal operation for highly exothermal reactions.
- A reaction quench should be applied and tested in order to monitor the reaction progress under nonequilibrium conditions.
- Variations in residence time should include different reactor lengths as well as different flow rates, in order to distinguish influences of fluid distribution from kinetic effects.

2) **Mixing Requirements**
After kinetic measurements have been performed under intensive mixing conditions, the mixing intensity should be reduced in order to observe the effect on selectivity and yield providing hints on potential mixing limitations. This can be done via reducing the energy dissipation or a reduction of a number of fluid lamella in ML mixers. Scale-up parameter for the mixing intensity can be the specific energy dissipation in the mixing zone or the fluid layer thickness that should be kept constant well above the identified limitation during scale-up. For residence time elements, constant local energy dissipation rate and back-mixing behavior are important scale-up factors.

3) **Heat of Reaction**
With kinetic and thermal information at hand the maximum allowable channel diameter to allow for thermally safe operation can be estimated using the heat transfer performance of the desired reactor setup. This requires detailed understanding of the flow patterns in the desired reactor and also includes the assessment of thermal stability and other thermal risks in the process [109].

In order to achieve stable and robust operation, different process concepts should be considered. Besides miniaturization of the channel geometry other concepts like loop reactors or multi-injection concepts for one starting material might be feasible to overcome hotspot formation. Also, a serial combination of different technologies might lead to improved results. For example, exothermic reactions of high order might be efficiently carried out in a combination of tube reactors for initial mixing and heat exchange followed by a larger continuous stirred tank reactor to reach high conversion. A more detailed analysis of complex process limitations as recently shown by Commenge and Falk [110] can help to identify a suitable process design.

The long-term stability of a desired reactor concept should be carefully monitored. Fouling on heat transfer surfaces can significantly reduce the heat exchange capacity of the reactor leading to hotspot formation and selectivity issues. In many cases, fouling can lead to plugging of the reactor channels, drastically limiting the maximum operation time between cleaning cycles. Also, a parallelized numbering-up of systems can lead to fluidic maldistributions, especially for systems with viscosity changes and solid formation. Therefore, the usage of parallelized channels should be critically evaluated. To avoid such limitations, a proof of concept for the later plant concept should be confirmed already at bench scale before further scale-up steps. This includes the testing of all critical technologies for inertization, purging, pumping technology (pulsation), redundant reactor setup with cleaning in place, and hot switching as well as process analytical technology for process monitoring and control.

The Role of Process Analytical Technology
PAT (process analytical technology) can be beneficially employed from early process development to full-scale manufacturing, be it in batch or continuous processing. For continuous processes, however, the use of online monitoring is often an integral part of the process control strategy and thus a prerequisite for achieving a stable process at the desired conditions [111]. Obviously, the implementation of control loops requires sensors with sufficiently short response times compared to the process dynamics. Also, it needs to be ensured that gradual fouling does not impair the measurements.

The strong interest of regulatory bodies, such as the FDA, to promote process development and implementation based on scientific knowledge and understanding has pushed the development of PAT. By allowing for (near) real-time measurements of quality attributes, PAT can be considered a facilitator of the Quality-by-Design paradigm ultimately leading to a more consistent product quality [112]. Recent reviews on the development of PAT have been published by different authors [111, 113]. As with most technologies, the question is whether the potential benefits and advantages justify the associated cost of implementation for a given application.

On lab scale the use of PAT to increase mechanistic and process understanding has been recognized for decades [111]. In combination with continuous processing it allows for a fast and efficient screening of reaction conditions. For instance, Schwolow *et al.* have reported on a process where more than 200 data points could be gathered through in-line Raman meansurements for a Micheal addition in a glass–metal microreactor in less than 1 h [114].

During scale-up of commercial manufacturing PAT helps to verify that the process behaves as expected. It has been used to ensure safe scale-up of hazardous processes [115, 116] and can replace traditional offline analysis where taking samples is difficult or where analyzing such samples would not be representative [114].

7.3.2
Flexible Implementation of Continuous Processes

It has already been pointed out that conventional batch processes often possess a higher degree of flexibility that can be an important factor when the market

Table 7.2 Drivers for modularization in small-scale continuous chemical production.

Time-to-market	Changeability in production	Mobility and risk mitigation
• Faster plant planning • Savings in procurement and construction • Early delivery of product samples	• Compensate lack of flexibility of continuous plants using exchangeable modules • Enable economies of scale through higher capacity usage in multipurpose plants	• Distributed production nearby the customer/feedstock • Flexible investment decisions/sequential numbering-up of small scale plants, following the market development

volatility is high and forecasting difficult or when multiple products are produced with the same plant. Potentially short product life cycles, the need for a stepwise expansion, or the opportunity for production near the final customer are factors that can influence the decision for a certain manufacturing technology. Therefore, in addition to substantial process optimization, the nature of plant engineering and construction needs to be adapted to fully leverage the benefits of continuous manufacturing in fine chemicals and specialties production. Especially, the speed of implementation and the versatility of the plants have to be addressed in order to be competitive with currently applied batch technology [18]. Given the fact that small-scale equipment is often available as standard equipment, individual design and equipment delivery times are much less of an issue than for large-scale plants. An approach to combine the advantages of continuous processes with the flexibility of batch manufacturing is modularization that is widely applied in other industries (with the automotive industry probably being the most prominent example), but has not yet found wider distribution in the fine chemical industry. However, two examples from the German chemical industry show that there is a strong interest in versatile and innovative small-scale production environments, which are described in the following paragraphs. Table 7.2 summarizes the main drivers to pursue such an approach.

Ecotrainer Concept for Rapid Process Development

The first example is the Ecotrainer developed by Evonik Industries AG that allows for rapid development and scale-up of continuous processes. The environment is housed inside an ISO transportation container divided into several segments. The main space is reserved for the installation of process equipment. A logistic room, adjacent to the process environment, provides storage for the required starting materials and products. Another segment which is encapsulated to be compatible with ATEX regulation provides space for process control systems and electrical installations. Furthermore, the container provides installations for fire and environmental protection [117]. Inside the process equipment area the flexible installation of equipment of various sizes and scales is possible.

This allows using the same infrastructure for bench top testing, piloting, and small-scale production [118]. It helps to reduce the efforts for the first scale-up steps and allows for faster testing and implementation. The container provides a moveable infrastructure that can be integrated into existing production environments.

Modularization for Changeable Production Systems
The second example is the F^3 Factory approach, developed in the corresponding, publicly funded EU project, which aims at combining the efficiency of continuous with the flexibility of batch operation. Since a static sequence of equipment is not well suited for multiproduct operation or short-term production campaigns, the plant is built from independent modules providing specific process functionality [6]. A module which is called PEA (Process Equipment Assembly) incorporates at least one main equipment element, providing the desired unit operation together with all needed peripheral components (e.g., pumps, heat exchangers, piping, and process control components). Near-field process control systems are provided in each PEA that can be connected to the overall process control system. To be compatible with each other additional geometrical and technical design guidelines regarding footprint as well as fluidic and electrical connections are established. The single PEAs are then combined to realize the desired process.

The advantages of this approach are twofold. First, the reuse of functional modules can help to speed-up plant engineering and construction. Here, the crucial point is the development of a module database where all PEAs are efficiently stored as fully documented engineering packages.

Second, the easily exchangeable modules can be used to realize flexible plant set-ups when frequent reconfiguration between production campaigns is required. Such a process can either be fully continuous or integrate individual PEAs into existing multipurpose/multiproduct batch plants to realize efficient hybrid production processes.

For stand-alone or decentralized production scenarios, the integration of PEAs into a process equipment container (PEC) can be a suitable solution. These modified freight containers provide a fully integrated infrastructure to build up a mobile and versatile production environment, requiring only basic utility supply on site. This PEA/PEC approach is summarized in Figure 7.2. An example for the successful application of this approach is described at the end of Section 7.4.

7.4
Selected Case Studies

The first example has been selected from a wide range of research activities at Bayer AG in the area of process intensification via continuous manufacturing of polymer, pharmaceutical, and agrochemical products. It illustrates the benefits when a tight control of residence time is required because the

Process Equipment Assembly (PEA)	Process Equipment Container (PEC)
	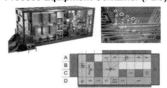

Description
• Modules containing unit operations and all supporting equipment
• Development based on technical and geometrical guidelines to ensure compatibility
	• Standardized infrastructure for:
	  • installation and fixation of PEAs
	  • connection and distribution of energy, utilities, and process media

Drivers ➤ Speed up development through reuse of engineering information ➤ Flexible process design through changeable and combinable functional units	➤ Mobile plant infrastructure ➤ Mobile and adaptable production units

Figure 7.2 Modularization concept for versatile continuous production.

desired product can be consumed in a consecutive reaction. The oxidation of thioethers, illustrated in Scheme 7.1, is highly exothermic and very often the one-step oxidation leading to the sulfoxide intermediate product is desired. Therefore, temperature control as well as the ability to rapidly quench the reaction after the optimum residence time were drivers to develop this process in flow.

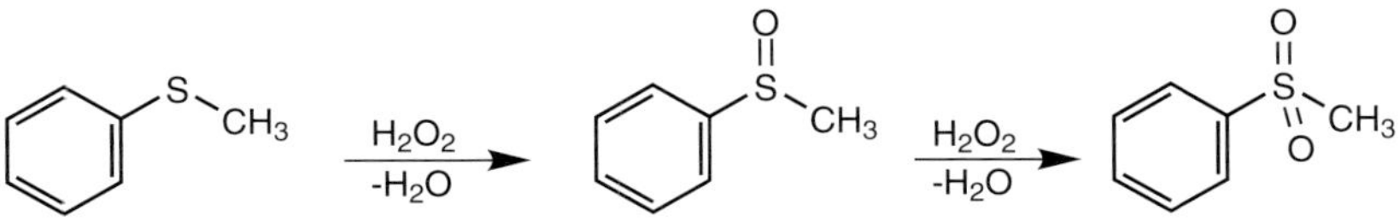

Scheme 7.1 Thioether oxidation (general reaction with model substrate).

The continuous process was developed using modular microstructured flow systems from Ehrfeld Mikrotechnik and applying H_2O_2 as an oxidative agent. The selectivity toward the desired sulfoxide was optimized and the process yield could be increased by approximately 6–8% compared to the batch process. The optimized process was successfully transferred to a fully automated pilot plant (scale-up factor 50) where the good laboratory performance could be fully reproduced [6].

While reaction performance was the most important factor in the previous example, safety is an important driver during the synthesis of 2-nitroethanol (Scheme 7.2), which is a raw material for the pharmaceutical ingredient aliskiren. The work of Roberge et al. nicely illustrates a systematic approach toward the development of an optimized continuous process. Experimentally gathered data were used to model the reaction under various conditions. Based on considerations concerning energy requirements for the workup section, safety constraints,

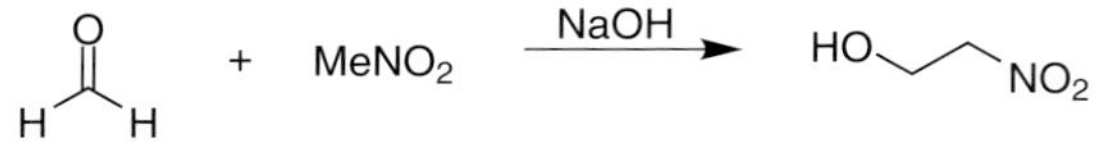

H−CHO + MeNO₂ —NaOH→ HO−CH₂−CH₂−NO₂ (implied)

Scheme 7.2 Synthesis of 2-nitro-ethanol from formaldehyde and nitromethane.

waste minimization, and reactor costs of the process conditions were selected to combine high conversion (>90%), moderate reaction temperature (<40 °C for safety reasons), and excess of $MeNO_2$ (preferably 60–80 equivalents) to suppress the formation of various nitroalcohols as by-products. This process optimization was the base for the design of a miniplant in two lab fume hoods, including solid separation followed by a distillation step for product removal. A long run study was performed resulting in a robust process and production of several kilograms of 2-nitro-ethanol with constant selectivity and product quality [119].

It has already been mentioned that over 90% of APIs accrue as solids [83] and that the handling of solids is a significant challenge in the design of continuous processes. A successful example has been published by researchers from Eli Lily [76]. The authors describe a process, including an asymmetric hydrogenation reaction at 70 bar with subsequent liquid–liquid extraction, solvent exchange distillation, crystallization, and filtration at lab and pilot scale (Scheme 7.3). In this process, a continuous process was particularly beneficial for the reaction and the crystallization. While the *in situ* yield and purity are comparable in batch and continuous mode, from a safety point of view the continuous process is beneficial because there is less hydrogen in the reactor at any given time. Furthermore, the capital investment for the continuous reactor was significantly lower than for a 70-bar hydrogenation batch autoclave. As no such capacity was already available, this investment would have had to be made at a stage where the fate of the given API as a commercial product was still uncertain, therefore, representing a significant financial risk. The second unit operation that benefitted from continuous processing is the 2-stage mixed suspension mixed product removal (MSMPR) crystallization. Continuous manufacturing enabled a kinetically controlled impurity rejection of the undesired enantiomer. This way an enantiomeric excess of 99.5% or higher could be achieved compared to 95% at thermodynamic equilibrium, although the authors point out that this is not (yet) a generally accepted strategy for impurity control.

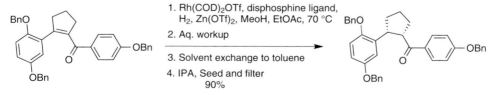

Scheme 7.3 Asymmetric enone hydrogenation.

Apart from the aforementioned benefits this example illustrates that it is not always required to design a fully and truly continuous process. In this case, the solvent exchange distillation was designed as automated repeating semibatch operation (about 20 turnovers per day) which, however, was also capable of reducing the required amount of solvent compared to conventional batch operation. Similarly, the slurry was fed to one of two parallel filters intermittently (once every 30 min in the final set-up). The slurry was then filtered and washed before the next charge of material was supplied. The authors point out that the identification of suitable parameters for filtering and washing is crucial for the formation of an evenly distributed filter cake and to avoid losses in the enantiomeric purity through crystallization of the impurity during filtration. Intermittent transport (about every 2 min), employing the already discussed pressure swing chambers (cf. Section 7.2.2), was also used out of the two crystallization vessels to avoid blocking of the tubing. Obviously, the combination of continuous and semicontinuous operations requires the use of additional buffer vessels. On the one hand, this increases the plant's footprint, hold-up, and time to achieve steady state; on the other hand, it also decouples the individual unit operations facilitating plant operation and control. Definitely, such options including the intelligent and possibly automated coupling of continuous and semicontinuous unit operations should be considered in order to develop an efficient overall process.

In total, the manufacture of 144 kg of product was accomplished in lab fume hoods and a lab hydrogenation bunker over two campaigns. The maximum employed vessel size was 22 l and the employed tubular reactor had a volume of 73 l. To achieve the same weekly throughput with traditional batch processing, a plant module with 400 l vessels would have been required.

The Novartis-MIT Center for Continuous Manufacturing has taken the development of integrated continuous process one step further. They have developed what they claim to be the first end-to-end continuous process for a pharmaceutical product (aliskiren) integrating two synthesis steps, purification, formulation, and tableting. In a plant with a footprint of $2.4 \times 7.3\,m^2$ and a nominal capacity of $45\,g\,h^{-1}$, they produced the final tablets that were reported to meet specifications for drug product quality [80, 120] (Figure 7.3). During the design of the process, many aspects of the existing batch processes had to be modified especially to avoid difficulties with the handling of solids and solvent changes, which shows that the optimum batch process is not necessarily equal to the optimum continuous process. For the particular product, the continuous process required 14 unit operations compared to 21 in batch, mostly due to integration of some previously separate downstream units into a single integrated device. Furthermore, some new pieces of small-scale equipment for continuous production, namely, for filtration and drying, were developed. Even though, equipment might not yet be available for all desired operations and at all required scales, this shows that it is possible to overcome some of the hurdles through in-house developments. Finally, the example also illustrates that significant efforts are required to develop a suitable process automation and control strategy for such

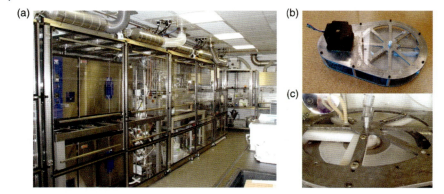

Figure 7.3 Process set up. (a) Continuous filter. (b) Filtration and wash of a wet cake (c) for end-to-end aliskiren process [80]. (Reprinted with permission from Ref. [80]. Copyright 2013, John Wiley & Sons, Inc.)

an integrated continuous process in order to ensure constant, in-spec production [121, 122].

While the previous example illustrates an end-to-end process for one particular product, Mleczko and Zhao from Bayer AG have described the use of continuous processes for a flexible platform technology capable of carrying out a range of similar reactions [123]. The different reactions are shown in Scheme 7.4. The first step, that is, the functionalization of an aromatic ring via deprotonation by a lithium base, is a fast, exothermic reaction that is typically mixing controlled. According to the frequently cited classification of Roberge *et al.* [8] this reaction would be classified as type A reaction and is an obvious target where microreactors can offer benefits. The subsequent reaction with different electrophiles allows for the synthesis of a variety of molecules. In this case,

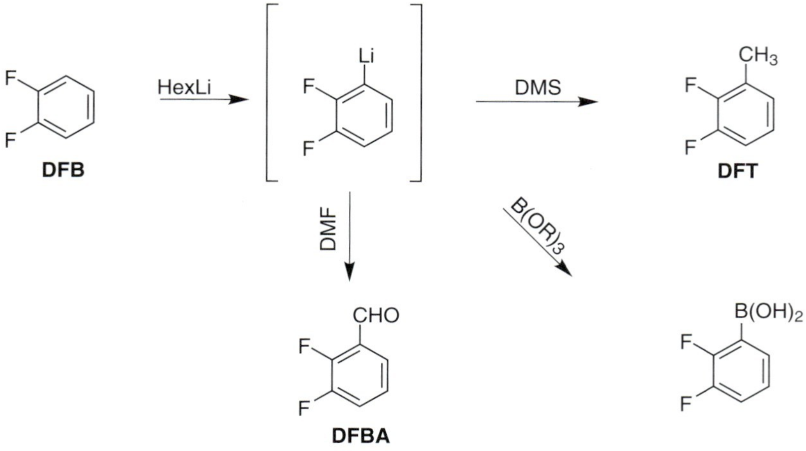

Scheme 7.4 Reaction network for DFT/DFBA synthesis.

the synthesis of 2,3-difluorotoluene (DFT), difluorobenzaldehyde (DFBA), and 2,3-difluorophenylboronic acid.

Optimization of the reaction parameters took place on lab scale using microstructured systems and the engineering concept for industrial production (up to 150 tonnes per year) was finalized. In addition, the formation of various side products was suppressed and the flexible technology developed was used to produce different products via one intermediate.

An even higher degree of flexibility can be realized through the previously discussed F^3 Factory modularization concept (Section 7.3.2). Within the F^3 Factory project, Bayer has demonstrated the technological viability of modular continuous production for an API intermediate at production scale. A three-step synthesis of an oncology drug candidate was successfully transformed from batch processing to an intensified, fully continuous process on lab scale. Subsequently, two synthesis steps and adjacent downstream operations (liquid/liquid extraction, solvent exchange), as shown in Figure 7.4, were scaled up and implemented in a systematically modularized container-based demonstration plant and operated at the INVITE backbone facility. The process container is mobile and can be operated with minimum requirements for energy and utilities onsite. These are provided via a standardized backbone infrastructure. The production container was operated at a scale of $1\,\mathrm{kg\,h^{-1}}$ API intermediate in a production near environment. Product specification of the API was met in a use test with the intermediate, inline analytics successfully applied, and technical feasibility of the intermediate production in a continuous process container was shown.

During the F^3 Factory project further case studies were also investigated. As a result, the broad applicability of the approach and its potential benefits were

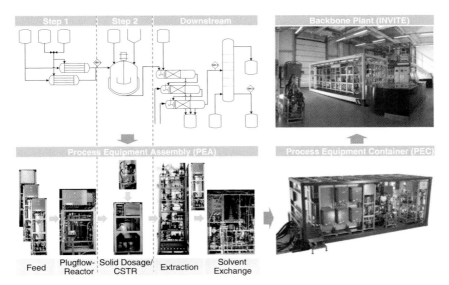

Figure 7.4 Modular design approach and INVITE platform.

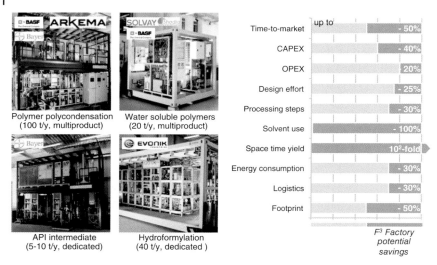

Figure 7.5 F³ Factory case studies and demonstration results.

demonstrated. Figure 7.5 shows all industrial case studies as well as the demonstration results reported in F³ Factory project.

7.5
Conclusion and Outlook

From a technological point of view, many unit operations can be carried out continuously at different scales and often these processes can be more efficient than conventional batch manufacturing. However, particularly on small scale, it might also be necessary to develop specialized solutions for some required steps to implement given processes. While some technologies, for example, static mixers, have been widely employed others have a lower technology readiness level and still need to prove their robustness. During the technology life cycle this is a crucial stage as there is a gap between the capabilities of academic research and the industrially expected level of development to pick up a technology at their own risk. Especially, challenging remain those steps involving the feeding or processing of slurries with the inherent risk of fouling and clogging. The development of technological solutions as well as models for predicting these processes are important fields of research to further develop continuous manufacturing in the fine chemical industry. Further potential for equipment design lies in the combination of simulation with innovative equipment manufacturing technologies such as 3D printing, which allows for the computer-supported geometry optimization with very few limitations imposed by manufacturing constraints.

An underestimated hurdle in the way of the wider acceptance of continuous manufacturing can probably be the experience and knowledge that has been acquired with conventional equipment. Companies have experts for different unit operations who know how to select, design, and operate suitable equipment. Such knowledge is limited for new technologies and mostly exists at the equipment suppliers, but not necessarily where the equipment is used and the process is developed. Even if people are aware of a certain technology, the implementation requires the cooperation and exchange of information across company boarders. These barriers can cause technologies not be routinely considered for quick proof-of-concept experiments during process development. The development of a thorough understanding of the devices' capabilities and operating windows is therefore important.

Based on a sound knowledge of different available technologies, process development should combine (experimental) data gathering and subsequent process modeling, simulation, optimization, and implementation in a structured approach. Here, it is important to keep the initial drivers in mind. Efficiency, time-to-market, and flexibility are partly contradictory aims and the optimal trade-off depends on the given process as well as the boundary conditions. On the one hand, an automated end-to-end process can be very efficient. However, the higher possible degree of automation and a decreased response time to market changes need to outweigh the higher technical complexity and the risk of single component failures stopping the entire process. On the other hand, for processes where time-to-market or flexibility is critical, the use of standardized modules or conventional batch manufacturing might be the better approach even at the expense of decreased process efficiency.

Additional implications on logistics and supply chain also need to be considered. Multipurpose batch plants are capable of producing a large variety of products. Producing the same product in a dedicated end-to-end continuous plant removes the need for intermediate isolations leading to reduced inventory. Apart from the fact that the intermediate isolation often also accomplishes the removal of impurities, the production cannot be easily transferred to another site in case of problems with the process thus requiring additional stocks to avoid shortages and delayed deliveries. In fact, flexibility of batch manufacturing is one of the biggest challenges for continuous manufacturing. Flexibility should therefore be kept in mind during equipment development and selection as well as process design and implementation in order to realize continuous processes competitive with the standard batch approach.

Given the presented technological advantages, continuous processing is sometimes described as panacea of chemical manufacturing. Although we believe in the discussed opportunities, we do not share this opinion. We do, however, think that continuous processing should be considered an equal alternative to conventional batch operation that can – depending on the context – enable more sustainable production. Likewise, hybrid processes in which some continuous steps (e.g., highly exothermic reactions) are combined with batch operations can constitute a viable option. What is important is the detailed analysis of a given process and the crucial drivers during implementation. Based on this analysis the most suitable process option should be selected be it batch, continuous, or any sort of hybrid form in between. From our perspective, one of the main challenges is the limited understanding and awareness when continuous processes should at least be considered as manufacturing option. As the optimum continuous process often is not a one-to-one transfer of the batch variant, such consideration need to be made already at the early stages of process development. To enable this, a stronger integration in the chemistry curricula and a stronger interdisciplinary connection between chemists and engineers would be highly desirable, even more so because the development of integrated continuous processes intrinsically needs to take a holistic approach instead of separately optimizing individual process steps. Furthermore, the companies' organizational structures should support the continuity from early process development to final production and foster the cooperation between all involved departments.

In order to increase the acceptance of continuous manufacturing in a traditionally conservative environment, it is probably easier to start picking the low hanging fruits and demonstrate the potential benefits for those syntheses or workup steps where a continuous process is particularly advantageous. Given the still existing technological (e.g., equipment robustness, fouling, clogging) and, in the pharmaceutical industry, regulatory hurdles (e.g., batch definition, material traceability[1)]), the industrial use of continuous end-to-end processes in small- to medium-scale chemical production is expected to be limited in the midterm. However, due to the significant interest and effort devoted to the subject, further progress in this direction can be expected steadily broadening the scope of continuous manufacturing from individual process steps toward more integrated processes. Eventually, this will hopefully enable to leverage the technology's full potential that is not limited to the mostly technical aspects such as operating windows and scale-up, but also includes the development of optimized supply chain and logistic concepts.

References

1 Stankiewicz, A. and Moulijn, J.A. (2000) Process intensification: transforming chemical engineering. *Chemical Engineering and Processing*, **96** (1), 22–34.

2 Hessel, V. (2009) Novel process windows – gate to maximizing process intensification via flow chemistry. *Chemical Engineering and Technology*, **32** (11), 1655–1681.

3 Hartwig, J., Metternich, J.B., Nikbin, N., Kirschning, A., and Ley, S.V. (2014) Continuous flow chemistry: a discovery tool for new chemical reactivity patterns. *Organic and Biomolecular Chemistry*, **12** (22), 3611–3615.

4 Hessel, V., Kralisch, D., and Kockmann, N. (2015) *Novel Process Windows*, Wiley-VCH Verlag GmbH, Weinheim.

5 Poechlauer, P., Colberg, J., Fisher, E., Jansen, M., Johnson, M.D., Koenig, S.G., Lawler, M., Laporte, T., Manley, J., Martin, B., and O'Kearney-McMullan, A. (2013) Pharmaceutical roundtable study demonstrates the value of continuous manufacturing in the design of greener

---

1) The regulatory implications of continuous manufacturing are discussed in more detail in Ref. [124].

processes. *Organic Process Research and Development*, **17** (12), 1472–1478.
6 Bieringer, T., Buchholz, S., and Kockmann, N. (2013) Future production concepts in the chemical industry: modular – small-scale – continuous. *Chemical Engineering and Technology*, **36** (6), 900–910.
7 Roberge, D.M., Zimmermann, B., Rainone, F., Gottsponer, M., Eyholzer, M., and Kockmann, N. (2008) Microreactor technology and continuous processes in the fine chemical and pharmaceutical industry: is the revolution underway? *Organic Process Research and Development*, **12** (5), 905–910.
8 Roberge, D.M., Ducry, L., Bieler, N., Cretton, P., and Zimmermann, B. (2005) Microreactor technology: a revolution for the fine chemical and pharmaceutical industries? *Chemical Engineering and Technology*, **28** (3), 318–323.
9 am Ende, D.J. (2011) Chemical engineering in the pharmaceutical industry: an introduction. *Chemical Engineering in the Pharmaceutical Industry*, John Wiley & Sons, Inc., Hoboken, N.J, pp. 3–20.
10 Holvey, C.P., Roberge, D.M., Gottsponer, M., Kockmann, N., and Macchi, A. (2011) Pressure drop and mixing in single phase microreactors: simplified designs of micromixers. *Chemical Engineering and Processing*, **50** (10), 1069–1075.
11 Stankiewicz, A. (2006) Energy matters. *Chemical Engineering Research and Design*, **84** (7), 511–521.
12 Morschhäuser, R., Krull, M., Kayser, C., Boberski, C., Bierbaum, R., Püschner, P.A., Glasnov, T.N., and Kappe, C.O. (2012) Microwave-assisted continuous flow synthesis on industrial scale. *Green Processing and Synthesis*, **1** (3), 281–290.
13 Oelgemöller, M. and Shvydkiv, O. (2011) Recent advances in microflow photochemistry. *Molecules*, **16** (9), 7522–7550.
14 Mason, T.J. and Peters, D. (2002) *Practical Sonochemistry: Power Ultrasound Uses and Applications*, Woodhead Publishing Ltd., Sawston, Cambridge.
15 Jiménez-González, C., Poechlauer, P., Broxterman, Q.B., Yang, B.-S., am Ende, D.J., Baird, J., Bertsch, C., Hannah, R.E., Dell'Orco, P., Noorman, H., Yee, S., Reintjens, R., Wells, A., Massonneau, V., and Manley, J. (2011) Key green engineering research areas for sustainable manufacturing: a perspective from pharmaceutical and fine chemicals manufacturers. *Organic Process Research and Development*, **15** (4), 900–911.
16 Baxendale, I.R., Braatz, R.D., Hodnett, B.K., Jensen, K.F., Johnson, M.D., Sharratt, P., Sherlock, J.-P., and Florence, A.J. (2014) Achieving Continuous Manufacturing: Technologies and Approaches for Synthesis, Work-Up and Isolation of Drug Substance. Available at https://iscmp.mit.edu/white-papers/introductory-white-paper (accessed February 24, 2015).
17 Kirschneck, D. (2008) Process development innovation. Specialty Chemicals Magazine, 28 (10).
18 Poechlauer, P., Manley, J., Broxterman, R., Gregertsen, B., and Ridemark, M. (2012) Continuous processing in the manufacture of active pharmaceutical ingredients and finished dosage forms: an industry perspective. *Organic Process Research and Development*, **16** (10), 1586–1590.
19 Thayer, A.M. (2013) Fine chemicals diversify. *Chemical and Engineering News*, **91** (21), 13–19.
20 Abboud, L. and Hensley, S. (2003) New prescription for drug makers: update the plants. The Wall Street Journal.
21 Nicholson Price, W. II (2014) Making do in making drugs. *Boston College Law Review*, **55** (2), 491.
22 FDA (2004) Innovation and Continuous Improvement in Pharmaceutical Manufacturing. Available at http://www.fda.gov/ohrms/dockets/ac/04/briefing/2004-4080b1_01_manufSciWP.pdf (accessed November 2, 2014).
23 Blacker J. and Williams M.T. Future trends and challenges. *Pharmaceutical Process Development: Current Chemical and Engineering Challenges*, Royal Society of Chemistry, Cambridge, pp. 331–341 (Watts, C. (2006) FDA and

continuous processing. Presentation held at the ISPE Congress, Vienna, 21 September 2006, cited after Blacker, A.J. and Williams, M.T. (2011)).

24 PhRMA (2014) Biopharmaceuticals in Perspective (Chart Pack). Available at http://www.phrma.org/sites/default/files/pdf/ChartPack_4%200_FINAL_2014MAR25.pdf (accessed November 10, 2014).

25 Grabowski, H.G. and Kyle, M. (2007) Generic competition and market exclusivity periods in pharmaceuticals. *Managerial and Decision Economics*, **28** (4–5), 491–502.

26 Strickland, I. (2012) World preview 2018: embracing the patent cliff. Available at http://www.evaluategroup.com/public/EvaluatePharma-World-Preview-2018-Embracing-the-Patent-Cliff.aspx.

27 Dhankhar, A., Evers, M., and Møller, M. (2012) Escaping the sword of Damocles: toward a new future for pharmaceutical R&D. *Evolution or Revolution?*, McKinsey, pp. 2–11.

28 Gonce, A. and Schrader, U. (2012) Plantopia? A mandate for innovation in pharma manufacturing. Available at http://www.mckinsey.com/insights/health_systems_and_services/pharma_manufacturing_for_a_new_era (accessed November 10, 2014).

29 Adams, C.P. and van Brantner, V. (2006) Estimating the cost of new drug development: is it really $802 million? *Health Affairs*, **25** (2), 420–428.

30 Adams, C.P. and van Brantner, V. (2010) Spending on new drug development. *Health Economics*, **19** (2), 130–141.

31 DiMasi, J.A., Hansen, R.W., and Grabowski, H.G. (2003) The price of innovation: new estimates of drug development costs. *Journal of Health Economics*, **22** (2), 151–185.

32 Paul, S.M., Mytelka, D.S., Dunwiddie, C.T., Persinger, C.C., Munos, B.H., Lindborg, S.R., and Schacht, A.L. (2010) How to improve R&D productivity: the pharmaceutical industry's grand challenge. *Nature Reviews. Drug Discovery*, **9** (3), 203–214.

33 Scannell, J.W., Blanckley, A., Boldon, H., and Warrington, B. (2012) Diagnosing the decline in pharmaceutical R&D efficiency. *Nature Reviews. Drug Discovery*, **11** (3), 191–200.

34 Congressional Budget Office (2006) Research and Development in the Pharmaceutical Industry. Available at http://www.cbo.gov/publication/18176 (accessed November 10, 2014).

35 Vernon, J.A., Golec, J.H., and DiMasi, J.A. (2010) Drug development costs when financial risk is measured using the fama-french three-factor model. *Health Economics*, **19** (8), 1002–1005.

36 Outterson, K. (2005) Pharmaceutical arbitrage: balancing access and innovation in international prescription drug markets. *Yale Journal of Health Policy, Law, and Ethics*, **5** (1), 193–292.

37 Berndt, E.R., Bui, L., Reiley, D.R., and Urban, G.L. (1995) Information, marketing, and pricing in the U.S. antiulcer drug market. *American Economic Review*, **85** (2), 100–105.

38 Basu, P., Joglekar, G., Rai, S., Suresh, P., and Vernon, J. (2008) Analysis of manufacturing costs in pharmaceutical companies. *Journal of Pharmaceutical Innovation*, **3** (1), 30–40.

39 Federsel, H.-J. (2003) Logistics of process R&D: transforming laboratory methods to manufacturing scale. *Nature Reviews. Drug Discovery*, **2** (8), 654–664.

40 Ahlawat, H., Chierchia, G., and van Arkel, P. (2014) The secret of successful drug launches. Available at http://www.mckinsey.com/insights/health_systems_and_services/the_secret_of_successful_drug_launches (accessed November 10, 2014).

41 Stirling, W. (2013) http://www.themanufacturer.com/articles/pharmas-new-love-in-can-build-a-c21st-british-supply-chain/ (accessed November 10, 2014).

42 Chemistry Innovation (2014 https://sbri.innovateuk.org/documents/3255552/3747706/16+-+Genzyme+and+NiTech+collaboration+delivers+the+world%E2%80%99s+largest+patent+protected+continuous+API+manufacturing+plant.pdf/bc5fa2ed-40e5-481b-8862-94a72d2bf4c9 (accessed November 5, 2014).

43 Heriott Watt University (2013) http://nitechsolutions.co.uk/wp-content/uploads/2014/03/HW-case-study-Nov13.pdf (accessed November 5, 2014).

44 Richards, P. (2007) http://newsoffice.mit.edu/2007/novartis-0928 (accessed October 21, 2014).

45 Montheit, M.J. and Mitchell, M.B. (2011) Development enabling technologies, in *Pharmaceutical Process Development: Current Chemical and Engineering Challenges* (eds J. Blacker and M.T. Williams), Royal Society of Chemistry, Cambridge, pp. 238–259.

46 Foster Wheeler (2009) http://www.fwc.com/Who-We-Are/Publications-Videos/Magazines.aspx (accessed November 11, 2014).

47 Sapien (2009) http://sapienprocesstechnologies.wordpress.com/tag/innovative-manufacturing-iniative/ (accessed November 11, 2014).

48 Vashisth, S., Kumar, V., and Nigam, K. (2008) A review on the potential applications of curved geometries in process industry. *Industrial and Engineering Chemistry Research*, **47** (10), 3291–3337.

49 Saxena, A.K. and Nigam, K. (1984) Coiled configuration for flow inversion and its effect on residence time distribution. *AIChE Journal*, **30** (3), 363–368.

50 Klutz, S., Kurt, S.K., Lobedann, M., and Kockmann, N. (2015) Narrow residence time distribution in tubular reactor concept for Reynolds number range of 10–100. *Chemical Engineering Research and Design*, **95**, 22–33.

51 Ghanem, A., Lemenand, T., Della Valle, D., and Peerhossaini, H. (2014) Static mixers: mechanisms, applications, and characterization methods – a review. *Chemical Engineering Research and Design*, **92** (2), 205–228.

52 Falk, L. and Commenge, J.-M. (2010) Performance comparison of micromixers. *Chemical Engineering Science*, **65** (1), 405–411.

53 Ghaini, A., Balon-Burger, M., Bogdan, A., Krtschil, U., and Löb, P. (2015) Modular microstructured reactors for pilot- and production scale chemistry. *Chemical Engineering and Technology*, **38** (1), 33–43.

54 Reay, D.A., Ramshaw, C., and Harvey, A.P. (2013) *Process Intensification*, Butterworth-Heinemann, Oxford.

55 Oxley, P., Brechtelsbauer, C., Ricard, F., Lewis, N., and Ramshaw, C. (2000) Evaluation of spinning disk reactor technology for the manufacture of pharmaceuticals. *Industrial and Engineering Chemistry Research*, **39** (7), 2175–2182.

56 Boswell, C. (2004) http://www.icis.com/resources/news/2004/06/04/587031/technology-watch-kreido-s-spinning-tube-in-tube-reactor-takes-on-batch-production/ (accessed January 9, 2015).

57 Kotrba, R. (2008) Putting the squeeze on convention. *Biodiesel Magazine*, **5** (2), 67–72.

58 Marui, T. (2013) An introduction to micro/nano-bubbles and their applications. *Journal on Systemics, Cybernetics and Informatics*, **11** (4), 68–73.

59 Chemistry Innovation (2014) https://connect.innovateuk.org/documents/3255552/3747706/12+-+NiTech+-+Continuous+Oscillatory+Baffled+Reactor%E2%84%A2.pdf/ad122d0a-184d-4531-b16d-e0209ab51c71 (accessed November 11, 2014).

60 Lawton, S., Steele, G., Shering, P., Zhao, L., Laird, I., and Ni, X.-W. (2009) Continuous crystallization of pharmaceuticals using a continuous oscillatory baffled crystallizer. *Organic Process Research and Development*, **13** (6), 1357–1363.

61 AM Technology (2014) http://www.amtechuk.com/Products.aspx (accessed October 22, 2014).

62 Browne, D.L., Deadman, B.J., Ashe, R., Baxendale, I.R., and Ley, S.V. (2011) Continuous flow processing of slurries: evaluation of an agitated cell reactor. *Organic Process Research and Development*, **15** (3), 693–697.

63 Salice, P., Fenaroli, D., de Filippo, C.C., Menna, E., Gasparini, G., and Maggini, M. (2012) Efficient functionalization of carbon nanotubes: an opportunity

enabled by flow chemistry. *Chemistry Today*, **60** (6), 37–39.
64 Strube, J., Grote, F., Josch, J.P., and Ditz, R. (2011) Process development and design of downstream processes. *Chemie Ingenieur Technik*, **83** (7), 1044–1065.
65 Schuur, B., Hallett, A.J., Winkelman, J.G.M., de Vries, J.G., and Heeres, H.J. (2009) Scalable enantioseparation of amino acid derivatives using continuous liquid–liquid extraction in a cascade of centrifugal contactor separators. *Organic Process Research and Development*, **13** (5), 911–914.
66 Trent, D.L. (2004) Chemical processing in high-gravity fields, in *Re-Engineering the Chemical Processing Plant: Process Intensification* (eds A. Stankiewicz and J.A. Moulijn), M. Dekker, New York, pp. 29–61.
67 Sudhoff, D., Neumann, K., and Lutze, P. (2014) An integrated design method for rotating packed beds for distillation. Proceedings of the 24th European Symposium on Computer Aided Process Engineering, 2014.
68 Bisschops, M.A.T., van Hateren, S.H., Luyben, K.C.A.M., and van der Wielen, L.A.M. (2000) Mass transfer performance of centrifugal adsorption technology. *Industrial and Engineering Chemistry Research*, **39** (11), 4376–4382.
69 Kenig, E.Y., Su, Y., Lautenschleger, A., Chasanis, P., and Grünewald, M. (2013) Micro-separation of fluid systems: a state-of-the-art review. *Separation and Purification Technology*, **120**, 245–264.
70 Helling, C., Fröhlich, H., Eggersglüss, J., and Strube, J. (2012) Fundamentals towards a modular microstructured production plant. *Chemie Ingenieur Technik*, **84** (6), 892–904.
71 Adamo, A., Heider, P.L., Weeranoppanant, N., and Jensen, K.F. (2013) Membrane-based, liquid–liquid separator with integrated pressure control. *Industrial and Engineering Chemistry Research*, **52** (31), 10802–10808.
72 Kumar, S., Kumar, B., Sampath, M., Sivakumar, D., Kamachi Mudali, U., and Natarajan, R. (2012) Development of a micro-mixer-settler for nuclear solvent extraction. *Journal of Radioanalytical and Nuclear Chemistry*, **291** (3), 797–800.
73 Nightingale, A.M. and Demello, J.C. (2013) Segmented flow reactors for nanocrystal synthesis. *Advanced Materials*, **25** (13), 1813–1821.
74 Geddert, T., Bialuch, I., Augustin, W., and Scholl, S. (2009) Extending the induction period of crystallization fouling through surface coating. *Heat Transfer Engineering.*, **30** (10–11), 868–875.
75 Narducci, O., Jones, A.G., and Kougoulos, E. (2011) Continuous crystallization of adipic acid with ultrasound. *Chemical Engineering Science*, **66** (6), 1069–1076.
76 Johnson, M.D., May, S.A., Calvin, J.R., Remacle, J., Stout, J.R., Diseroad, W.D., Zaborenko, N., Haeberle, B.D., Sun, W.-M., Miller, M.T., and Brennan, J. (2012) Development and scale-up of a continuous, high-pressure, asymmetric hydrogenation reaction, workup, and isolation. *Organic Process Research and Development*, **16** (5), 1017–1038.
77 Quon, J.L., Zhang, H., Alvarez, A., Evans, J., Myerson, A.S., and Trout, B.L. (2012) Continuous crystallization of aliskiren hemifumarate. *Crystal Growth and Design*, **12** (6), 3036–3044.
78 White, T.D., Alt, C.A., Cole, K.P., Groh, J.M., Johnson, M.D., and Miller, R.D. (2014) How to convert a walk-in hood into a manufacturing facility: demonstration of a continuous, high-temperature cyclization to process solids in flow. *Organic Process Research and Development*, **18** (11), 1482–1491.
79 Braden, T.M., Gonzalez, M.A., Jines, A.M., Johnson, M.D., and Sun, W.-M. (2008) Reactors and methods for processing reactants therein. WO/2009/023515 (A3), filed Aug. 7, 2008 and published April 30, 2009.
80 Mascia, S., Heider, P.L., Zhang, H., Lakerveld, R., Benyahia, B., Barton, P.I., Braatz, R.D., Cooney, C.L., Evans, J.M.B., Jamison, T.F., Jensen, K.F., Myerson, A.S., and Trout, B.L. (2013) End-to-end continuous manufacturing of pharmaceuticals: integrated synthesis, purification, and final dosage formation.

Angewandte Chemie, International Edition, **52** (47), 12359–12363.

81 Steadfast Equipment (2015) http://www.steadfastequipment.com/ (March 20, 2015).

82 Jones, A.G. (2002) *Crystallization Process Systems*, Butterworth-Heinemann, Oxford.

83 Alvarez, A.J. and Myerson, A.S. (2010) Continuous plug flow crystallization of pharmaceutical compounds. *Crystal Growth & Design*, **10** (5), 2219–2228.

84 GEA (2015) http://www.gea.com/de/de/products/consigma-ctl.jsp (December 8, 2015).

85 Lin, Y.-M., Liu, S.-L., Chuang, C.-H., and Chu, Y.-T. (2003) Effect of incipient removal of hydrogen through palladium membrane on the conversion of methane steam reforming: experimental and modeling. *Catalysis Today*, **82** (1–4), 127–139.

86 Harrison, D.P. (2008) Sorption-enhanced hydrogen production: a review. *Industrial and Engineering Chemistry Research*, **47** (17), 6486–6501.

87 Stankiewicz, A. (2003) Reactive separations for process intensification: an industrial perspective. *Chemical Engineering and Processing: Process Intensification*, **42** (3), 137–144.

88 Keller, T. (2015) Reactive distillation, in *Handbook of Distillation*, vol. **2** (eds A. Górak and Z. Olujic), Elsevier, pp. 261–294.

89 Wasewar, K.L., Yawalkar, A.A., Moulijn, J.A., and Pangarkar, V.G. (2004) Fermentation of glucose to lactic acid coupled with reactive extraction: a review. *Industrial and Engineering Chemistry Research*, **43** (19), 5969–5982.

90 Schmidt-Traub, H. and Górak, A. (2006) *Integrated Reaction and Separation Operations*, Springer, Berlin.

91 Giguere, R.J., Bray, T.L., Duncan, S.M., and Majetich, G. (1986) Application of commercial microwave ovens to organic synthesis. *Tetrahedron Letters*, **27** (41), 4945–4948.

92 Gedye, R., Smith, F., Westaway, K., Ali, H., Baldisera, L., Laberge, L., and Rousell, J. (1986) The use of microwave ovens for rapid organic synthesis. *Tetrahedron Letters*, **27** (3), 279–282.

93 Kappe, C.O., Stadler, A., Dallinger, D., Mannhold, R., Kubinyi, H., and Folkers, G. (2013) *Microwaves in Organic and Medicinal Chemistry*, Wiley-VCH Verlag GmbH, Weinheim, Germany.

94 Patil, N.G., Benaskar, F., Rebrov, E.V., Meuldijk, J., Hulshof, L.A., Hessel, V., and Schouten, J.C. (2014) Scale-up of microwave assisted flow synthesis by transient processing through monomode cavities in series. *Organic Process Research and Development*, **18** (11), 1400–1407.

95 Cablewski, T., Faux, A.F., and Strauss, C.R. (1994) Development and application of a continuous microwave reactor for organic synthesis. *Journal of Organic Chemistry*, **59** (12), 3408–3412.

96 Howarth, P. and Lockwood, M. (2004) Come of age. The Chemical Engineer, 29–31.

97 Hartman, R.L., Naber, J.R., Zaborenko, N., Buchwald, S.L., and Jensen, K.F. (2010) Overcoming the challenges of solid bridging and constriction during Pd-catalyzed C–N bond formation in microreactors. *Organic Process Research and Development*, **14** (6), 1347–1357.

98 Horie, T., Sumino, M., Tanaka, T., Matsushita, Y., Ichimura, T., and Yoshida, J.-i. (2010) Photodimerization of maleic anhydride in a microreactor without clogging. *Organic Process Research and Development*, **14** (2), 405–410.

99 Sedelmeier, J., Ley, S.V., Baxendale, I.R., and Baumann, M. (2010) KMnO(4)-Mediated oxidation as a continuous flow process. *Organic Letters*, **12** (16), 3618–3621.

100 Leonelli, C. and Mason, T.J. (2010) Microwave and ultrasonic processing: now a realistic option for industry. *Chemical Engineering and Processing: Process Intensification*, **49** (9), 885–900.

101 van Gerven, T., Mul, G., Moulijn, J., and Stankiewicz, A. (2007) A review of intensification of photocatalytic processes. *Chemical Engineering and Processing: Process Intensification*, **46** (9), 781–789.

102 Loponov, K.N., Lopes, J., Barlog, M., Astrova, E.V., Malkov, A.V., and Lapkin, A.A. (2014) Optimization of a scalable photochemical reactor for reactions with singlet oxygen. *Organic Process Research and Development*, **18** (11), 1443–1454.

103 Lévesque, F. and Seeberger, P.H. (2012) Continuous-flow synthesis of the antimalaria drug artemisinin. *Angewandte Chemie (International ed. in English)*, **51** (7), 1706–1709.

104 Wiles, C. and Watts, P. (2011) *Micro Reaction Technology in Organic Synthesis*, CRC Press, Boca Raton, FL.

105 Fekete, M. and Glasnov, T. (2014) Technology overview/overview of the devices, in *Flow Chemistry – Volume 1: Fundamentals* (eds F. Darvas, D. György, and V. Hessel), De Gruyter, Berlin, pp. 95–140.

106 Syrris (2015) http://syrris.com/flow-products/asia-flow-chemistry (accessed August 31, 2015).

107 ThalesNano (2015) http://thalesnano.com/ (accessed February 25, 2015).

108 Vapourtec (2015) http://www.vapourtec.co.uk/ (accessed February 25, 2015).

109 Krasberg, N., Hohmann, L., Bieringer, T., Bramsiepe, C., and Kockmann, N. (2014) Selection of technical reactor equipment for modular, continuous small-scale plants. *Processes*, **2** (1), 265–292.

110 Commenge, J.-M. and Falk, L. (2014) Methodological framework for choice of intensified equipment and development of innovative technologies. *Chemical Engineering and Processing*, **84**, 109–127.

111 Simon, L.L., Pataki, H., Marosi, G., Meemken, F., Hungerbühler, K., Baiker, A., Tummala, S., Glennon, B., Kuentz, M., Steele, G., Kramer, H.J.M., Rydzak, J.W., Chen, Z., Morris, J., Kjell, F., Singh, R., Gani, R., Gernaey, K.V., Louhi-Kultanen, M., O'Reilly, J., Sandler, N., Antikainen, O., Yliruusi, J., Frohberg, P., Ulrich, J., Braatz, R.D., Leyssens, T., von Stosch, M., Oliveira, R., Tan, R.B.H., Wu, H., Khan, M., O'Grady, D., Pandey, A., Westra, R., Delle-Case, E., Pape, D., Angelosante, D., Maret, Y., Steiger, O., Lenner, M., Abbou-Oucherif, K., Nagy, Z.K., Litster, J.D., Kamaraju, V.K., and Chiu, M.-S. (2015) Assessment of recent process analytical technology (PAT) trends: a multiauthor review. *Organic Process Research and Development*, **19** (1), 3–62.

112 Rathore, A.S. (2009) Roadmap for implementation of quality by design (QbD) for biotechnology products. *Trends in Biotechnology*, **27** (9), 546–553.

113 Chew, W. and Sharratt, P. (2010) Trends in process analytical technology. *Analytical Methods*, **2** (10), 1412.

114 Schwolow, S., Braun, F., Rädle, M., Kockmann, N., and Röder, T. (2015) Fast and efficient acquisition of kinetic data in microreactors using in-line raman analysis. *Organic Process Research and Development*, **19** (9), 1286–1292.

115 Wiss, J., Länzlinger, M., and Wermuth, M. (2005) Safety improvement of a Grignard reaction using on-line NIR monitoring. *Organic Process Research and Development*, **9** (3), 365–371.

116 Barrios Sosa, A.C., Conway, R., Williamson, R.T., Suchy, J.P., Edwards, W., and Cleary, T. (2011) Application of PAT tools for the safe and reliable production of a dihydro-1 H -imidazole. *Organic Process Research and Development*, **15** (6), 1458–1463.

117 Lang, J.E., Hoppe, C.-F., Rauleder, H., and Müh, E. (2008) Universal infrastructure for chemical processes, DE 10 2008 041 950 (A1), filed Sep. 10, 2008 and published March 11, 2010.

118 Hüser, T. (2014) http://www.process.vogel.de/anlagenbau_effizienz/articles/470931/ (accessed February 10, 2015).

119 Roberge, D.M., Noti, C., Irle, E., Eyholzer, M., Rittiner, B., Penn, G., Sedelmeier, G., and Schenkel, B. (2014) Control of hazardous processes in flow: synthesis of 2-nitroethanol. *Journal of Flow Chemistry*, **4** (1), 26–34.

120 Heider, P.L., Born, S.C., Basak, S., Benyahia, B., Lakerveld, R., Zhang, H., Hogan, R., Buchbinder, L., Wolfe, A., Mascia, S., Evans, J.M.B., Jamison, T.F., and Jensen, K.F. (2014) Development of a multi-step synthesis and workup sequence for an integrated, continuous manufacturing process of a pharmaceutical. *Organic Process Research and Development*, **18** (3), 402–409.

121 Lakerveld, R., Benyahia, B., Braatz, R.D., and Barton, P.I. (2013) Model-based design of a plant-wide control strategy for a continuous pharmaceutical plant. *AIChE Journal*, **59** (10), 3671–3685.

122 Zhang, H., Lakerveld, R., Heider, P.L., Tao, M., Su, M., Testa, C.J., D'Antonio, A.N., Barton, P.I., Braatz, R.D., Trout, B.L., Myerson, A.S., Jensen, K.F., and Evans, J.M.B. (2014) Application of continuous crystallization in an integrated continuous pharmaceutical pilot plant. *Crystal Growth & Design*, **14** (5), 2148–2157.

123 Mleczko, L. and Zhao, D. (2014) Technology for continuous production of fine chemicals, in *Managing Hazardous Reactions and Compounds in Process Chemistry* (eds J.A. Pesti and A.F. Abdel-Magid), American Chemical Society, Washington, DC, pp. 403–440.

124 Allison, G., Cain, Y.T., Cooney, C., Garcia, T., Bizjak, T.G., Holte, O., Jagota, N., Komas, B., Korakianiti, E., Kourti, D., Madurawe, R., Morefield, E., Montgomery, F., Nasr, M., Randolph, W., Robert, J.-L., Rudd, D., and Zezza, D. (2015) Regulatory and quality considerations for continuous manufacturing. May 20–21, 2014 continuous manufacturing symposium. *Journal of Pharmaceutical Sciences*, **104** (3), 803–812.

8
How Technical Innovation in Manufacturing Is Fostered through Business Innovation

Nicolas Eghbali, Marianne Hoppenbrouwers, Steven Lemain, Gert De Bruyn, and Bart Vander Velpen

8.1
General Introduction

Manufacturing is defined by a range of activities that transform raw materials into finished products or finished goods. It includes, for instance, the production of textiles, automobiles, steel, chemicals, electronics, or wood products. Over the past century, the manufacturing industry has undergone transformations from both technological and organizational innovation points of view; one of the most famous changes being the introduction of lean manufacturing [1]. These changes are often driven by evolving customer's needs and changes in the business and legal environment.

Like many other industries, the manufacturing industry is constantly under pressure to adapt. Companies are now facing major challenges ranging from resources scarcity, energy price volatility, and global supply chain management to regulatory challenges. The business environment has become increasingly complex and difficult to predict, and so innovation is necessary for companies to remain resilient.

Besides lean production, one of the transformations that occurred relatively recently is the embedment of the manufacturing industry with the so-called service industry. This was particularly marked with the rise of outsourcing. This integration of service components was observed at all levels within the companies and represents a transition from traditional manufacturing practices [2]. It even reached a new level with the development of the "functional economy" and the introduction of product-service models [3].

As it is the case in the manufacturing of many goods, industries have to deal with noncore activities (less valued) or secondary activities, which are part of the company's value chain but are not an essential part of the value proposition. In fact, it is this very differentiation between valued and less-valued activities that led to the bigger concept of lean production (also called lean manufacturing). Lean manufacturing is a philosophy derived from the Japanese auto industry

(Toyota industries), which focuses on the reduction of the less-valued activities (maximum efficiency and flexibility) [1]. It is an attempt to optimize, reduce, or eliminate every step, situation, and/or activities that do not contribute enough to the value creation process. It is seeking the maximum optimization. When these secondary activities are perceived as increasing cost, it raises the question as whether they should or could not be best outsourced and handled by specialized suppliers (for which they become core activities – more efficient), henceforth introducing the notion of services as part of the manufacturing process.

Although the introduction of lean manufacturing and the subsequent outsourcing practices can be seen as a positive managerial transformation, it did not necessarily translate into an actual technological innovation, especially for those activities that are perceived by one of the parties as cost. This situation could even be seen as a barrier to radically improving or changing the noncore processes. Introducing services in the form of outsourcing may have therefore limited the extent to which cost reduction and optimization can be achieved.

One way to overcome this barrier and allow the introduction of technological innovation within secondary processes is to consider the optimization of the supply chain (learner supply chain). Product-service systems (PSS) [3, 4] are examples of business models used to optimize a supply chain, for achieving not only a more efficient production but also a more sustainable production–consumption pattern.

These types of business models rely on the concept that users seek a desired result, outcome, or function and not *per se* a specific product. It is a business model in which a supplier provides a service supported by a product. This has the effect to partially decouple the profit from the product itself and allows a better optimization of the cost and resources. The model differentiates itself from outsourcing precisely because it encourages the use of technological innovation to achieve the desired outcome.

Given the constantly changing economic situation, one way to support the sustainability of the industry in its very definition [5] would be to further transition from traditional practices to the more recent service-oriented models such as PSS. This chapter focuses on one example of PSS: Take Back Chemicals (shorten to TaBaChem), a result-oriented business model inspired by Chemical Leasing [6]. The chapter is written to familiarize the readers with the concept of TaBaChem and to show how the model can renew the competitive advantage of the industry by providing greater resources optimization and by stimulating technological innovations through an improved production–consumption pattern.

8.2
Concept of Chemical Leasing and Take Back Chemicals

The concept of Chemical Leasing seems to find its root in the 2002 World Summit on Sustainable Development that sets up ambitious goals for the sound management of chemicals and hazardous waste. The same year, the Austrian Ministry of Agriculture, Forestry, Environment and Water Management launched a national program to promote the business model. Chemical Leasing is presented as new tool to reconcile policymakers and industries by creating both economy and environment benefits. In 2004, the United Nations Industrial Development Organization (UNIDO) adopted this model with the support of the Austrian government bringing an international perspective. The international development of Chemical Leasing can be replaced into the context of UNIDO sustainable resource management program [6]. Based on the official definition of UNIDO, Chemical Leasing is described as a *service-oriented business model that shifts the focus from increasing the sales volume of chemicals toward a value-added approach in which the producer mainly sells the functions performed by the chemical.* Chemical Leasing is a typical example of product-service system in which the earning model is changed, as explained in the definition, from selling chemicals to a service supported by the product. The focus is shifted from the product to its function, and therefore the sale of product's service.

A chemicals supplier applying the Chemical Leasing model is paid for the function that the chemical substance performs instead of the volume sales of that same chemical. As seen in Figure 8.1, this approach has the advantages of realigning the economic incentives of both the supplier and the user, ergo optimizing cost and consumption of material within supply chain.

In the traditional business model based on volume sales, a chemical supplier has no interest in reducing its volume sales because it reduces its profit. On the other hand, the user has no interest in purchasing more of the same product, especially if this chemical is used in a noncore process that is purely seen as a cost. The economic incentives of the supplier and the user are in conflict. In the Chemical Leasing business model, which is marked by the introduction of a service, the supplier no longer sells the chemical but rather keeps the ownership of the product during its entire life cycle and takes over the responsibilities of the use and the end-of-life treatment. Higher volume will translate into higher cost for the supplier. It then becomes the interest of both the user and the supplier to reduce the volumes of chemicals used and find new ways to optimize the entire

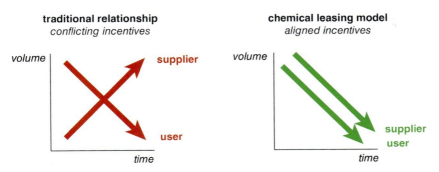

Figure 8.1 Traditional sales business model versus Chemical Leasing business model.

process that would otherwise not be as efficient. Chemical Leasing has been underscored as a more sustainable chemical management practice [6, 7].

8.2.1
The Concept of Take Back Chemicals

Take Back Chemicals is inspired by the Chemical Leasing concept and therefore follows the same overarching principles. Where Chemical Leasing can be seen as a continuum between rudimentary take back schemes and a fully closed life cycle, TaBaChem specifically focuses on improving environmental performance and on strictly closing the life cycle. TaBaChem especially aims at creating this closed loop: The same valuable materials will be reused after the take back step. In addition, it introduces stricter requirements for sustainable development and environmental protection. TaBaChem is therefore a result-oriented business model for resource-efficient and environment-friendly manufacturing. Originally developed in Belgium by a consortium of companies, governmental agencies, and knowledge institutes under MIP support (Milieu- en energietechnologie Innovatie Platform), such a model aims at improving the competitiveness of the industry by reducing the consumption of chemicals, further protection of the environment, stimulation of the introduction of nontoxic chemicals as substitutes, and increase of economic development.

Just as it is the case in the traditional Chemical Leasing model, it is the purpose of the TaBaChem model to realign the economic incentives within a supply chain (supplier–user) and bring benefits for both the supplier and the user of chemical substances and/or mixtures (Figure 8.2). This approach requires a concerted effort in finding the best cost-effective and environment-friendly way to deliver value.

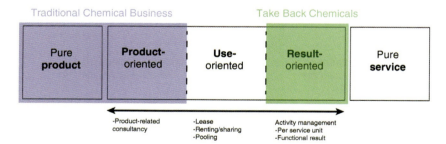

Figure 8.2 Continuum of a product-service system (as defined by Tukker [4]). The "traditional sale" business model and the TaBaChem model are mapped on the figure. Though variants of leasing or renting concepts are seen in the chemical industry, the majority is focused around the product itself and the directly related services (stocking, transportation, consultancy, etc.). Take Back Chemicals is fundamentally different by focusing on the result (including the waste prevention part) rather than the product.

Table 8.1 Business models in the chemical industry.

Business models in chemical industry	External expertise in process optimization	Chemical production and service within one firm	Shared goals and know-how	Smart pricing: innovation pressure and perspective for investments	Environmental impact and closed loop
Volume based					
Outsourcing	X				
Traditional Leasing	X	X/0			
Chemical Management Services	X	x/0	X		
Chemical Leasing	X	x	X	X	
Take Back Chemicals	X	x	X	X	X

Gray: no involvement; light blue: possible involvement; dark blue: high involvement of the indicated activities per business model.
X: applies, X/0: applies occasionally, blanc: does not apply.

1) *Volume based:* Typical sales based on amount of volume or mass (kg).
2) *Outsourcing:* It is the less complex step in involving external parties in the value chain, simply by having a third party take over a specific step.
3) *Traditional Leasing:* May go further (hence the semishading) by integrating production and services especially in the case where the producer is also the supplier.
4) *Chemical Management Services:* Shifts from use-oriented to result-oriented, and goals and know-how are shared.
5) *Chemical Leasing:* Goes even further by introducing "smart pricing": reallocation of gains and investments in a business model where the supplier is paid per servicing unit.
6) *Take Back Chemicals:* Includes additional environmental impact goals and aims at a fully closed loop by not only taking back the chemicals but also repurposing them in a highest valuable way.

Table 8.1 outlines the differences between various existing and emerging business models in the industry. This topic has been reviewed elsewhere [6, 8]. The table shows an evolution toward integration of incentives for the user, the producer, and the environment.

8.2.2
Advantages and Challenges of the Take Back Chemicals Model

As it is the case for every model, the TaBaChem model presents several advantages and challenges, which are summarized in Table 8.2. Knowing these drivers and hurdles is essential in determining where the model can be best applied, and

Table 8.2 Advantages and challenges of implementing TaBaChem.

Advantages	Challenges
Continuous innovation for flexibility and business continuity	Quality assurance
Lower production cost and distribution of investment cost	Pricing
Environmental impact	Legal
Safety	Organization
Long-term partnership	Single supplier dependency

in reaching a successful implementation. A business model innovation will require commitment from several departments per company. The advantages and challenges are described in Sections 8.2.2.1 and 8.2.2.2, respectively. Some aspects are given special attention further in the chapter: pricing is elaborated in Section 8.3.1.2; the main organizational challenges are discussed in Section 8.3.3; and Section 8.4 is dedicated to the legal aspect of implementation.

8.2.2.1 What Are the Advantages of Implementing TaBaChem

Continuous Innovation

As stated above, the introduction of a model such as the TaBaChem model leads to an economic dematerialization, that is, the switch from the sale of a specific product to providing a product-service system (functional economy). By switching to a product-service system, the supplier is technically no longer bound to use one specific chemical. In consultation with the user and following the agreed specifications for the function to be performed, the supplier may then decide to switch to a different product that may be more adapted to the user's situation or may generate more cost savings for the partners while keeping to the agreed functional specifications. This dematerialization step clearly encourages the introduction of technical innovation and the exchange of knowledge as the supplier can investigate other products for the same function. This leads to more flexibility for the existing process and help with business continuity.

Production and Investment Cost

Implementing the new business model clearly leads to process improvements, namely, the reduction of material and process water used, energy savings, the reduction of greenhouse gas emissions, and the reduction of waste production. These process improvements subsequently translate into cost reduction for both parties and contribute to strengthen the market position of the partners. This means a reduction in operating costs and to some extent an indirect reduction of operational expenses (OPEX) related, for instance, to the cost of inventory and/or licenses. By gaining a better understanding of the client's needs, the supplier also gains a competitive advantage (product leadership) over other suppliers.

The cost reduction can also be linked to the reduction in the number of administrative steps necessary to ensure compliance, including permits for waste treatment (permits are required for all aspects of waste management from handling, processing, and transporting to recycling/disposing of waste; transport and handling of hazardous waste can be particularly costly, cumbersome, and time-consuming). The cost of compliance is perceived by industry as rather high, especially in Europe with the introduction of increasingly stricter environmental legislation. The legal aspects including the waste legislation are further detailed in Section 8.4.

In addition to the decrease of operating costs and in some situations the decrease of OPEX, one net advantage of the model is the redistribution of the investment cost between the parties involved. Section 8.3 further elaborates on the economic gains and on the redistribution of the investment cost.

Environmental Impact and Safety
Another advantage of such a dematerialization is the reduction in both resources and energy consumption. Reductions of material and energy losses, reductions of emissions during production, and significant decreases in waste generation have all a beneficial impact on the environment.

In addition to a reduced environmental footprint, this business model – through its potential for innovation at both organizational and operational levels – is a solution to reducing the usage or stimulating the phasing-out of hazardous chemicals. The direct and indirect costs of continuing operation with potentially highly regulated chemicals become a major incentive for both the supplier and the user to seek a safer and less costly alternative.

Long-Term Partnership
The success of TaBaChem relies on mutual trust between all the stakeholders involved and especially between the supplier and customer who are engaged in an economic cooperation. An already existing business relationship (preferred supplier) does facilitate the implementation of the model as trust already exists between some of the partners. Once established, such a relationship based on trust will foster loyalty, closer collaboration, and commitment to continuity of supply – which facilitate to some extent supply chain risk management.

8.2.2.2 What Are the Impediments in Implementing the New Business Models?
Implementing innovative solutions always requires overcoming certain barriers. The following section discusses the specific challenges of a single supplier dependency and the importance of quality assurance (QA). As the entire chapter itself reflects on the implementation of the business model, some of the challenges identified in Table 8.2 (negotiation on pricing, legal, organization) are given special attention and further discussed in other sections.

Single Supplier Dependency
Although there is a great benefit in working hand-in-hand with a trusted supplier, the implementation of business models such as TaBaChem can raise additional questions, especially on how to deal with other suppliers when they do not provide the same service. By developing a close collaboration with a partner, a

company may become a lot more dependent on its partner and may not have the same freedom to remodel its supply chain. Such a greater economic dependence may be perceived as a risk or even as a liability if one of the partners is economically weak. Contractual agreement could also make it longer and more costly for a user to change supplier.

Quality Assurance
The most obvious technical challenge is to retain the minimum quality standards for the reuse of the chemical product. The reused material must obviously meet the proper specifications of a use (whether it is the same use or whether it is a different use). In some applications, the presence of impurities even when only in trace amount can really be an issue. To make this business model work, it should therefore be technically and economically feasible to produce a suitable quality of the material after it has been used in the process. Proper QA is defined as one of the *sine qua none* conditions.

8.3
General Economic, Technical, and Management Aspects

8.3.1
Economic Aspects

8.3.1.1 Direct Gains, Indirect Gains, and Investments
As briefly described in the previous section, the economic gains of implementing TaBaChem are linked to the three domains: an increase in leanness, agility, and adaptability (business continuity). A summary of the gains are given in Table 8.3.

When considering the benefits, it is important to take into account both *direct* and *indirect* benefits. The direct benefits are related to the improvement made on the so-called *operating costs*, for example, the cost of purchasing substances or the cost of energy consumption, while the indirect benefits are related to the improvement made on flexibility, on business continuity, and on sharing of investment cost.

Elaborating on the example of the direct benefits linked to the cost of purchasing substances, it should be noted that this does not necessarily mean that the price per volume of the substance purchased decreases. In essence, it is the volume used that decreases. In the case of a product substitution, the new substance selected could potentially have a higher price per volume when TaBaChem is applied. For example, this could be the case when the producer and user agree to use a more expensive but more powerful detergent in a cleaning process. Since the chemical used is taken back and recycled, the use of more expensive chemicals may still result in overall direct financial gains.

Implementing TaBaChem means not only a change in the business model but also a change in the logistics between the partners involved, and in the way the materials are processed. This often requires an investment in order to support

Table 8.3 Example of gains profile in a TaBaChem business case in which the producer is handling the material recovery.

Gain	Producer (including recovery)	User
Leanness of the existing process	• Reduction of cost linked to material and energy consumption in production • Process optimization (installation for recovery/adaptation to existing production line for greater efficiency)	• Reduction of cost linked to resource material and energy consumption in use • Reduction of cost linked to waste management and/or valorization of side streams • Reduction of cost linked to EHS compliance (time of EHS personal, consultancy cost, and compliance cost) • Knowledge on use optimization
Flexibility/agility of the existing process	The partners (producer, user) put in place a product service system. Their activities become more independent of the product itself.	
Adaptability/business continuity	• Sharing investment cost. Producer provides the CAPEX, while user provides some form guarantee to secure the investment • Through aligned incentive, it is easier to create joint innovation to tackle challenging issue from business continuity. For instance, using knowledge from both parties can help find alternative chemicals in case the chemical product used during the leasing is threatened (regulations, supply, etc.)	

the necessary changes to the existing equipment. Although the largest investment may mostly be done by the producer/supplier of the material, it is mirrored by some form of guarantee from the user. Hence, the cost of investment is shared between the parties involved, ultimately resulting in a reduced risk profile. This is a strong advantage of the model as it is a way to secure investment for the purpose of reducing a process footprint. Minor investments may be required by the user – either for QA before the material can re-enter the process or for allowing suitable transportation in the case where the material is to be transported. In a traditional business case, such differences in investment would hamper innovation. In the business model TaBaChem, contracts allow shared investments and shared gains, which lower the investment barrier to innovation.

8.3.1.2 Pricing

The price for a chemical is traditionally expressed in unit per volume and is linked to its cost of production, which in its turn is influenced by the price of raw material and energy resources. TaBaChem includes a price setting defined by a result sought rather than a volume and is therefore linked to the total added value. This added value is defined by comparing the existing situation (situation as it is) to a prospect (situation when TaBaChem is implemented). This should however not result in long-term fixed pricing based on a single moment in time.

Table 8.4 Contribution and volatility of pricing in a traditional model versus a TaBaChem situation.

	Traditional sales	Take Back Chemicals
Contribution to pricing	Volume of substance Cost of production: • Material resources • Energy resources • Other • Margin	Result Cost of production: • Material resources • Energy resources • Other • Margin Extra added value: linked to economic gains
Volatility of pricing	Largely dependent on (global) raw material and energy price	Reused raw material is a smaller share of the total production cost

Though a certain level of price fixing is desired to stimulate continuous innovation, the price of the service cannot be fully decoupled from the cost of production of the substance used.

As seen in Table 8.4, the cost of production still influences the pricing; however, this influence of the cost of production is reduced compared to the traditional sale and other financial elements must be taken into consideration. The "extra added value" as given in Table 8.4 shall include a "fair" cost and profit sharing. As discussed in Section 8.3.1.1, the (direct and indirect) gains including investment costs are distributed over the stakeholders involved. An unequal distribution of the gains is not rare, and is often a show-stopper in innovation in general and in TaBaChem in particular. A typical example is found in the situation where major investments are to be made by one party and the direct gains are taken by the other. To bridge the unequal distribution of the gains and investments, a "fair" cost and profit sharing is introduced. It is generally perceived as fair that the involved parties who incur the costs/investments required to introduce the new business get compensated. The profit that remains is distributed so as to ensure that each company makes a profit that is in line with the investments made and the risks taken in comparison with the as-is situation. Fair profit sharing – including the required transparency to achieve it – is a *sine qua non* for the success of this cooperation.

8.3.1.3 Conclusion on the Economic Aspects

The above section has outlined how the implementation of TaBaChem can be challenged by an unequal distribution of investments and direct and indirect gains (leanness, flexibility, and adaptability). Ensuring that the party making the major investment does get a fair return is critical to the success of the model. Proper pricing should take into consideration a fair share of the costs and benefits among the partners.

8.3.2
Technical Aspects

8.3.2.1 Reuse of Chemicals

The TaBaChem business model is designed to foster material efficiency. The implication of reusing chemicals ranges from the implementation of a relatively easy-to-feed step to the necessity of major investments in new equipment and rather radical technology.

As the implementation of new business models such as TaBaChem leads to multiple level of innovation (not only technical but also economical, organizational, and legal), the technical innovation may remain rather modest, or be performed through a tiered approach. The immediate introduction of radical technical innovation would introduce too many risk factors simultaneously, leading to a higher risk of project failure.

The type of investment is obviously linked with the feasibility of the project finance scheme and is in practice mostly linked to the potential size and economic value of the stream to be taken back relative to the traditional resources. When the material stream to be taken back is relatively small, it could possibly be recycled by incorporating the material as a feed in existing virgin production. This has the preference in terms of ease regarding logistics, organization, and investments. When the material stream is relatively large – or potentially large when scaled up to more than a single user – the option of investing in new equipment (dedicated, if required) becomes viable.

8.3.2.2 Process Optimization

The TaBaChem model does not only reduce the amount of material used (or reused) but also foster process optimization. Shared goals and knowledge between the parties involved will also accelerate optimization of the use of the chemical, that is, the production process of the user. This can result in both increased *efficiency* and *effectiveness*. Increased efficiency means that fewer resources (time, material, energy, water, etc.) will be used for the same output. Increased effectiveness means that the technical specifications of endpoints in processes quality (e.g., better cleanliness or higher surface quality) are increased with the same amount of resources. Optimization in both efficiency and effectiveness often requires adjustment to the existing equipment of the user.

8.3.2.3 Conclusion on the Technical Aspects

Technical innovation is to be expected when implementing a new business model such as TaBaChem. This will concern the recycling of substances and the optimization of use of the chemical. As the innovation touches many levels within the company at the same time (not only technical but also economical, organizational, and legal), the technical innovation is expected to be first modest. The size of investments for new equipment will in practice be mostly defined by the potential size of the take-back stream relative to the size of traditional resources.

8.3.3
Organizational/Managerial Aspects

8.3.3.1 Sales
Collaborative business models such as TaBaChem require the establishment of long-term relationships. From a sales perspective, long-term relationships are mostly beneficial, as they bring certainty on future incomes. The definition of sales – possibly including monitoring and rewarding of KPIs – should in this case be changed to providing a service.

8.3.3.2 Quality Assurance
Quality assurance should be perceived from both the producer and the users perspectives. As the user's process is optimized with additional knowledge from the producer, the potential of TaBaChem is realized and quality increased. On the other hand, both the producer and the user will be handling a purified product and should therefore be aware of the specifications for acceptance of impurities.

8.3.3.3 Tendering and Rewarding
Procurement in a free market is designed based on the volatility of price of resources per volume and ensures availability of resources. According to the traditional way of procurement, these criteria are fulfilled if there is the ability to change supplier. This procurement design includes tender specification and monitoring, and rewarding systems for personnel. A change of this rather short-term-oriented model is needed to allow long-term leasing-like models to be a success.

Long-term tenders include a certain fixing of pricing and commitment to a single supplier. The focus of procurement will shift from a short-term deal for the best price-per-volume to a partnership where the total cost of ownership is crucial.

Redefining tender documents, structures, and rewarding systems for procurement is key to allow the full organization – including its procurement department – to move ahead with lease-like business models. Having such long-term contract in place is necessary for the implementation of the model (investment) and its sustainability.

8.3.3.4 Knowledge Sharing
The producer of chemicals is usually thought to have a certain knowledge of its own products, which is a valuable information for the proper use in downstream processes. On the other hand, the user of the chemicals may have indications of requirements for improved performance that are valuable for product (chemical) optimization. Knowledge sharing beyond usual business practices has in this context a mutual benefit.

This exchange of knowledge mostly occurs on plant or installation level. Though each equipment is designed for optimal usage on paper, operators and engineers often know best how equipment really works. This means that the role of the operator/engineer is crucial for adequate process optimization. Sharing of such knowledge may lead to competition and intellectual properties issues. The legal consideration under which this knowledge sharing can be achieved is extensively described in Section 8.4.

8.3.3.5 Logistics

A change in process design, especially between different sites, requires attention to logistics. How will material streams be produced and/or received? Will it be batch or continuous? In which state? How will it be transported? Is additional storage needed? Just a few questions that need answering to achieve smooth material handling. When it comes to treatment and storage of material, adequate equipment needs to be at hand. This often results in modifications of existing equipment.

8.3.3.6 Conclusion on the Organizational/Managerial Aspects

This section underscored that several managerial aspects need reviewing for a leasing model such as TaBaChem to be successful. Starting a TaBaChem construction without considerations to procurement, operations, and logistics is therefore strongly discouraged.

8.4
Compatibility of the Service Model with the Actual Legislation: Some Important Aspects[1]

In many innovative projects, statutory and regulatory norms are not considered until the end. Unwelcomed surprises are then no exception. Instead of running into legal obstacles, considering the legal framework from the beginning ensures the feasibility and economic viability of an innovative project. The implementation of the TaBaChem business model is no different. It requires a timely consideration of different regulatory aspects.

In the following sections, some important points are discussed. The approach is pragmatic and generic. In Section 8.4.1, the transition from selling chemicals to providing a service through chemical substances is analyzed. Section 8.4.2 assesses the impact of the closure of the life cycle as it relates to sustainability and waste. The model with its particular close relations between parties involved also requires some thoughts on business confidentiality and (European) competition law. These are found in Section 8.4.3.

---

1) This information is of a general nature only and is not intended to address the specific circumstances of any particular individual or entity. This information is based on the status of the legislation in October 2014.

8.4.1
Transition from Sales to Providing a Service to the Customer

An essential aspect of the TaBaChem model is that the supplier of the chemical no longer sells the substance to his customer. Instead, he delivers a service based on the functionality of the chemical substance. The supplier thereby commits to delivering an end product following the standards agreed upon with his customer.

8.4.1.1 The Supplier Retains Ownership of the Chemical

When a chemical substance is used for its function, it is in itself not sold. Consequently, the substance remains the property of the supplier (producer, importer, or distributor), who is now the provider of a service. As an owner, the provider has the full authority of its property. If the property of the substance were to be sold, the buying customer could fully decide on the use of the substance: He can resell, use, dispose, and destroy it. In this case, the potential of the TaBaChem model concerning optimization of process would be lost. Additionally, the focus on an efficient and effective use of chemicals would disappear. It is thus essential for a successful implementation of the business model that the property of the chemical(s) remains with the provider of the service.

Strictly speaking, it is not necessary to contractually describe the ownership in the contract, although it would be useful. It is more important that factual circumstances support it: The provider of the service should act as the owner of the substance used. If it is decided to mention ownership in a contract, the formulation should be carefully drafted in order to avoid incompleteness or incorrect stipulations, which create some doubt on the ownership.

Ownership also has an impact on responsibilities and liability. Concerning REACH, the supplier also becomes, for example, a downstream user [9]. They will have to comply with the regulations, including the obligations that formerly belonged to their customers. Other responsibilities, such as environmental permits, reporting, and safety management, will or can shift to the suppliers. Because each TaBaChem project is different, national and international laws regulating these aspects should each time be checked.

8.4.1.2 Result-Oriented Services Lead to Different Pricing of a Chemical

The optimization of the TaBaChem model is further encouraged by an adequate price setting. The supplier must be paid for the result and no longer for the volume of the chemical. The calculation of fees has to be based on the specifications of the final product as it is contractual, defining quality, function, material, and so on.

The price setting is thus different from a traditional sales relationship. For example, a suitable formula is calculating the supplier's fee on the basis of the cleaned surface if the service is degreasing objects. In practice, the finding of a suitable algorithm that incorporates the decisive aspects of a successful completion of the agreed end result has proved to be quite challenging. The customer has to define clear and measurable quality standards, indicators, and functionalities of the end product. The options depend on the individual case, but reference to the volume of chemicals used should never be included. Reference to volume would turn the project into a (partial) sale. Consequently, the particular benefits of the business model will be to a great extent lost.

8.4.1.3 A Transparent and Elaborated Contract Is Necessary

No standard contract is available as each project is specific. Stipulations need to be customized and elaborated in line with the specificity of delivering the service.

Specific focus points are assigning authority of and responsibility to the respective project partners, quality requirements concerning the end result/product, communication and reporting rules between parties and with third parties, procedures for the sharing of knowledge, rules on cooperation and working with each others' employees, access to premises, and administrative duties. Confidentiality and intellectual property rights should be strictly regulated.

Providing a product-service system implies that the supplier has the authority and independence to determine and/or influence the choice of the substance and eventually the process used to achieve the result imposed by the customer. In practice, this will most often be discussed and agreed upon with the customer. However, since the supplier is knowledgeable on chemicals and their properties, it is within his competency to optimize the use. The knowledge and the relative freedom to redesign the process to achieve the agreed result are the basis for the realization of the advantages for the parties involved. When including conditions in relation to the freedom of the provider of the product-service system, care should be taken not to jeopardize essential elements of the business model, like ownership, choice of the substance, and input in the process.

Other essential elements in a TaBaChem agreement are contingency plans and end-of-contract rules. In view of the closeness in the project, relations will be more long term than in traditional supplier–customers dealings. Moreover, procedures to deal with unexpected breakdowns, loss of quality, and other difficulties have to be agreed upon at the start of the project. This must avoid major problems if any calamities should occur.

To keep the contracts workable, it is recommended to structure the contractual stipulations into a master agreement with the framework conditions and stipulations and subagreements. The latter elaborates details and conditions that should be more flexible or are more likely to change. This approach focuses negotiations on the issue at hand and avoids rediscussing the whole contract.

8.4.2
Closing the Life Cycle and Preventing Waste

The core characteristic of the TaBaChem model is that the use of chemicals becomes more controllable and more efficient since one is no longer fixed on volume. Consequently, the sustainable use of chemical substances, through improved life cycle management, becomes economically interesting and feasible.

With TaBaChem, partial and even full closing of the life cycle of chemical substances is possible. In any case, waste production is positively influenced.

Prevention of waste, as the ultimate objective of sustainable materials management, has thus become an achievable target. Prevention has to be understood as "measures taken before a substance, material, or product has become waste," thereby reducing the quantity of waste, including the reuse of products or the extension of the life span of products, the adverse impacts of the generated waste on the environment and human health, or the content of harmful substances in materials and products [10].

Prevention means that there is no waste. Fully closing a life cycle skips the waste phase. Government authorities responsible for waste management are suspicious when processes that used to produce waste now no longer do so. Waste is a highly regulated material (e.g., Directive 2008/98/EC on waste (Waste Framework Directive)). The experience with TaBaChem projects has however proved that full closure of a life cycle with avoidance of the waste phase is legally possible. The elimination of a waste phase was consequently recognized by the waste authorities. The following paragraph briefly explains how this was achieved in the first TaBaChem projects.

Waste is legally defined as any substance or object that the holder discards or intends or is required to discard [11]. When it comes to the closing of the life cycle, the owner as the holder of the substance is important (see supra – importance of the ownership). As already indicated, supplier of the product-service system as the owner of the chemical has the most complete authority to decide on the use of the substance. If that owner of a chemical substance does not want nor intends to discard the chemical after use, then the substance does not become waste.

The intention is important. An intention is however difficult to prove and as such not observable. The interpretations of the conduct and of the actions of the owner are necessary. It is this decisive element that leads to discussion. A lot of case law exist on the topic. Some criteria and guidelines can be distilled on the basis of litigation. The criteria are however mainly negative, that is, they refer to conditions that are not sufficient for the avoidance or elimination of the waste status. But the elements can be used *a contrario*. First of all, each situation has to be considered on its individual facts and circumstances. Second, the use, on the basis of the economic value of the substance and commercial objectives of the owner, can provide strong arguments supporting that the intention is not to discard the substance.

If the owner/holder of a used chemical does not intend to discard it because he reuses that substance to perform an agreed upon product-service system, then the substance does not meet the criteria for being considered as waste.

In addition, since the supplier of the product-service system retains the responsibility for complying with the relevant statutory norms and the administrative implementation of obligations, the argument that the substance is not waste becomes only stronger.

All above conditions can be met in the TaBaChem model because the supplier delivers a product-service system while retaining ownership of the chemical.

Concrete cases proved that the life cycle can be fully closed and controlled when the circumstances are right.

In view of the innovative aspect of fully avoiding the waste phase, it is advisable to consolidate the prevention of waste in an agreement with the local waste authorities. Although, strictly speaking, the waste authorities should not be involved because there is no waste, such an agreement avoids lengthy discussions. A clear understanding of the TaBaChem business model and the proof of the closure of the life cycle is necessary.

In concreto, the local waste authorities (e.g., Flanders, Belgium) can agree with the closing of the life cycle and the avoidance of the waste phase if the following conditions are met and controlled:

- the ownership of substance does not change;
- service agreement is concluded between the owner–supplier of the chemical and his customer;
- the owner–supplier is knowledgeable about the properties and the safe use of the chemical substance;
- the substance is taken back and further use of the chemical is guaranteed;
- the chemical is during the agreement not changed into a substance with more hazardous characteristics;
- the residual fraction after eventual treatment/purification of the used chemical is treated as waste if the owner–supplier intends to discard of these residues.

The fact that the substances are still covered by the REACH regulation secures the control and the risk management of the chemicals involved. It is important to note that the above does not work when the owner or holder of the substance is a waste handler, since substances and products have always in such cases the waste status.

Undoubtedly, similar results can be achieved in other countries (as stated above, the agreement was reached in Belgium by the Flemish authorities). Concerning Europe, it is necessary to keep in mind that the European framework for waste is a directive (Waste Framework Directive). A directive needs to be translated into national rules. These "local" rules differ among the member states. Consequently, the national waste legislation should be considered, mainly on the practical translation into national norms of the directive's guidelines. Given this situation, an international implementation of closed life cycle without waste phase could be difficult because of the differences in legislation. However, the European general definition of waste is to be taken into account as it is based on directives and regulations on waste. An essential element of that definition is the intention to discard. Consequently, when an owner of a substance does not want to discard it, that is sufficient to avoid the waste status of that substance. It is clear that a unified applicability and enforcement of a regulation instead of a directive would facilitate the use of models that avoid waste creation.

Last but not least, if the substance after use needs some treatment, like the removal of impurities, the extracted substances are waste if the owner/holder wants to discard these. This waste fraction must follow all relevant legal norms.

On the other hand, even when the life cycle cannot be fully closed, waste reduction is possible and above reasoning can be applied *mutatis mutandis*.

8.4.3
Business Confidentiality and the Protection of Competition

It is clear that the relationship between the partners in a TaBaChem model is close. The question is thus (i) if parties can work a mutually required standard of business confidentiality and (ii) if the model is acceptable in relation to the European and local competition legislation. The sharing of information and mutual trust are essential in TaBaChem projects, since parties agree to achieve a joint result, namely, the end product of the process caught by the business model. Two aspects are explained in the following sections: the operation of the process and the sharing of (innovative) knowledge.

8.4.3.1 Intellectual Property Rights
The business model TaBaChem aims at not only supporting sustainable material management but also stimulating innovative thinking. The responsibilities of the product-service system provider together with the change in the economic income source motivate the optimization of the process, of the choice, and use of chemicals. Logically, issues could arise concerning intellectual property rights.

The general principle should be that intellectual rights remain the property of the developer of the item. Other arrangements might be necessary. Especially the opportunity provided by the customer to develop new methods and approaches makes it difficult to assign unique rights to one party.

On the other hand, the investments to be made by the product-service system provider are only economically viable if the provider can use his acquired knowledge and expertise for other customers, potentially competitors of the original customer. Usually newly acquired knowledge on the basis of the cooperation is the property of the involved parties. Deviations of this principle are of course possible. Contractual arrangement should clarify the limits on the use of ideas, concepts, knowledge, methods, technologies, and so on.

8.4.3.2 Competition
Doing business involves concluding agreements. Agreements can lead to concerted practices and/or affect competition. Both are forbidden by national and European Community competition rules. Each enterprise should determine independently its business policies and market conduct. Meaning that influencing of the market affecting the competition is not allowed. That is clear for direct effects, but indirect effects can also be considered restrictive. This is the case when as a side effect of an activity, competition or the structure of the market is negatively influenced [12]. The key distinction between harm to competitors and harm to competition has to do with the following question: Is the contract under review likely to reduce the competitive pressure exerted by the dominant

company's rivals? [13]. In other words and in order to breach the competition law, an act should impact the relevant market. Influencing the market position of an individual competitor is not sufficient.

In order to avoid problems with the anticompetition rules, agreements between undertakings, decisions by associations of undertakings, and concerted practices should not or cannot affect trade. Neither should they have as their objective or affect the prevention, restriction, or distortion of competition. Agreements that contribute to improving the production or distribution of goods or promote technical or economic progress, while allowing consumers a fair share of the resulting benefit, also fall under the exclusion [14].

The burden of proving an infringement of competition law rests on the party or on the authority alleging the infringement. This does not imply that a "wait and see" approach is the best practice. It is always useful to counter any impression of affecting competition negatively. An impression of an attempt to affect the market is most frequently linked to price, output/turnover, product quality, product variety, or innovation. These elements should be treated with care when working in a close business relationship.

TaBaChem is a business model requiring such close contact between supplier and customer, resulting in long-term relations and knowledge of the respective production processes. On the other hand, the TaBaChem's objective is efficient and innovative production, sustainable material use, environmental benefits, and safer use of chemicals. These objectives are accepted as a reason to influence the market, but only up to necessity.

The following points should at least be considered when implementing the TaBaChem model.

1) The effect on the market should be mainly assessed for the consortium of the TaBaChem parties and not on the level of the individual participant.

    The fact that the product-service systems provided by the supplier concern (generally) noncore processes of the customer makes it less probable that the consortium breaches competition rules. Benefits of creating economies of scale and scope, cost reductions, qualitative efficiencies, such as improvements in product quality or safety, and the production of goods or services not otherwise possible are acceptable under competition law up to their point of necessity. A similar exception is made for innovation and efficiency measures.

2) Market-sensitive information should never be shared. Parties should not share data/information on production, import volumes, pricing volumes, marketing conditions, strategies, relationships to suppliers, and so on. The basic principle is that each undertaking should independently perform and manage its business. It is very important to realize that a common supplier should also not share sensitive information acquired from his customers, regardless if this happens willingly, unconsciously, or negligently [12]. To reduce risk, it is strongly advised to use a third party for the handling of the

information of parties and for the communication between the parties to avoid issues with confidentiality and competition.

3) Parties must ensure that the rights to information furnished by the individual members remain with them entirely and that information and knowledge jointly obtained is owned by all parties.
4) In the contract, the data to be shared in the project must be defined clearly.
5) Long-term supply agreements between a supplier and a customer, as nearly certain will be concluded in a TaBaChem structure, need special attention. Contracts that are exclusive, that cover almost the entirety of the buyers' needs, and that take up the full capacity of the supplier are of concern [13]. Contracts that are concluded for long durations (typically longer than 5 years) are also considered as negatively impacting the market. In practice, the problems with the duration of the contracts mainly occurred in contracts with energy providers.

   There should be no exclusion or foreclosure of similar (respectively) suppliers/customers in the contract.

   Preventing the purchaser in practice from switching to other suppliers and thereby foreclose customers for other suppliers is a breach of competition law. This means that if another supplier can deliver the same service to the customer as the actual supplier, the former should be able to acquire the customer. A TaBaChem contract should not make this impossible.
6) The structure of the project must reflect the level at which its members cooperate. In order to make a TaBaChem model viable for the supplier, namely, the provider of the product-service system, an upscale of the product-service model to other customers might be desirable. Foreclosing the supplier from acquiring other customers on the basis of the TaBaChem relation is forbidden (see former paragraphs). However, the supplier should always respect the confidentiality of the knowledge and information acquired in the relationship with his customers.

Clear arrangements and agreements should be made between the parties. Some elements should be explicitly mentioned in a contract: the identity of each party, a declaration of intent explaining the overall purpose of the relationship between them, list of tasks on which parties are planning to work is also helpful. Provisions on liability, sanctions in case of breach, and confidentiality are also to be included. Generally, it is considered safer to use an independent third party as a go-between when working in such close relationships.

Finally, it is important to note that the TaBaChem is in itself not a treat to competition. On the contrary, it could bring competition to another level, where sustainability and safety are key.

In case a competition complaint is filed, the situation will be assessed on the merits of the individual case. Parties can prepare by calculating the value of efficiencies, substantiating projections about when the efficiencies will become operational and have market data readily at their disposal. But the most important is a clear contract regulating what activities will be performed and to consider working with an independent third party when sensitive information is involved. It is therefore advised to clearly describe in the contract what activities will be performed within the lines of the (long-term) contract and to foresee possibility to appeal to the assistance of an independent third party for sensitive aspects.

8.5
General Conclusion

This chapter gives an overview of how technical innovation in manufacturing can be fostered through business innovation and focuses on one concrete example: the Take Back Chemicals business model. Outlining the major advantages of the business model – next to fostering technical innovation itself – it provides handholds on how to best implement the model, whether acting as a chemicals supplier or user. It also addresses the challenges that may be faced during the implementation phase, with an emphasis on the economic, organizational, and especially legal impediments. It is important to consider each of these aspects before going in depth on the technical aspects of process optimization.

When addressed in the right way, this type of business model will not only improve profitability but will also *invite* technological innovation. Whereas innovation in traditional processes today is mostly incremental, the introduction of new business models such as the Take Back Chemicals model may induce rather radical innovation. Although such a business model requires a shift of mindset in key positions throughout the value chain – often not an easy process – the solution provided can be rather powerful on the long term and renew the competitive advantage of the industry.

References

1 (a) Krafcik, J.F. (1988) Triumph of the lean production system. *Sloan Management Review*, **30** (1), 41–52; Womack, J.P., Jones, D.T., and Roos, D. (1990) *The Machine That Changed the World*, Rawson Associates, New York; Holweg, M. (2007) The genealogy of lean production. *Journal of Operations Management*, **25** (2), 420–437.

2 This is a reference to the transition from industrial to postindustrial economy: Bell, D. (1973) *The Coming of Post-Industrial Society*, Basic Books, New York. It is not the purpose of this chapter to review outsourcing practices that are widely covered elsewhere. For further reading, we suggest McIvor, R. (2005) *The Outsourcing Process: Strategies for Evaluation and Management*, Cambridge University Press, New York; Tompkins, J.A., Simonson, S.W., and Upchurch, B.E. (2005) *Logistics and Manufacturing Outsourcing: Harness Your Core Competencies*, Tompkins Press, Raleigh, NC.

3 According to M.W. Toffel, Contracting for servicizing, HBS working papers 08–063 (2008), servicizing, functionalization, and product-service systems refer to the same concept. Stahel, W. (1994) The utilisation-focused service economy: resource efficiency and product-life extension, in *The Greening of*

Industrial Ecosystems (eds B.R. Allenby and D.J. Richards), National Academy Press, Washington, DC, pp. 178–190; Goedkoop, M.J., van Halen, C.J.G., te Riele, H.R.M., and Rommens, P.J.M. (1999) Product service systems: ecological and economic basics; Van Halen, C., Vezzoli, C., and Wimmer, R. (2005) *Methodology for Product Service System Innovation*, Koninklijke Van Gorcum, Assen, The Netherlands; Cooka, M.B., Bhamrab, T.A., and Lemonc, M. (2006) The transfer and application of product service systems: from academia to UK manufacturing firms. *Journal of Cleaner Production*, **14** (17), 1455–1465; Sakao, T. and Lindahl, M. (eds) (2009). *Introduction to Product/Service-System Design*, Springer, London; Towards the Circular Economy, Part 1: an economic and business rationale for an accelerated transition, Ellen Macarthur Foundation report (2012).

4 Tukker, A. (2004) Eight types of product-service system: eight ways to sustainability? Experiences from SusProNet. *Business Strategy and Environment*, **13**, 246–260.

5 UNO (1987) Report of the World Commission on Environment and Development: Our Common Future, United Nations General Assembly Document A/42/427.

6 Jakl, T. and Schwager, P. (eds) (2008) *Chemical Leasing Goes Global – Selling Services Instead of Barrels: A Win-Win Business Model for Environment and Industry*, Springer, New York; Geldermann, J., Daub, A., and Hesse, M. (2009) Chemical Leasing as a model for sustainable development, Research Paper No. 9. Lozanoa, R., Carpenterb, A., and Lozanod, F.J. (2014) Critical reflections on the chemical leasing concept. *Resources, Conservation and Recycling*, **86**, 53–60; Moser, F. and Jakl, T. (2015) Chemical Leasing: a review of implementation in the past decade. *Environmental Science and Pollution Research*, **22** (8), 6325–6348; Global Promotion and Implementation of Chemical Leasing Business Models in Industry, 10 Years Outlook, UNIDO Report, 2016.

7 Moser, F., Karavezyris, V., and Blum, C. (2014) Chemical leasing in the context of sustainable chemistry. *Environmental Science and Pollution Research*, **22** (9), 6968–6988.

8 Eder, P. and Delgado, L. (2006) Chemical Product Services in the European Union, JRC IPTS technical report series; van der Velpen, B. and Hoppenbrouwers, M. (2013) Implementing service-based chemical supply relationship: Chemical Leasing® – potential in EU? in *Management Principles of Sustainable Industrial Chemistry: Theories, Concepts and Industrial Examples for Achieving Sustainable Chemical Products and Processes from a Non-Technological Viewpoint* (eds G.L.L. Reniers, K. Sorensen, and K. Vrancken), Wiley-VCH Verlag GmbH, Weinheim, Germany, pp. 181–198.

9 Regulation (EC) No. 1907/2006 of the European Parliament and of the Council of 18 December 2006 concerning the Registration, Evaluation, Authorisation and Restriction of Chemicals (REACH), Official Journal of European Union, L396, December 30, 2006.

10 Article 3, 12°, Directive 2008/98/EC of the European Parliament and of the Council of 19 November 2008 on Waste and Repealing Certain Directives (Text with EEA Relevance), Official Journal of European Union, L312, November 22, 2008. This directive is called the Waste Framework Directive. See also Regulation (EC) No. 1013/2006 of the European Parliament and of the Council of 14 June 2006 on shipments of waste, Official Journal of European Union, L190, July 12, 2006.

11 Article 3, 1°, Directive 2008/98/EC of the European Parliament and of the Council of 19 November 2008 on waste and repealing certain Directives (Text with EEA relevance), Official Journal of European Union, L312, November 22, 2008.

12 European Commission, Communication: Guidelines on the Applicability of Article 101 of the Treaty on the Functioning of the European Union to Horizontal Co-Operation Agreements, January 14,

2011, Official Journal of European Union, C11/1.

13 Spector, D. (2014) The European Commission's approach to long-term contracts: an economist's view. *Journal of European Competition Law & Practice*, **5** (7), 492–497.

14 Article 101 of the Treaty on the Functioning of the European Union and Commission, Guidelines on the Application of Article 101(3) TFEU (formerly Article 81(3) TEC), Official Journal of European Union, C101, April 27, 2004.

9
Applications of 3D Printing in Synthetic Process and Analytical Chemistry

Victor Sans, Vincenza Dragone, and Leroy Cronin

9.1 Introduction

3D printing is a developing field that groups a set of emerging technologies aiming to fabricate 3D physical objects from a digital model [1, 3]. The latter is designed using 3D computer-aided drawing programs (3D CAD) and converted into a specific file format (.stl), which can be read by a 3D printer. This digital model is then sliced into layers of a defined thickness, which are physically formed by a 3D printer machine in a layer-by-layer fashion. Recently, 3D printing has been employed in manufacturing industries to produce a wide range of devices, including design prototypes [4], tissue scaffolds [5], biology systems [6], biomimetic microvascular systems [7], microfluidics and other functional devices for laboratory environment [8], electronics, and pneumatics [9]. This technique is becoming extremely popular both in industry and academia, counting nearly 3000 publications in 2016 (Figure 9.1).

Before moving to the discussion of the current and potential benefits of using 3D printing in chemistry, it is important to give a brief overview of the different techniques employed to fabricate 3D objects. The classification used here is based on the printing mechanism and materials employed. Indeed, these two aspects have a significant impact on the characteristics and the final application area of the printed devices (Table 9.1). Three different classifications for printing have been proposed, depending on the physical method employed to generate the layers.[1] G-I corresponds to photopolymerization techniques, G-II to the use of a binding agent, and G-III are parts built through melting of the materials.

---

1) https://www.additively.com/en/learn-about/3d-printing-technologies (accessed November 5, 2017).

Handbook of Green Chemistry Volume 12: Green Chemical Engineering, First Edition. Edited by Alexei A. Lapkin.
© 2018 Wiley-VCH Verlag GmbH & Co. KGaA. Published 2018 by Wiley-VCH Verlag GmbH & Co. KGaA.

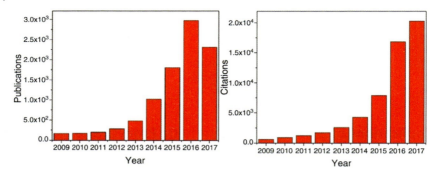

Figure 9.1 Results from the literature search for 3D printing keyword in terms of the number of publications and citations. (Web of knowledge, October 2017).

9.1.1
Polymerization-Based Additive Manufacturing (AM)

This technique generates a physical object employing liquid photosensitive inks that undergo a polymerization process when activated by UV light. In this case,

Table 9.1 A summary of the main 3D printing processes available and their key characteristics.

3D printing techniques			Minimum layer thickness (mm)	Minimum feature size (mm)	Common materials
G-I	SLA		0.025	0.1	Photopolymers
	PJ		0.016	0.016	Photopolymers and rubbers
G-II	BJ		0.09	0.1	Powders
G-III	SLM		0.02	0.05	Powders
	EBM		0.05	0.1	Powders
	FDM		0.1	0.178	Plastic filaments
	LS		0.1	0.2	Powders and plastic filaments
	MJ		0.016	0.016	Wax-like materials

SLA: stereolithography, PJ: photopolymer jetting, BJ: binder jetting, LM: laser melting, EBM: electron beam melting, LS: laser sintering, FDM: fused deposition modeling, and MJ: material jetting.
Source: Adapted from Refs [10].[1]

the printed objects are not durable over time and have poor mechanical properties, as they are made of a photopolymer. Two different techniques belong to this group and mainly differ in the type of UV source used for the activation of the polymerization process.

9.1.1.1 Stereolithography (SLA)
Stereolithography machines can print physical parts in 3D using a laser to initiate the polymerization of the liquid photosensitive inks. This technique enables building large objects with high degree of accuracy, good details, and high quality surface finish. The minimum layer thickness varies, but the low limit is around 0.025 mm with the minimum accuracy of the final print ±0.1 mm. The most common materials used are photoreactive acrylates [10]. The printing process consists of a build support platform with a vessel filled with the liquid photopolymer. The printing uses lasers to solidify a predefined pattern corresponding to a sliced layer of the product designed by initiating the photopolymer to react and to solidify on the build platform. The build platform is moved up and down to allow replenishing the build bed with fresh ink. This process is repeated until it finally yields the device built layer-by-layer.

9.1.1.2 Photopolymer Jetting (PJ)
Photopolymer jetting modeling machines can print physical objects in 3D employing inkjet printheads and a UV lamp to rapidly photopolymerize the jetted monomers on a supported structure. This AM technology is similar to stereolithography. The minimum layer thickness is 0.016 mm and the minimum feature size is 0.15 mm (with tolerance of ±0.025 mm), which is better than stereolithography. The most common photopolymers utilized are also in this case acrylate derivatives and acrylonitrile butadiene styrene (ABS) polymers.[1] The main advantage of this process is the possibility of loading multiple inkjet heads with different materials. This is ideal for printing multimaterials (and/or also multicolor) objects or mixing materials in different proportions in order to create functional graded objects. An additional advantage is the possibility of printing objects using transparent materials that can readily find application in visual and fit testing.

9.1.1.3 Physical Binding
3D printers can also be loaded with powders. These types of machines deposit layers of powder that are then glued together and solidified in a compact and homogeneous structure. With this technique, the printed parts are nonmechanically resistant and are often fragile. Nevertheless, good resolutions can be achieved, the minimum layer thickness is 0.09 mm and the minimum feature size is 0.1 mm (with tolerance of ±0.13 mm). This method is relatively simple, fast, and cheap, as well as more versatile, compared to the stereolithography and photopolymer jetting. For example, a wide range of materials can be used with this technology, such as metals, ceramics, and polymers; these materials are widely used for prototyping and to build moulds for casting (e.g., glass,

aluminum, silver, stainless steel, sand, and wax and among the plastics: ABS, polyamides (PA), polycarbonate (PC), polymethyl methacrylate (PMMA), and other photoactive polymers). One of the most employed machine of this category is the binder jetting (BJ) printer.[1] The powder in the BJ reservoir is distributed as layers on the tray of the printer. A liquid adhesive agent is then applied to the layer of powder through inkjet printheads. At this point, the build platform is moved down to continue the printing process layer-by-layer. In this case, a support structure is not necessary and several different objects can be printed at the same time as separated pieces of the same printing process.

9.1.2 Melting-Based Techniques

These additive manufacturing technologies employ materials that can be melted, shaped, and then solidified at room temperature layer-by-layer such as metal powders or thermopolymers. Several different techniques belong to this group, the objects printed with this methodology acquire improved mechanical properties than objects printed using the polymerization-based techniques.

9.1.2.1 Selective Laser Melting (SLM)

Laser melting machines can build objects using powders. In this case, a layer of material is loaded in the print hopper and features corresponding to the design pattern of a CAD file are melted using one or more lasers. The unmelted powder remains in the hopper. The print hopper is then lowered by the depth of a layer in order to load another powder layer, the process is repeated until all digital layers are printed. In this case, print hoppers are filled up to 99.9%. The unmelted powder then serves as a support to the structure built and to reduce thermal stress and prevent warping during the printing. This process is more expensive and slow than the physical binding, and has limited tolerance and surface finishing; minimum layer thickness is 0.02 mm and the minimum feature size is 0.05 mm, with ±0.1–0.2% of accuracy. However, this process is a suitable choice to build robust metallic devices as it offers a larger selection of materials (e.g., aluminum, copper, gold, stainless steel, nickel, silver, and titanium).[1]

9.1.2.2 Electron Beam Melting (EBM)

This method is similar to the SLM but uses electron beams instead of lasers during the melting process. Thus, electron beam melting (EBM) uses a milder heating process, making it possible to reduce the required support structure of the object. Other differences are in the lower quality of the surface finishes of the metallic objects printed than the LM – minimum layer thickness is 0.05 mm and the minimum feature size is 0.1 mm, with ±0.2 mm of tolerance. However, the EBM printing process is faster than the LM.[1]

9.1.2.3 Fused Deposition Modeling (FDM)

Fused filament fabrication machines create physical objects by melting plastic filaments. The filament is extruded through heated nozzles and deposited on the printing bed where upon cooling layers solidify. Each nozzle is loaded with a filament and therefore parts made of different materials can be printed simultaneously. The primary requirement of FDM is that the printing materials need to adhere to each other and to the support structure, which can also be made of a different material than the object. Once a layer has been created, the bed is lowered and the next sheet can be formed. This technique employs thermoplastic filaments, which can be melted at high temperature and molded in the desired shape before hardening, typically at room temperature. The most common thermoplastic polymers employed are nylon, polylactic acid (PLA), polyethilenimine (PEI), ABS, polycarbonate (PC), and polypropylene (PP). Many of these polymeric materials are not chemically resistant to organic solvents. Thus, devices obtained using this process have many constraints for their application in synthetic chemistry. Other limitations of this technique are anisotropy in the z-direction (vertical direction) of the printed parts and the resolution of the objects printed (minimum layer thickness is 0.35–0.1 mm with a tolerance of ±0.25 mm). However, the latter can be improved through postprinting processes [10].

9.1.2.4 Laser Sintering (LS)

Laser sintering (LS) AM printers consist of a layer-by-layer printing technique, where printer hopper is covered with a layer of powder that is subsequently sintered with one or more lasers. Afterward, the printer hopper is lowered by one-layer thickness, a second layer of powder is added to which the next layer is created. Machines of this type can be considered a middle choice between SLM and the FDM printers in terms of printer and material prices, and of the quality/properties of the 3D printed objects. This technique does not possess the finishing of the FDM – minimum layer thickness is 0.1 mm and the minimum feature size is 0.15 mm, with ±0.25 mm of tolerance, although this can be improved through postprinting processes. However, the printing process is support-free and can employ a huge variety of materials, including sand and metals, such as titanium, stainless steel (SS), and aluminum.[1] However, the most employed materials with this technique remain plastics such as PP, ABS, PEI, and polyether ether ketone (PEEK) in powder form.

9.1.2.5 Material Jetting (MJ)

Multijet modeling machines are the best of this group to achieve good accuracy and surface finishes – minimum layer thickness is 0.016 mm and the minimum feature size is 0.1 mm, with ±0.025 mm of tolerance. However, they can work only with wax-like materials, which limits their use to few materials, such as PMMA. The molten wax is pushed through an inkjet head to the supported structure.[1] The development of new materials with finely tuneable properties is a topic of intense research.

9.2
Chemical Reactors Manufacturing by Additive Manufacturing Techniques

In the last few years 3D printing started to attract the attention of chemists and chemical engineers both for the possibility of creating reaction vessels (reactionware) and for the creation of new materials to be printed. The significant advantage of this technique is that the architecture of the objects can be easily and precisely controlled without the need of additional expensive equipment, as compared to other techniques employed to manufacture microfluidic devices, such as MEMS technology or glass processing. Indeed, 3D printing allows chemists to build devices with high precision, including complex geometries and intricate internal structures, such as channels with well-defined size dimensions. Furthermore, understanding the kinetics of reaction processes can allow to rapidly develop and build additional knowledge, for example, details of mixing behavior, reaction kinetics or phase behavior, generated experimentally via reactor design. Moreover, the additive manufacturing process of generating the devices is very quick and represents a cost-efficient methodology for the fabrication of fluidic devices [10]. All this is important in chemistry and engineering, and in particular for the manufacture of micro- and meso-scale devices. AM technology also starts to populate synthetic chemistry laboratories and promises to revolutionize synthetic methodology [2] because of the ease and low costs of fabricating devices applicable in chemistry [11].

9.2.1
3D Printing Technologies in Chemistry

Reactor design and manufacture represent a bottleneck for the development of continuous flow synthetic chemical processes. 3D printing is an innovative technique that can enhance the development of chemical research and processing by fabricating rapidly configurable miniaturized devices tailored to each individual chemical challenge. There is a clear synergy between flow chemistry and 3D printing (Figure 9.2).

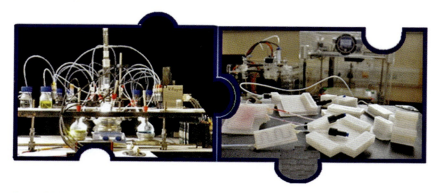

Figure 9.2 Representation of the development of flow chemistry and 3D printed reactionware.

The possibility of rapidly designing and building reactors coupled with liquid handling robots enables the combination of vessel and reaction design (Scheme 9.1). In this way, each reactor can be printed with tailored catalysts and reagents and controlled volumes, adjustable to the time required for each chemical reaction. Hence, this leads to a new term "reactionware." This means that the reactor is not anymore a mere container of the chemical reagents, but an active part of the process, influencing the chemistry by its design [2]. The possibility of rapidly adding chemical information feedback (in terms of yield, conversion, and selectivity) into the reactor design is only possible with this approach. This concept opens a new field of interdisciplinary research for chemists, engineers, physicists, and mathematicians.

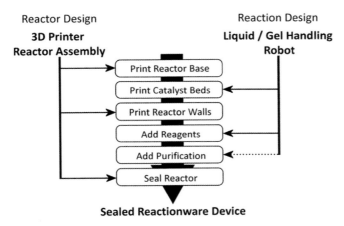

Scheme 9.1 Fabrication scheme for the integration of 3D printing techniques with automated liquid handling to produce sealed reactionware for multistep syntheses. (From Ref. [12]. Copyright 2013, The Royal Society of Chemistry.)

The first example of the use of 3D printing in chemistry is illustrated in Figure 9.3. A basic reactor made of acetoxysilicone polymer was created employing a Fab@Home Version 0.24 RC6 3D printer [2]. The reactor consists of a chamber sealed at the bottom with a microscope coverslip in order to allow optical monitoring of reactions and crystallization processes. A glass frit is utilized to improve the mixing of the reagents that are introduced on the top of the reactor and allowed to flow by gravity. Specifically designed solution holding chambers were developed to achieve this flow regime.

Reactor volumes can be readily tailored by simple modification of a CAD file, and thus rapidly made available for testing. In this way, reactor manufacturing is no longer a bottleneck for the design of chemical processes. Figure 9.4 shows an example of how a simple modification in the reactor geometry can lead to a completely different chemical output. Different mixing profiles and stoichiometry can be achieved by simply changing the reactor volume. For example, the heterocycle 1-(4-methoxyphenyl)-2,3-dihydro-1H-imidazo[1,2-f]phenanthridin-

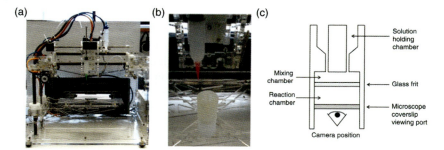

Figure 9.3 An example of a 3D printer employed to manufacture a basic chemical reactor. The Fab@Home version 0.24 RC6 3D printer. (a) The printer viewed from the front, with a single syringe of acetoxysilicone polymer loaded in the printing head. The white area below the printing head is a square of ordinary paper onto which the reactionware was printed. (b) The printing process of one device. (c) A schematic of the as-printed multipurpose reactionware. (Reprinted with permission from Ref. [2]. Copyright 2012, Macmillan Publishers Ltd: Nature Chemistry.)

4-ylium bromide (Figure 9.4, compound **3**) is obtained in high yield, when the reactor volume is large enough to accommodate a 1 : 3 molar ratio of 4-methoxyaniline and 5-(2-bromoethyl)phenanthridinium bromide (Figure 9.4, compound **1**) in d_6-DMSO in the presence of excess Et_3N. Again, the heterocycle 1-(4-methoxyphenyl)-1,2,3,12b-tetrahydroimidazo[1,2-f]phenanthridine (Figure 9.4, compound **4**) is obtained in good yield, but when the reactor volume is restricted to 2 ml, as only a 1 : 1 molar ratio of the same reagents mixture is allowed into the reactor, this changed the selectivity of the reaction.

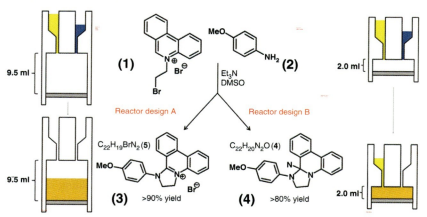

Figure 9.4 An example of a combined reactor and reaction design enabled by 3D printing. The modification of the reactor geometry influences the reaction stoichiometry. Consequently, the outcome of the reaction can be switched between these two products. In Reactor A, there is enough reactor volume to allow both reagents completely into the reaction chamber. In Reactor B, the lower reactor volume prevents a controlled amount of 5-(2-bromoethyl)phenanthridinium bromide (**1**) to enter the reactor, thus changing the reaction stoichiometry. (Reprinted with permission from Ref. [2]. Copyright 2012, Macmillan Publishers Ltd: Nature Chemistry.)

Another opportunity that 3D printing can offer over traditional fabrication process is the possibility of redefining reactor engineering to produce self-contained (stand-alone) reactors, where reagents and catalysts are added into the 3D printed vessel in a predefined sequence (Scheme 9.1). In this way, multistep chemical syntheses can be performed safely by reducing liquid handling and exposure to noxious compounds. In addition, the full automation of the synthetic process enables performing chemical syntheses outside specialized laboratory settings and by those without extensive chemical knowledge.

A monolithic multichamber block with two different 3D printed catalysts and integrated purification was designed to perform consecutive reactions. As a proof of principle, a sealed three-chamber reactionware was designed and manufactured to perform three sequential reactions: a Diels–Alder cycloaddition, an imine synthesis, and a hydrogenation (see Figure 9.5). These are very important chemical transformations, employed in a wide range of chemical syntheses. This multistep reaction sequence was possible by simply rotating the device; operation that enables the reaction mixture to flow from one chamber to the other. A channel filled with silica was designed and printed at the end of the reaction in order to trap impurities. This meant to improve the purity of the final product, which could be potentially used directly in a solution.

Figure 9.6 shows the design of a bespoke 3D printed device for a multistep synthesis. The reaction chambers were designed as cubes of side 20 mm, with circular passages connecting the chambers of diameter 4 mm (Figure 9.6a). During the printing process of this device, starting materials were deposited in each separate chamber and afterward sealed off in it to avoid chemical spills. The purification column space printed into the sealed reactor design was 24 mm in length with a square cross section of 49 mm^2. The fabrication of the reactors was achieved using two 3D printing machines: a Fab@Home version 0.24 RC6 3D printer to print the catalysts, alongside a Bits from Bytes 3DTouchTM 3D printer to manufacture the reactor body. The catalysts were deposited in a paste formulation of montmorillonite and palladium on carbon (Pd/C), both suspended in a silicone polymer. The

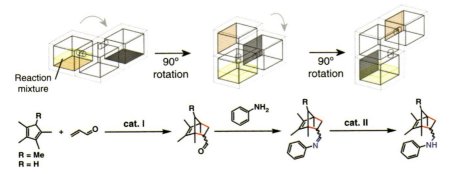

Figure 9.5 Schematic diagram of the multistep reaction sequence in the three main (reaction) chambers by simple rotation of the reactor. The reagents can be transferred into subsequent reaction chambers by rotation of the device through 90° intervals. (From Ref. [12]. Copyright 2013, The Royal Society of Chemistry.)

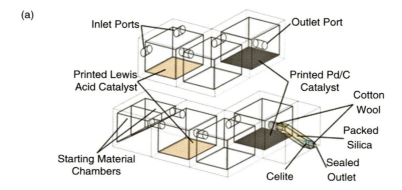

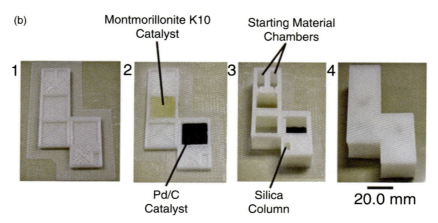

Figure 9.6 (a) A schematic diagram of the 3D printed sealed sequential reactors with starting material reservoirs, incorporating packed silica purification column. (b) 1. Reactor base with purification column before printing of catalyst regions. 2. Reactor base with purification column after printing of catalyst regions. 3. The fabricated reactor with purification column after addition of starting materials, reagents, and packing of silica. 4. The final sealed reactor. (Adapted from Ref. [12]. Copyright 2013, The Royal Society of Chemistry.)

reactor body was printed in polypropylene (PP), which is chemically resistant to solvents, reagents, and catalysts employed (see Figure 9.6b).

Despite the obvious limitations of this approach in terms of mixing, mass, and heat transfer, comparable yields to a traditional batch protocol were attained. ^{1}H NMR spectroscopy to make this analysis was used for the monitoring of the reaction products at different stages (Figure 9.7). Hence, this proves that it is possible to reproduce specialist laborious tasks with simple, inexpensive, and commercially available 3D printers.

The application of the reactionware concept to more challenging reaction conditions, like hydrothermal synthesis in complex and unknown reaction landscapes, showcases the potential of AM techniques to contribute to chemical research. The work of Ayers *et al.* [13] shows the possibility of using 3D printing to overcome

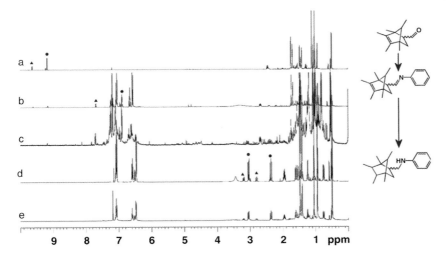

Figure 9.7 ¹H NMR (400 MHz, 298 K, CDCl$_3$) spectra of the reaction mixtures extracted from the 3D printed reactors at various stages of the reaction sequence. (a) After Diels–Alder cyclization. (b) After imine formation with aniline. (c) Crude reaction mixture after final reaction. (d) Final product after full purification using traditional column chromatography from the open reactor. (e) Final product after built-in purification using the sealed reactor. Peaks corresponding to aldehyde, imine, and newly formed aliphatic protons are highlighted in spectra (a), (b), and (e) with circles indicating the major product signals and triangles indicating minor product signals. (From Ref. [12]. Copyright 2013, The Royal Society of Chemistry.)

technological limitations in a more affordable and fast way than traditional methodologies. In this work a versatile alternative to conventional hydrothermal synthesis devices is presented, which results in the design and fabrication of different sizes of 3D printed hydrothermal devices as sealed plastic monolithic vessels that can be loaded with the reagents during the printing process (Table 9.2).

The printed reactors allow increasing the experimental throughput while considerably reducing the costs. A new reactor design was developed, consisting of 5 × 5 arrays of 1 ml capacity reaction chambers, producing devices suitable for high-throughput screening (HTS) of 25 simultaneous, discovery-scale reactions per device. This 3D printed HTS array was employed to investigate the synthetic parameter space of a number of typical diacid (L1–L5) linkers and bis(pyridine)-based pillars (P1–P5) with different metal cations, resulting in the discovery of two new metal–organic frameworks (MOFs), [Cu(L5-2H)(P3)]$_n$ and [Cd(L5-2H)(P3)]$_n$. The 5 × 5 array reactors were then used to investigate the synthetic conditions used in the synthesis of [Cu(L5-2H)(P3)]$_n$ allowing the scale-up of the synthesis to preparative quantities, with no loss of purity compared to traditional hydrothermal apparatus. To summarize, with this HTS approach it was possible to discover two new coordination polymers, optimize the synthesis of one of those, and then increase the scale of the synthesis using larger monolithic reactors manufactured with the same 3D printer, see Figure 9.8 for the summary of this study.

Table 9.2 Internal volumes and reaction mixture volumes of the reactors.

	Reactor			
	R1	R2	R3	R4
Internal volume (ml)	3	20	25 × 1	1.05
Liquid fraction (%)	66	50	60	47.6
Total printing time (h)	5	7	18	5

3D printing functional devices also find applications in energy devices manufacturing. Chisolm et al. [14] recently reported the use of 3D printing to fabricate flow plates for an electrolyzer. These parts were made of PP, employing a 3DTouch from Bits from Bytes 3D printer. Figure 9.9a shows a schematic representation of an electrolytic cell employed for proton exchange membranes. Two flow plates are employed to seal the device. These are generally manufactured in metals, typically stainless steel, titanium, or graphite [15]. Hence, the fabrication of flow plates requires specialist machining techniques, accounting for around 40% of the overall cost of the devices [13]. By employing plastic 3D printed plates, coated with silver to increase the conductivity (see Figure 9.9c), cheap and efficient proton exchange membrane (PEM) devices were manufactured. The weight and cost of the PEM devices is largely reduced with this technique. Current densities of 1.04–1.09 A cm^{-2} were achieved and the current obtained was stable for 100 h of continuous operation. These results open the door to manufacture tailored designs for electrolysers at a greatly reduced cost than the state of the art.

9.3
3D Printing Applied to Flow Chemistry

9.3.1
Mesoscale Reactors

Bespoke devices compatible with a wide range of organic solvents and reagents are usually made of silicon or glass [16]. This requires expensive specialized manufacturing techniques [17]. There is a growing interest in the use of

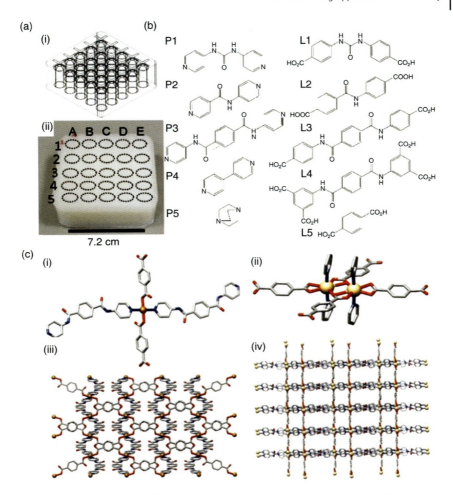

Figure 9.8 (a) (i) CAD model of the reactor R3. (ii) Photograph of the completed reactor depicting the placement of ligands into rows and columns. (b) Pyridyl (P1–P5) and carboxylic acid (L1–L5)-based ligands. (c) Solid-state structures of two coordination polymers discovered in this work: (i) Primary coordination environment of Cu^{II} in $[Cu(L5\text{-}2\,H)(P3)]_n$. (ii) Packing of $[Cu(L5\text{-}2\,H)(P3)]_n$. (iii) Primary coordination environment of the dimeric Cd^{II} subunit in $[Cd(L5\text{-}2\,H)(P3)]_n$. (iv) Packing of $[Cd(L5\text{-}2\,H)(P3)]_n$. Cu orange, Cd yellow, N blue, O red, and C gray. (Adapted with permission from Ref. [13]. Copyright 2012, John Wiley & Sons, Inc.)

polymers that can be employed to fabricate devices in a rapid and affordable fashion. One of the most commonly utilized polymers is PDMS due to its low cost and due to the possibility of being employed for rapid prototyping [18]. Nevertheless, it is not suitable for organic reactions as it can react with the chemicals and swell in most nonaqueous solvents [19]. Laboratory-scale flow reactors can generally be

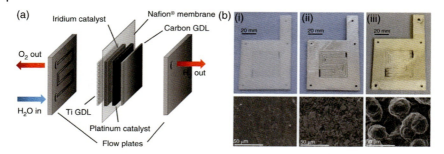

Figure 9.9 FDM printed plates for an electrolytic cell. (a) Schematic representation of an electrolysis cell components. (b) Photographs and scanning electron microscopy (SEM) images of 3D printed PP flow plates in different coating stages: (i) Uncoated plate. (ii) After two layers of silver coating. (iii) After electrodeposition of silver. (Reference [14], Published by The Royal Society of Chemistry.)

divided into two broad classes on the basis of channels' size and volume: micro- and mesoscale reactors. In general, microflow reactors present channels with diameters ranging from 10 to 1000 μm, whereas mesoflow reactors are characterized by larger channels with diameter up to 1000 μm [20] (Figure 9.10).

A wide range of materials can be chosen for the fabrication of objects by 3D printing. There is an increasing number of new materials being developed to increase the chemical compatibility of the 3D printed objects [21]. Indeed, research for the optimal fabrication of functional structural materials and devices by 3D printing is a field continuing in growth. At the moment, thermoplastics are the most employed materials, in particular PLA. Polypropylene (PP) is a better choice to fabricate micro- and millifluidic devices, as it is chemically

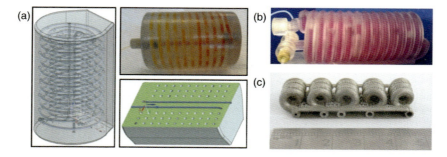

Figure 9.10 Examples of 3D printed flow reactors employing different AM techniques. (a) A SLA-based flow chip made from Accura 60 consisting of a tube with length 33 cm, diameter 3 mm, and a total volume of 23 ml, using a 3D System Viper si2 SLA 3d printer. The reactor was applied for a selective oxidation of aldehyde to methylesters. (b) A split-and-recombine mixer for the reaction of elemental iodine and potassium iodide. (c) An SLM-based flow device demonstrating the potential build resolution possible. (Reproduced with permission from Ref. [10]. Copyright 2013,The Royal Society of Chemistry.)

stable in a wide range of organic solvents and organic compounds, cheaper than PDMS, and compatible with commercial 3D printers. This plastic possesses the required characteristics to perform a chemical reaction: thermostability up to 150 °C, high chemical compatibility, and low cost. PLA is widely used in medicinal chemistry because of its biocompatibility. From a chemical point of view, its use is rather limited to a few solvents and organic compounds. In addition, operating reactor temperatures cannot exceed 66 °C, in order to preserve the 3D printed objects integrity [22]. Polyacrylates consist of a vast group of polymers with different physical and chemical properties, but their chemical compatibility is also low, like in the case of PLA. In fact, they are not generally recommended for exposure to alcohol, glycols, alkalis, brake fluids, or chlorinated or aromatic hydrocarbons [23]. Therefore, PP is currently the best choice thermoplastic for the 3D printed object to use for organic reactions.

The previous examples show the potential of 3D printing to the manufacturing of reactor devices tailored to bespoke chemical processes. Efficient reactor engineering based on continuous flow processes helps reducing waste and energy needs, and is known to contribute to the green chemistry principles [24]. Nevertheless, the manufacture of continuous flow reactors is costly, requires specialized equipment, and is generally limited to simple geometries, due to the available manufacturing techniques. A range of commercial platforms has been developed, but these are all restricted to design only standard reactor architectures, generally tubular reactors with reduced mixing efficiency under laminar flow conditions [25]. Another general shortcoming is the gap among reactor design, manufacturing, and application in chemistry. 3D printing is a perfect tool to overcome these limitations, enabling tailored continuous flow chemistry in nonspecialist laboratories and industry. An example of this is shown in Figure 9.11.

As reported by Kitson *et al.* [26], a 3DTouch™ 3D printer was used to print fluidic devices made of PP in a cheap, fast, and precise way. This is a FDM

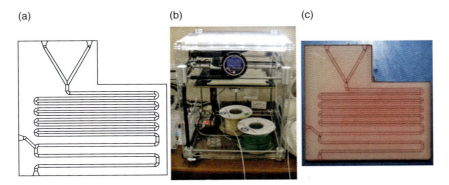

Figure 9.11 Examples of rapid prototyping of millifluidic devices employing 3D printing. (a) CAD design of a continuous flow reactor with multiple inlets and outlets. (b) 3D-Touch 3D printer employed to manufacture the reactor devices. (c) Manufactured reactor in polypropylene. A solution of Rhodamine C was employed as a tracer to visually observe the morphology of the channels. (Reproduced with permission from Ref. [26]. Copyright 2012, The Royal Society of Chemistry.)

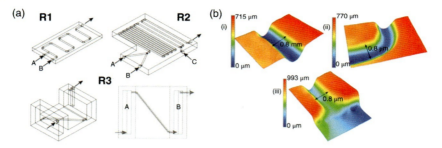

Figure 9.12 (a) CAD profiles of 3 reactors with different geometries. (b) 3D profiles of reactors obtained by optical microscopy of half-printed channels (i) a typical straight channel, (ii) 90° bend in a channel, and (iii) opening of a channel at the wall of a device. (Reproduced with permission from Ref. [26]. Copyright 2012, The Royal Society of Chemistry.)

printer and as such, it heats a thermopolymer through the extruder, depositing the material in a layer-by-layer fashion, converting a digital model into its corresponding object. In particular, this 3D printer has a tolerance of ±0.2 mm and a Z-axis resolution of 0.125 mm (±0.06 mm). The channels of the devices herein printed were approximately circular in cross section and with diameter of approximately 0.8 ± 0.1 mm, as determined by optical microscopy of half-printed channels (see Figure 9.12b). Different geometries were successfully printed (Figure 9.12a) and tested for organic, inorganic, and nanomaterial synthesis under continuous flow.

Table 9.3 has a summary of features corresponding to the manufacturing of the chemical reactors. All the devices can be printed in few hours and for less than $1.

The printed chips were connected with epoxy resins to standard tubing and integrated into continuous flow synthetic platforms, equipped with programmable syringe pumps (Tricontinent, C3000) and inline analytics (UV-vis and ATR-IR) to monitor reaction progress in real time (Figure 9.13) [26].

In subsequent publications, the device design was improved to accommodate standard fitting connectors [27]. This facilitated the integration of the chips in bespoke processing platforms with multiple devices also connected in series with inline analytics. In the work of Dragone *et al.* [27a], two devices are

Table 9.3 A summary of main features for the 3D printing of millifluidic devices.

	R1	R2	R3
Printing time (h)	2	4	4
Reactor weight (g)	3.9	8.8	13.4
Materials cost (US$)	0.09	0.20	0.31
Dimensions (mm)	25 × 50 × 3	56 × 52 × 6	40 × 30 × 20

Source: Reproduced with permission from Ref. [26], The Royal Society of Chemistry.

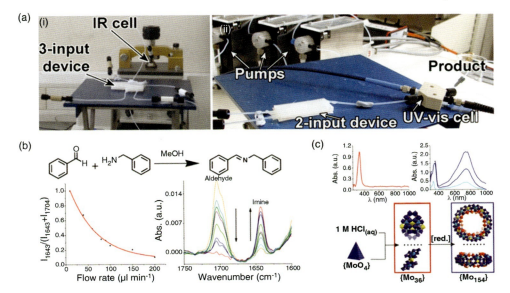

Figure 9.13 3D printed continuous flow millifluidic devices applied in organic and inorganic synthesis. (a) A picture corresponding to a 3-input device (i) connected to a ATR-IR flow-cell and a 2-input device (ii) connected to a UV-vis flow cell. (b) An example of an imine synthesis monitored by ATR-IR. Increasing residence time results in an increase in conversion, reflected in a reduction in the aldehyde band and an increase in the imine band. (c) Synthesis of two complex inorganic clusters under continuous flow monitored by UV-vis. Mixing a molybdenum source with concentrated HCl (1M) leads to the formation of a molybdenum cluster with a characteristic band at 335 nm. The addition of a mild reducing agent yields the cluster with a characteristic blue color. (Adapted with permission from Ref. [26]. Copyright 2012, The Royal Society of Chemistry.)

connected with each other to perform imine synthesis in the first reactor and imine reduction in the second one (Figure 9.14).

The experiments were conducted using 2 M methanolic solutions of the different substrates as reported in Table 9.4. The use of highly concentrated reagents favors the kinetics of the transformation, thus reducing reaction times and resulting in greener processes. Furthermore, the compatibility of the materials with environment-friendly solvents, like methanol and ethanol, is another added value of this case study. The reactor output was connected with a 0.1 ml tubing to the ATR-IR flow cell; hence, the total flow reactor volume (V_R) was 0.5 ml. The syntheses of the imines were monitored by an inline ATR-IR flow cell and conducted at different flow rates, using all times equimolar solutions. The disappearance of the aldehyde band and the simultaneous appearance of the corresponding imine band with increasing total residence time (t_R) were indicative of the reaction progress. The second reactor connected in series (R2″ in Figure 9.14) was employed for the reduction of the imines with a solution of cyanoborohydride ($NaBH_3CN$) in MeOH (1M) as a reducing agent. The two

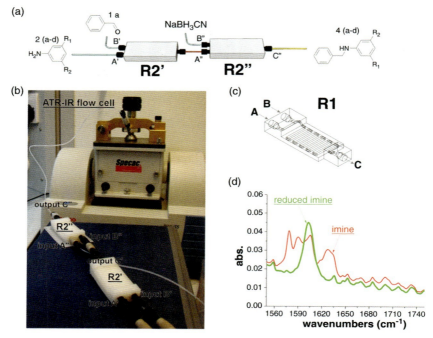

Figure 9.14 Integrated modular setup for multistep organic transformations employing 3D printed flow reactors. (a) Reaction scheme employed as benchmark. In the first step an amine and an aldehyde are mixed together to produce the corresponding imine. The product stream is directly fed into a second reactor, where it is mixed with a mild reducing agent to yield the corresponding substituted amine. (b) Picture of the assembled setup connected inline to an ATR-IR flow cell. This allows real-time monitoring of the reaction mixtures. (c) CAD design of a millifluidic reactor with two inlets, a tubular reactor, and an outlet. The design accommodates standard fitting connectors. (d) Detail of the IR spectra after the synthesis of the imine (red line) and its subsequent reduction (green line). (Reproduced from Ref. [27a].)

solutions were pumped through R2″ at the same flow rate. The molar and volumetric ratios hydride : imine were kept constant (1 : 1) to produce the corresponding amines with a t_R of 7 min in the second reactor. The reducing agent was selected because it is mild but effective. Additionally, it prevents the undesired formation of bubbles or problems related to overreduction, which could be expected in this range of concentrations when using conventional reducing agents, such as $NaBH_4$. Using this methodology, a range of imines were reduced to produce the corresponding secondary amines shown in Table 9.4.

Another similar example is the work of Mathieson *et al.* [27b] in which the possibility of 3D printing flow reactors that can be easily integrated in a flow setup and directly linked to a high-resolution electrospray ionization mass spectrometer (ESI-MS) for real-time, inline observations is demonstrated. In this

Table 9.4 Table of the compounds used to study the imine reduction.

Entry	R1	R2	Product	Yield (%)[a]
1	H	H	(PhCH2-HN-Ph)	78
2	CF$_3$	H	(PhCH2-NH-C6H4-CF3)	99
3	Cl	H	(PhCH2-NH-C6H4-Cl)	96
4	CH$_3$	CH$_3$	(PhCH2-HN-C6H3(CH3)2)	97

a) Combined yield after two stages determined by ^{1}H NMR.
Source: Reproduced from Ref. [27a].

work, the 3D printed reactionware inputs were connected to the desired pumps and the output to a stream splitting for interfacing directly the outlet stream with inline ESI-MS. Different supramolecular experiments were carried out to emphasize the versatility of the reactionware setup. The first set of experiments consisted in screening by inline ESI-MS a reaction using different reagent flow rates in order to change the product stoichiometry. The same reactionware setup was then used to switch between two complexes formed by employing cis,trans-1,3,5-tris(pyridine-2-ylmethylene)cyclohexane-1,3,5-triamine (top) ligands with two metals. Indeed, Cu and Ni salts were alternately pumped through the device, with top remaining constant. Real-time, continuous ESI-MS was successfully used to observe the reaction and the oscillation between the two salt complexes with a simultaneous collection of the products, see Figure 9.15 for a summary of the setup and results of this work.

Very recently, the possibility of 3D printing advanced reactor geometries has been further demonstrated by Okafor et al. by manufacturing miniaturized continuous flow oscillatory baffled reactors (COBRs) [28] employing SLA to generate bespoke mesoscale baffled reactors (mCOBRs) for the efficient synthesis of nanostructured materials (Figure 9.16) [29]. The reactors were printed in commercially available acrylate-based resins, which demonstrated good solvent resistance in a number of organic solvents, including acetonitrile (MeCN), ethanol, and water. Residence time distribution studies showed the optimal flow regimes easily achievable under oscillatory flow conditions, with high Peclet numbers at relatively low flow rates (and therefore Reynolds numbers).

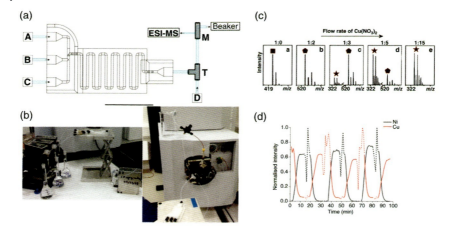

Figure 9.15 (a) Schematic overview of the device setup. The three inlets were each connected to a syringe pump, which were connected to stock solutions of the required starting materials or MeOH. (b) Photograph of the device setup and connection to the mass spectrometer. (c) ESI-MS isotope patterns of top + Na, m/z 419.2 (square), [Cu(C$_{24}$H$_{24}$N$_6$)(NO$_3$)]$^+$, m/z 521.1 (pentagon), [Cu$_2$(C$_{24}$H$_{24}$N$_6$)(NO$_3$)$_2$]$^{2+}$, m/z 323.0 (star), showing how an increase in flow rate from the pump containing Cu(NO$_3$)$_2$·6H$_2$O changes the stoichiometry of the complex from 1:1 to 1:2 top: Cu(NO$_3$)$_2$·6H$_2$O. (d) Normalized intensity plotter against time showing the change of [Ni(C$_{24}$H$_{24}$N$_6$)(NO$_3$)]$^+$, m/z 516.1 (black) and [Cu(C$_{24}$H$_{24}$N$_6$)(NO$_3$)]$^+$, m/z 521.1 (red). (Reproduced from Ref. [27b].)

The chemical modification of the surface properties of 3D printed continuous flow reactors has been employed to efficiently immobilize enzymes, resulting in efficient continuous flow bioreactors for transaminations [30]. An FDM 3D printer was employed to manufacture simple fluidic devices employing commercially available Nylon 6 (Taulman 618 and 645). A protocol to functionalize the surface with aldehyde groups was applied in batch and flow devices. The modified surfaces proved to be very efficient to support ω-transaminases. The simple protocol allows printing well plates for rapid parallel screening of multiple conditions (Figure 9.17).

A key advantage of this novel methodology is the direct translation of the optimal properties into continuous flow devices. This extent was demonstrated by employing the optimal conditions established in the well screening (see Figure 9.17e). It was found that surfaces functionalized with glutaraldehyde were particularly efficient in stabilizing the enzymes. The same functionalization approach was employed to functionalize a continuous flow reactor. The walls of a 0.8 ml reactor were functionalized with 170 µg of enzyme ATA117. The resulting bioreactor was tested under continuous flow for the kinetic resolution of rac-methylbenzylamine, showing a remarkable activity and stability for about 100 h on stream (Figure 9.18).

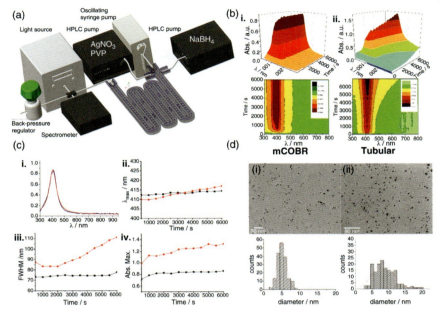

Figure 9.16 (a) Automated platform employed to synthesize silver nanoparticles under continuous flow conditions employing 3D printed mCOBRs. (b) Comparison of the time resolved local surface plasmon of resonance (LSPR) for (i) a mCOBR and (ii) tubular reactor. Spectra collected every 10 s. (c) Data analysis from (i) fitting the UV-vis extinction spectra (blue line) to a Lorentzian function for the results obtained in the mCOBR and in a tubular reactor in regular intervals (500 s). The results obtained in both reactor configurations for (ii) λ_{max}, (iii) full width at half maximum (FWHM), and (iv) maximum absorption plotted as a function of time. (d) Particle size distribution of the Ag-NPs synthesized in a mCOBR (i) compared to a tubular reactor (ii) at 90 min. (Reproduced with permission from Ref. [29b]. Copyright 2017, The Royal Society of Chemistry.)

9.3.2
3D Printed Membranes

Membranes play a fundamental role in advanced chemical reactor engineering [31]. Membrane technology enables a broad range of applications for selective separations relevant to grand challenges like climate change and water purification, including osmosis for wastewater treatment [32], CO_2 capture and separation [33], and advanced features like the integration of reaction and separation as a single unit operation [34].

The possibility of 3D printing membranes has been demonstrated by Femmer et al. [35] by employing a digital light rapid prototype (Perfactory Minimultilense from EnvisionTEC). A mixture of copolymer (methacryloxypropyl methylsiloxane- domethylsiloxane), a photoinitiator (ethyl(2,4,6-trimethylbenzoyl)phenyl

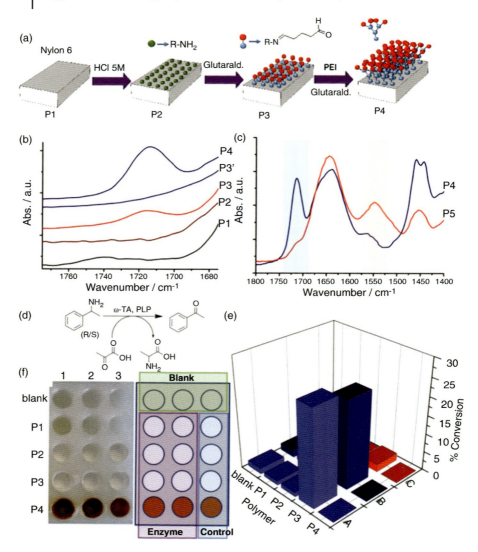

Figure 9.17 Chemical modification of nylon-based 3D printed materials for enzyme immobilization. (a) Sequential protocol to modify nylon surfaces, including (P1) HCl 5M; (P2) glutaraldehyde; (P3) polyethyleneimine, and (P4) glutaraldehyde. (b) Details of the aldehyde region of the ATR-IR spectra of the subsequent modified surfaces. (c) ATR-IR spectra corresponding to the modified nylon surface before and after enzyme immobilization. A strong reduction in the aldehyde band (blue region) and an increase in the amide-II region (brown region) were observed. (d) Scheme of the reaction employed for the screening of enzymatic activity, more concretelythe transformation of methylbenzylamine to acetophenone employing a commercially available ATA117 enzyme. (e) Results of the screening of biocatalytic activity of the different wells. (f) 3D printed wells with modified surfaces and design strategy. (Adapted from Ref. [30] with permission of The Royal Society of Chemistry.)

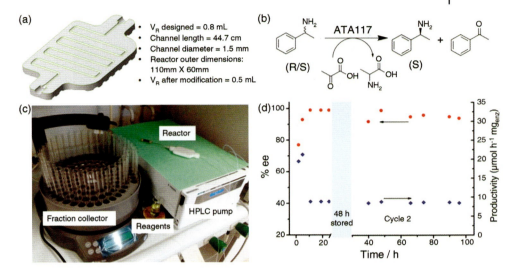

Figure 9.18 Setup employed for the kinetic resolution of rac-methylbenzylamine with 3D printed bioreactors under continuous flow. (a) CAD drawing of the reactor employed in this work with the main reactor parameters. (b) Reaction scheme of the kinetic reaction studied as benchmark for the performance of the bioreactors. (c) Picture corresponding to the experimental setup with labels. (d) Productivity and enantiomeric excess observed as a function of time on stream. First two data points correspond to 25 µl min^{-1}, then the flow rate was reduced to 10 µl min^{-1}. A first run of 24 h was followed by 48 h of storage in fresh buffer. Afterward another run of 78 h was held with remarkably high activity and stability of the immobilized enzyme. (Reproduced with permission from Ref. [30]. Copyright 2017, The Royal Society of Chemistry.)

phosphinate), and a small amount of a dye (Orasol Orange) for increased resolution was cured layer-by-layer (Figure 9.19) employing UVlight irradiation with a 180 W Hg vapor lamp illuminating 1400 × 1050 pixels (30 × 30 µm/pixel). A curing time of 12 s was applied. Features with size 150–300 µm were well resolved employing this approach.

The performance of the 3D printed membranes was tested by fabricating a cross-flow gas–liquid contactor based on a Schwarz-P triple periodic structure [36]. Figure 9.20 shows a CAD design and the actual device connected with standard tubing to a CO_2 stream that crosses with H_2O stream. The change in color from the bromothymol pH indicator (blue above pH 7.6 and yellow below pH 6.0) indicates the transfer of CO_2 to the aqueous phase, thus forming H_2CO_3.

Pure gas permeability of 3D printed flat sheet membranes with a thickness of 840 µm was determined for nitrogen, oxygen, and carbon dioxide. The results show the selectivity of the membrane toward CO_2 compared to nitrogen. The values of permeability are marginally lower than reported in the literature. This was attributed to a higher degree of cross-linking (Table 9.5).

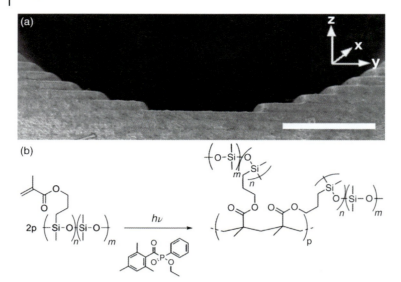

Figure 9.19 (a) SEM image of a 3D printed PDMS layer. The membrane is cured layer by layer, subsequent layers are added in the z-axis. Scale bar = 1 mm. (b) Scheme of the cross-linking reaction. (Reproduced with permission from Ref. [35]. Copyright 2014, The Royal Society of Chemistry.)

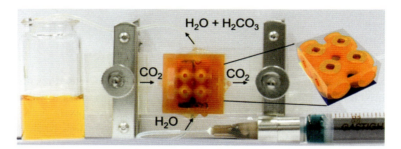

Figure 9.20 Membrane contactor manufactured with 3D printing in PDMS and artistic rendering of the respective TPMS structure (inset right). Gas flows horizontally from left to right and the bromothymol blue pH indicator is pumped vertically from the bottom to the top through the contactor. The color change of the pH indicator from blue to yellow indicates the CO_2 transport over the membrane. (Reproduced with permission from Ref. [35]. Copyright 2014, The Royal Society of Chemistry.)

Table 9.5 Permeability and selectivity data of different gases in 3D printed PDMS photoresist compared to the literature.

Entry	Gas	Permeability[a]	Ideal selectivity over N_2	Young's modulus[b]
1	N_2	235 (280)	—	—
2	O_2	523 (600)	2.3 (2)	11.45 ± 2.20
3	CO_2	2611 (3200)	11.1 (12)	(1.44 ± 0.20)

a) Time lag permeabilities measured at 35 °C, values in parentheses from literature [37] for PDMS RTV 615 at 35 °C, in barrer (1 barrer = 1×10–10 cm3(STP) cm^{-1} s^{-1} cm Hg^{-1}).
b) Young's modulus expressed in MPa, in parenthesis values for Sylgard 184 with a ratio of 10:1, base to cross-linker.
Source: Reproduced with permission from Ref. [35], The Royal Society of Chemistry.

9.4
Applications of 3D Printed Flow Devices in Analytical Chemistry

9.4.1
3D Printing of Valves, Pumps and Actuators

A step forward in the development of advanced reactor architectures based on 3D printing consists of integrating actuators, such as valves and switches, in the device design, which enables complex reactions and analytical sequences to be seamlessly automated. Au et al. employed stereolithography to produce well defined and characterized valves based on Watershed XC 11122 resin (Figure 9.21) [38]. This resin is practically transparent, does not swell in water and is biocompatible. This is limited for most chemical processing, but viable for biological applications. The critical part of the valve is a membrane with a minimal target thickness (t) of 100 μm. The membrane separates two chambers (Figure 9.21b and c) that can move by applying pressurized air through the "control line." In this way the membrane opens and closes. The membrane must cover the 200 μm distance to contact the nozzle and apply an extra pressure to ensure a good seal is achieved.

The membrane radius (r) necessary to close the gap under a controlled applied pressure can be calculated from Eq. (9.1), where E is the Young's modulus, ν is the Poisson's ratio, t is the membrane thickness, and y is the deflection distance.

$$\frac{Pr^4}{Et^4} = \frac{5.33}{1-\nu^2} \cdot \frac{y}{t} + \frac{2.6}{1-\nu^2} \cdot \left(\frac{y}{t}\right)^3 \quad (9.1)$$

It was calculated that for a Watershed resin ($E = 2700$ MPa, $\nu = 0.30$) with a value of $r = 5$ mm, a distance of 204 μm was deflected at a $P = 2.9$ psi. Experimentally, a variation between 1 and 6 psi (3.30 ± 1.78 psi) was observed due to the variability in the printing process. A basic peristaltic pump was generated employing three valves in series achieving a maximum flow rate of 0.68 ml min^{-1} at a frequency of 6.7 Hz for a single valve. The valves can also be employed as switches in complex

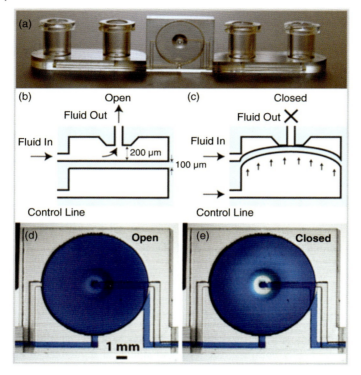

Figure 9.21 Images corresponding to the valve design. (a) Photograph of a single-valve device. Schematics of a valve unit in its open (b) and closed (c) states. Micrographs of a valve unit in its open (d) and closed (e) states. (Reproduced with permission from Ref. [38]. Copyright 2015, The Royal Society of Chemistry.)

fluid environments, in a configuration of four component mixing valve, connected with printed luer connectors (Figure 9.22a and b). Food-coloring dyes were employed to characterize the switch efficiency. Two measuring points were defined, one close to the mixing point (switch) and the other close to a 3D printed cell culture chamber (Figure 9.22c) and the time necessary to switch fluids (shown in Figure 9.22d) was determined. It was found that the difference in time to achieve 10% (t_{10}) and 90% (t_{90}) of fluorescence could determine two parameters $t_{on} = t_{90} - t_{10}$ and $t_{off} = t_{10} - t_{90}$. In the junction, $t_{on} = 2.53 \pm 0.14$ s and $t_{off} = 4.93 \pm 0.05$ s, and in the chamber $t_{on} = 4.93 \pm 0.15$ s and $t_{on} = 4.86 \pm 0.33$ s (mean ± SD). Furthermore, the variability between the three valves regulating the dyes was below 5.8%, thus indicating the reproducibility of the operations.

The 3D printed valve system was employed to control stimulation of cells in a culture chamber. A cell culture chamber was printed without floor and then bonded to a PDMS slab that ensured gas exchange by diffusion. Besides, it enables microscopy imaging and cell attachment. CHK-O cells were stimulated with 20 s pulses of ATP solutions and their Ca^{2+} response was monitored with a

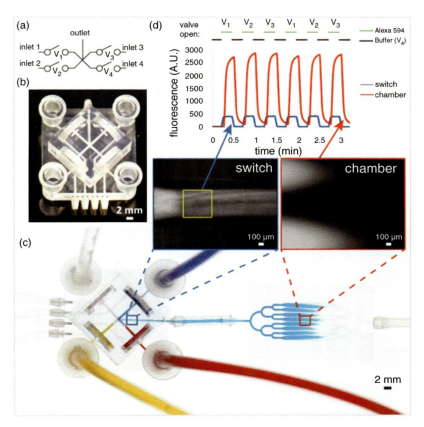

Figure 9.22 A 3D printed cell perfusion system based on four interconnected valves. (a) Circuit diagram. (b) Image of the four-valve switch. (c) Photograph of a dye-filled switch with four different dyes connected to a cell culture chamber. Insets near the mixing point and in the cell culture chamber are shown. (d) Time-dependent switch based on concentration profiles in the switch and chamber during fluid-switching events measured by fluorescence. Red and blue curves correspond to the fluorescence intensity in the cell chamber and the switching device regions, respectively. Valve actuation events are shown in green and black. (Reproduced with permission from Ref. [38]. Copyright 2015, The Royal Society of Chemistry.)

fluorescent dye [38]. The response of the cells over 30 min demonstrates the viability of this approach.

Very recently, Tsuda *et al.* have reported another example of 3D printed valves [39]. The design shown in Figure 9.23 is based on the combination of two FDM-based 3D printed modules (manufactured with an Ultimaker 2 3D printer). In the first module, two inlets were employed to pump the reagents that can be selectively delivered by the printed device to the outlet. The second module was employed to pump compressed air to open and close the valves. A PDMS-based flexible membrane was placed to connect both modules, thus allowing the actuation of the valve pneumatically. A treatment with plasma oxygen was employed to

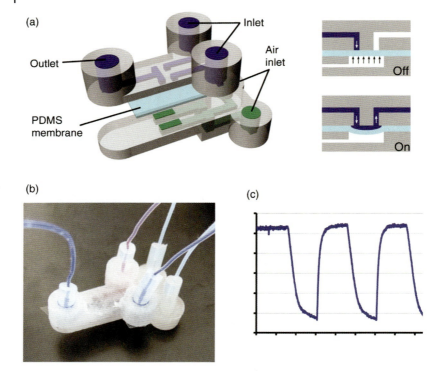

Figure 9.23 3D printed fluid selector employing a "plug and play" concept. (a) Scheme of the 3D printed flow selector consisting of two 3D printed modules with an intercalated PDMS membrane. Valves are selectively closed applying compressed air at a higher pressure than the liquid. (b) Image of the actual device. (c) Plot of the temporal evolution of a dye level measured at the outlet of the device through consecutive switching cycles every 15 s. (Reproduced from Ref. [39].)

bind both modules to the PDMS membrane. The fluid was pumped at low pressures (180 mbar), then compressed air (250 mbar pressure) was applied to close the membrane. An efficient actuation was achieved with this design, demonstrated with a sequence of opening and closing cycles lasting together for 15 s (Figure 9.23c). Indeed, on opening the valve 85% of the maximum level was recovered within 1 s. This device can be easily integrated into complex reactor architectures.

9.4.2
Modular Devices Based on SL

3D printing enables manufacturing of configurable blocks that can be connected to produce configurable and modular continuous flow reactors for the first time. Bhargava *et al.* employed stereolithography to manufacture a set of inter-connectable blocks that could be manually linked, enabling the manufacture of complex and configurable three-dimensional reactors [40]. Figure 9.24 represents an

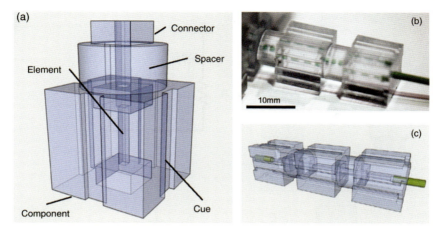

Figure 9.24 A CAD drawing of a flow module consisting of a 1-mm male–male connector aligned with female-type port terminating a 750-μm microfluidic straight channel. (Reproduced with permission from Ref. [40]. Copyright 2014, National Academy of Sciences.)

example of a 1 mm reactor module with a 750 μm microfluidic element, depicted with the CAD drawing and the assembly of two printed modules. The linking component was designed in a male shape that fits in a female port of the next module. A circular spacer was added to facilitate the manual connection between modules.

Once this concept has been established, it is possible to design, model, and construct chemical reactor circuits employing electrical engineering design concepts. As an example, Figure 9.25 shows a model of a reactor system with two modules in parallel with different channel morphologies. In this configuration, R_{ref} is a reference straight channel and R_{select} corresponds to a selected complex structure that determines the mixing between the channels. The remaining of the structure is identical and therefore shows a comparable resistance (R_{struct}). The system can be modeled as an equivalent internal circuit with internal segment resistances. The resistance from each module can be calculated from the solution of the Navier–Stokes equation for Poiseuille flow in straight square channels where μ is the dynamic viscosity of pure water at room temperature (1 mPa s), L is the length of the channel, and h is the height (of a square cross section) of the channel (Eq. (9.2)).

$$R_{hyd} = \frac{28.4\,\mu L}{h^4} \quad (9.2)$$

The experimental mixing ratios follow a similar trend to that acquired from the model, even though some deviations were observed (Figure 9.26). This was modeled employing Eq. (9.3).

$$m_0 = \frac{Q_y}{Q_b} = \frac{R_{struct} + R_{ref}}{R_{struct} + R_{select}} \quad (9.3)$$

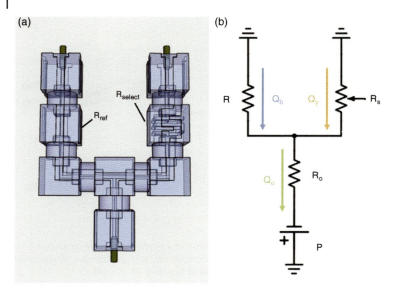

Figure 9.25 (a) A CAD drawing for a 2-input, 1-output device to generate concentration gradients in which a single branch "resistor" varies the mixing ratio according to selected R_{ref} and R_{select}. (b) Equivalent circuit diagram to analyze mixing ratios employing Kirchoff's laws. (Reproduced with permission from Ref. [40]. Copyright 2014, National Academy of Sciences.)

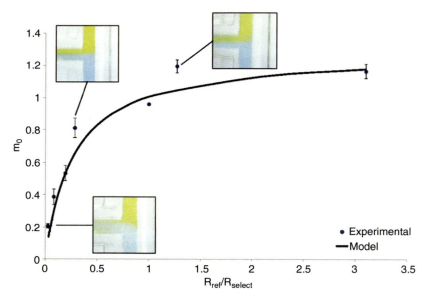

Figure 9.26 A comparison of experimental and theoretical mixing ratios for a variety of R_{select} values (error bars represent the SD over 12 measurements). (Reproduced from Ref. [40].)

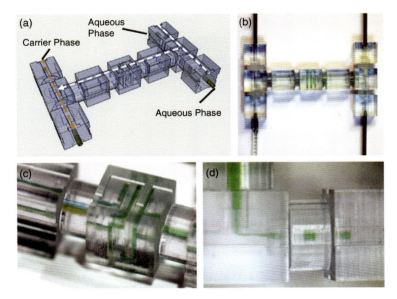

Figure 9.27 (a) Scheme of a CAD reactor configuration design and (b) the printed version. (c and d) Details of a complex fluidic module and a T-junction configuration for a droplet generator. (Reproduced with permission from Ref. [40]. Copyright 2014, National Academy of Sciences.)

Functional devices like droplet and Taylor flow generators can be easily realized employing this approach. A T-junction (see Figure 9.27d) and flow focus devices were assembled by conveniently connecting a set of modules.

The length of the dyed aqueous plugs of flow generated with both devices was determined as a function of the aqueous flow rate while the carrier flow rate was constant (Figure 9.28). In Figure 9.28, it can be observed that the length of the droplets increased with the flow rate and smaller droplets were achieved in the flow-focused device (600 μm droplet length at 1 ml h^{-1}).

In the same work, real-time analytics were integrated in the design of the reactor with the same modular approach, employing a 3D printed IR sensor block. The printed device containing the sensor consisted of a straight channel with 642.5 μm cross section that intersects a beam created by a near-infrared diode emitter and a phototransistor (Figure 9.29). The voltage signal was monitored across the photoresistor. The system was tested employing a Taylor flow of aqueous droplets in a fluorous oil phase. The water absorbs much more in the infrared, thus yielding a traceable signal. The flow rates were set to 5 and 2 ml h^{-1} for the carrier and aqueous phases, respectively. Average droplet lengths were in agreement with both NIR sensor and optical micrographs techniques, and were, respectively, 421.22 ± 27.54 and 416.90 ± 16.36 μm.

Another approach to produce modular and configurable flow configurations employing 3D printing was recently reported by Lee *et al.* [41]. A set of modules (30 × 30 × 5 mm W, H, L) were printed employing inkjet printing with a UV

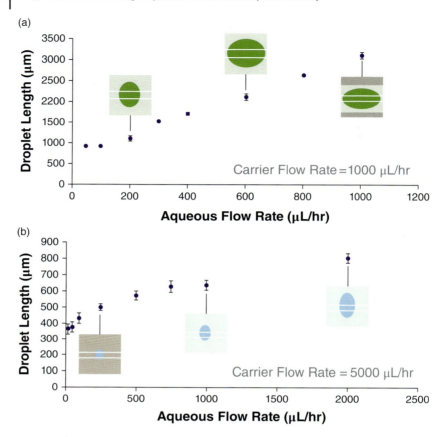

Figure 9.28 Droplet length as a function of the aqueous phase flow rate at constant carrier flow rate. The length was measured along the center axis of exit tubing for (a) T-junction and (b) flow-focus subsystems. Error bars represent SD over 12 measurements. (Reproduced with permission from Ref. [40]. Copyright 2014, National Academy of Sciences.)

curable polymer, employing wax as a filler of the voids preventing the collapse of the structures (Figure 9.30). Employing this methodology, microchannels of 100–1000 μm in width and 50–500 μm in height were printed. Below this limit the structures were found to be inaccurate and collapsed. Variability of 35 μm in width and 11 μm in height was observed. To ensure a good seal, rubber O-rings were employed, which showed a resistance of up to 200 kPa and up to 800 kPa with silicon greased O-rings.

The modules were locked with pins and O-rings and configured according to different experimental requirements. Figure 9.31 shows five examples of fluidic configurations where red dye is employed to highlight the channels. The reactor configuration shown in Figure 9.31a was employed for the AFP immunoreaction for liver cancer diagnosis [42].

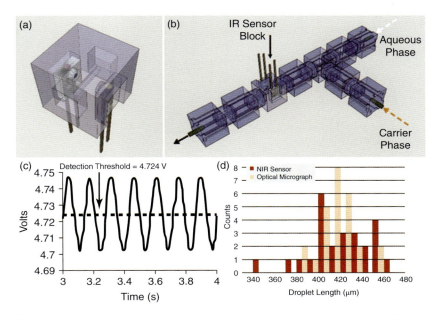

Figure 9.29 A CAD representation of a component with a straight pass channel of 642.5 μm cross-sectional side length intersecting the beam created between a discrete near-infrared (NIR) diode emitter and phototransistor receiver. (Reproduced with permission from Ref. [40]. Copyright 2014, National Academy of Sciences.)

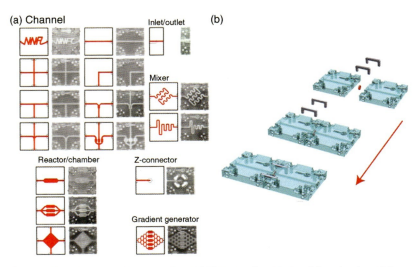

Figure 9.30 (a) A schematic illustration and photographic images of functional modules. (b) Illustration of block chip assembly employing O-rings and clips. (Reproduced with permission from Ref. [41]. Copyright 2014, The Royal Society of Chemistry.)

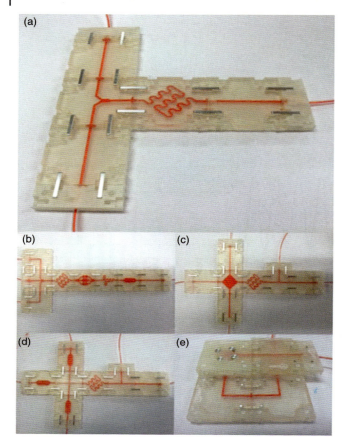

Figure 9.31 Image of assembled microfluidic device for (a) biosensing and (b–e) several possible applications. (Reproduced with permission from Ref. [41]. Copyright 2014, The Royal Society of Chemistry.)

9.5
Future Trends

The development of novel methodologies to 3D print devices with higher precision, lower cost, and in shorter timescales than current methodologies, is highly desirable for the development of the field. A new approach for 3D printing has recently been reported by Lee *et al.* [43], which employs scanning projection stereolithography (SPL). It is inspired by maskless lithography with a digital micromirror device (DMD) that is used to project patterns with resolution up to 10 μm onto a layer of photoresist. The printer was built employing low cost and off the shelf parts and materials. Compared to traditional stereolithography, it does not require expensive masks and thus reduces the dependence on

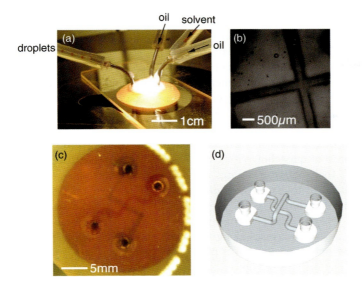

Figure 9.32 (a) and (b) A droplet device. The channels are 400 μm in diameter and the reactionware has around 1 cm² footprint and is 2.3 mm tall. In (a) there is a photograph of the device in operation, outputting droplets on the left. (b) A microscope image of the channels within the device. (c) A micrograph of a flyover channel, where red and blue aqueous dyes cross over one another. (d) 3D model of the flyover reactionware. (Reproduced from Ref. [43].)

specialized suppliers. Hence, this is a low-cost approach that offers an alternative to stereolithography methods for the production of large area objects (5 cm²).

A microfluidic device was manufactured employing SPL to generate droplets of water in a fluorinated carrier solvent. The device was created using a flow focusing geometry. It was designed and manufactured over a surface of 1 cm² with a channel width of 500 μm and employed to generate droplets of water in a fluorinated solvent (Figure 9.32).

9.5.1
Ultrafast Printing

Despite being called rapid prototyping, additive manufacturing can be regarded as a relatively slow and time-consuming layer-by-layer manufacturing technique. Tumbleston *et al.* [44] developed a novel method for the continuous liquid interface production (CLIP) of 3D objects. This approach is based on limiting oxygen inhibition reactions during the photopolymerization process. This was achieved by inverting the light source to the bottom of the polymer VAT, which was formed with an UV-vis transparent and gas permeable window made of Teflon AF-2400. This material has an excellent oxygen permeability (10^{-7} cm³ (standard temperature and pressure) s^{-1} cm Hg^{-1}), is UV transparent, and chemically inert. The irradiation of the polymer liquid resin under a controlled oxygen supply allows the polymerization at a distance from the window called "dead zone" (DZ). The build support platform moves up at a controlled speed, enabling a constant supply of polymer to the printing zone. The printing rate is determined by viscosity and photopolymerization kinetics and not by stepwise layer formation. Therefore, the printing times are reduced from hours to minutes and this approach has the potential to become a standard in polymer-based 3D printing.

The thickness of the DZ changes with the UV light irradiation. In the case of nitrogen, the absence of quenching effect was observed by a nonexisting layer. This demonstrates that oxygen generates the DZ. For air and oxygen, higher irradiation intensities led to a reduction of the DZ. The DZ generates an area where oxygen and free radicals coexist, which prevents the polymer from sticking to the bottom of the bath. The concentration of oxygen decreases as the distance from the bottom increases along the z-axis. When all the oxygen has been consumed the polymerization takes place. In a well synchronised fashion, the build support platform moves upward, thus allowing a continuous printing process.

9.5.2
Smart Materials through 4D Printing

The design for self-assembly concept [45] aims to build parts that will evolve over time as the fourth dimension to change shape. Stimuli-responsive polymers have been printed, taking into account different possible shapes that can be taken as a function of time and the stimuli applied. By 3D printing shape memory polymer fibers in an elastomeric matrix, Ge *et al.* expand this concept to yield temperature-dependent structures [46]. A 3D multimaterial polymer printer (Objet Connex 260, Stratasys, Edina, MN, USA) was employed to directly print an object from a CAD file with specific fiber architectures, in laminate shape embedded in an elastomeric matrix. The composite materials are designed to have shape memory effect. In fact, the fibers are thermoresponsive and present anisotropic shape memory behavior. Composite laminate (2 mm thickness) were printed with a single row of fibers (volume fraction = 0.28) that were oriented at different angles. The storage modulus (E_s) was determined with a dynamic mechanical analyzer (DMA) as a function of temperature. The results demonstrate that it is possible to control (E_s) by varying the temperature and the orientation of the fibers. Hence, differential behavior can be obtained by a controlled design and manufacture of polymeric composite materials.

The application of shape memory materials in 3D printing was demonstrated by printing a folding cube made of shape memory hinges combined with plates of inert matrix material (Figure 9.33). This methodology opens the door to the development of stimuli-responsive devices that can actively influence a range of processes, including chemical reactors and reactions, membrane separations, drug delivery, and so on.

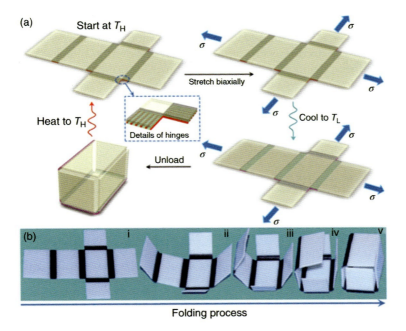

Figure 9.33 A self-folding and opening box fabricated by printing polymeric active composites as hinges connecting plates of an inert matrix. (a) Schematic of thermomechanical protocol to achieve the self-folding and opening box. (b) Photographs corresponding to the folding process from a flat to a closed box (i–vi). (Reprinted with permission from Ref. [46]. Copyright 2013, AIP Publishing.)

9.6 Conclusions

Additive manufacturing techniques based on a layer-by-layer deposition by 3D printing have the potential to revolutionize manufacturing by enabling the production of personalized designs that could not be possible by any other techniques. Furthermore, this can be achieved in many cases without the specialized equipment previously required in traditional manufacturing. The impact of this technique in research and scientific environments is apparent by the increasing number of publications.

The potential of this technique to support chemical research and manufacturing has been very recently realized. The untapped potential is being rapidly explored and developed. Indeed, this technique enables the facile integration of the design of the chemical reaction and the reactor media where the transformations are set to occur, thus leading to a novel concept called "reactionware." This is of particular relevance to develop continuous flow reactors with tailored features to the reaction studied. Applications in synthesis, catalysis, and analytics will be developed based on this concept.

New trends that will impact the future include the development of faster, more reliable techniques to manufacture devices that will facilitate the design of advanced reactor architectures with advanced features, such as membranes and baffles. Furthermore, the integration of stimuli-responsive materials will enable the next-generation devices to be able to adapt to changes in external factors, such as temperature or feedstock materials.

References

1 Dimitrov, D., Schreve, K., and de Beer, N. (2006) Advances in three dimensional printing – state of the art and future perspectives. *Rapid Prototyping Journal*, **12**, 136–147.
2 Symes, M.D., Kitson, P.J., Yan, J., Richmond, C.J., Cooper, G.J.T., Bowman, R.W., Vilbrandt, T., and Cronin, L. (2012) Integrated 3D-printed reactionware for chemical synthesis and analysis. *Nature Chemistry*, **4**, 349–354.
3 Geissler, M. and Xia, Y. (2004) Patterning: principles and some new developments. *Advanced Materials*, **16**, 1249–1269.
4 (a) Ahn, B.Y., Duoss, E.B., Motala, M.J., Guo, X., Park, S.-I., Xiong, Y., Yoon, J., Nuzzo, R.G., Rogers, J.A., and Lewis, J.A. (2009) Omnidirectional printing of flexible, stretchable, and spanning silver microelectrodes. *Science*, **323**, 1590–1593; (b) Stampfl, J. and Liska, R. (2005) New materials for rapid prototyping applications. *Macromolecular Chemistry and Physics*, **206**, 1253–1256.
5 (a) Connell, J.L., Ritschdorff, E.T., Whiteley, M., and Shear, J.B. (2013) 3D printing of microscopic bacterial communities. *Proceedings of the National Academy of Sciences*, **110**, 18380–18385; (b) Mironov, V., Boland, T., Trusk, T., Forgacs, G., and Markwald, R.R. (2003) Organ printing: computer-aided jet-based 3D tissue engineering. *Trends in Biotechnology*, **21**, 157–161; (c) Peltola, S.M., Melchels, F.P.W., Grijpma, D.W., and Kellomäki, M. (2008) A review of rapid prototyping techniques for tissue engineering purposes. *Annals of Medicine*, **40**, 268–280; (d) Seitz, H., Rieder, W., Irsen, S., Leukers, B., and Tille, C. (2005) Three-dimensional printing of porous ceramic scaffolds for bone tissue engineering. *Journal of Biomedical Materials Research Part B: Applied Biomaterials*, **74**, 782–788.
6 Hribar, K.C., Soman, P., Warner, J., Chung, P., and Chen, S. (2014) Light-assisted direct-write of 3D functional biomaterials. *Lab on a Chip*, **14**, 268–275.
7 (a) Pataky, K., Braschler, T., Negro, A., Renaud, P., Lutolf, M.P., and Brugger, J. (2012) Microdrop printing of hydrogel bioinks into 3D tissue-like geometries. *Advanced Materials*, **24**, 391–396; (b) Therriault, D., White, S.R. and Lewis, J.A. (2003) Chaotic mixing in three-dimensional microvascular networks fabricated by direct-write assembly. *Nature Materials*, **2**, 265–271.
8 (a) Anderson, K.B., Lockwood, S.Y., Martin, R.S., and Spence, D.M. (2013) A 3D printed fluidic device that enables integrated features. *Analytical Chemistry*, **85**, 5622–5626; (b) Erkal, J.L., Selimovic, A., Gross, B.C., Lockwood, S.Y., Walton, E.L., McNamara, S., Martin, R.S., and Spence, D.M. (2014) 3D printed microfluidic devices with integrated versatile and reusable electrodes. *Lab on a Chip*, **14**, 2023–2032; (c) Krejcova, L., Nejdl, L., Rodrigo, M.A.M., Zurek, M., Matousek, M., Hynek, D., Zitka, O., Kopel, P., Adam, V., and Kizek, R. (2014) 3D printed chip for electrochemical detection of influenza virus labeled with CdS quantum dots. *Biosensors and Bioelectronics*, **54**, 421–427; (d) Shallan, A.I., Smejkal, P., Corban, M., Guijt, R.M., and Breadmore, M.C. (2014) Cost-effective three-dimensional printing of visibly transparent microchips within minutes. *Analytical Chemistry*, **86**, 3124–3130.

9 (a) Ilievski, F., Mazzeo, A.D., Shepherd, R.F., Chen, X., and Whitesides, G.M. (2011) Soft robotics for chemists. *Angewandte Chemie, International Edition*, **50**, 1890–1895; (b) Morin, S.A., Shepherd, R.F., Kwok, S.W., Stokes, A.A., Nemiroski, A., and Whitesides, G.M. (2012) Camouflage and display for soft machines. *Science*, **337**, 828–832.

10 Capel, A.J., Edmondson, S., Christie, S.D.R., Goodridge, R.D., Bibb, R.J., and Thurstans, M. (2013) Design and additive manufacture for flow chemistry. *Lab on a Chip*, **13**, 4583–4590.

11 Causier, A., Carret, G., Boutin, C., Berthelot, T., and Berthault, P. (2015) 3D-printed system optimizing dissolution of hyperpolarized gaseous species for micro-sized NMR. *Lab on a Chip*, **15**, 2049–2054.

12 Kitson, P.J., Symes, M.D., Dragone, V., and Cronin, L. (2013) Combining 3D printing and liquid handling to produce user-friendly reactionware for chemical synthesis and purification. *Chemical Science*, **4**, 3099–3103.

13 Ayers, K.E., Capuano, C.B., and Anderson, E.B. (2012) Recent advances in cell cost and efficiency for PEM-based water electrolysis. *ECS Transactions*, **41**, 15–22.

14 Chisholm, G., Kitson, P.J., Kirkaldy, N.D., Bloor, L.G., and Cronin, L. (2014) 3D printed flow plates for the electrolysis of water: an economic and adaptable approach to device manufacture. *Energy & Environmental Science*, **7**, 3026–3032.

15 Carmo, M., Fritz, D.L., Mergel, J., and Stolten, D. (2013) A comprehensive review on PEM water electrolysis. *International Journal of Hydrogen Energy*, **38**, 4901–4934.

16 Whitesides, G.M. (2006) The origins and the future of microfluidics. *Nature*, **442**, 368–373.

17 Jahnisch, K., Hessel, V., Lowe, H., and Baerns, M. (2004) Chemistry in microstructured reactors. *Angewandte Chemie, International Edition*, **43**, 406–446.

18 McDonald, J.C., Duffy, D.C., Anderson, J.R., Chiu, D.T., Wu, H.K., Schueller, O.J.A., and Whitesides, G.M. (2000) Fabrication of microfluidic systems in poly (dimethylsiloxane). *Electrophoresis*, **21**, 27–40.

19 Lee, J.N., Park, C., and Whitesides, G.M. (2003) Solvent compatibility of poly (dimethylsiloxane)-based microfluidic devices. *Analytical Chemistry*, **75**, 6544–6554.

20 Krishna, K.S., Navin, C.V., Biswas, S., Singh, V., Ham, K., Bovenkamp, G.L., Theegala, C.S., Miller, J.T., Spivey, J.J., and Kumar, C. (2013) Millifluidics for time-resolved mapping of the growth of gold nanostructures. *Journal of the American Chemical Society*, **135**, 5450–5456.

21 (a) Peterson, G.I., Larsen, M.B., Ganter, M.A., Storti, D.W., and Boydston, A.J. (2015) 3D-printed mechanochromic materials. *ACS Applied Materials & Interfaces*, **7**, 577–583; (b) Wang, X., Guo, Q., Cai, X., Zhou, S., Kobe, B., and Yang, J. (2014) Initiator-integrated 3D printing enables the formation of complex metallic architectures. *ACS Applied Materials & Interfaces*, **6**, 2583–2587.

22 Giordano, R.A., Wu, B.M., Borland, S.W., Cima, L.G., Sachs, E.M., and Cima, M.J. (1997) Mechanical properties of dense polylactic acid structures fabricated by three dimensional printing. *Journal of Biomaterials Science, Polymer Edition*, **8**, 63–75.

23 Novosel, E.C., Meyer, W., Klechowitz, N., Krüger, H., Wegener, M., Walles, H., Tovar, G.E.M., Hirth, T., and Kluger, P.J. (2011) Evaluation of cell-material interactions on newly designed, printable polymers for tissue engineering applications. *Advanced Engineering Materials*, **13**, B467–B475.

24 Wiles, C. and Watts, P. (2012) Continuous flow reactors: a perspective. *Green Chemistry*, **14**, 38–54.

25 Hartman, R.L., McMullen, J.P., and Jensen, K.F. (2011) Deciding whether to go with the flow: evaluating the merits of flow reactors for synthesis. *Angewandte Chemie, International Edition*, **50**, 7502–7519.

26 Kitson, P.J., Rosnes, M.H., Sans, V., Dragone, V., and Cronin, L. (2012) Configurable 3D-printed millifluidic and microfluidic 'lab on a chip' reactionware devices. *Lab on a Chip*, **12**, 3267–3271.

27 (a) Dragone, V., Sans, V., Rosnes, M.H., Kitson, P.J., and Cronin, L. (2013) 3D-printed devices for continuous-flow organic chemistry. *Beilstein Journal of Organic Chemistry*, **9**, 951–959; (b) Mathieson, J.S., Rosnes, M.H., Sans, V., Kitson, P.J., and Cronin, L. (2013) Continuous parallel ESI-MS analysis of reactions carried out in a bespoke 3D printed device. *Beilstein Journal of Nanotechnology*, **4**, 285–291.

28 (a) Harvey, A.P., Mackley, M.R., and Seliger, T. (2003) Process intensification of biodiesel production using a continuous oscillatory flow reactor. *Journal of Chemical Technology and Biotechnology*, **78**, 338–341; (b) Ni, X., Gao, S., Cumming, R.H., and Pritchard, D.W. (1995) A comparative-study of mass-transfer in yeast for a batch pulsed baffled bioreactor and a stirred-tank fermenter. *Chemical Engineering Science*, **50**, 2127–2136.

29 (a) Okafor, O., Goodridge, R. and Sans, V. (2017) New trends in reactor engineering with additive manufacturing. *Chimica Oggi-Chemistry Today*, **35**, 4–6; (b) Okafor, O., Weilhard, A., Fernandes, J.A., Karjalainen, E., Goodridge, R., and Sans, V. (2017) Advanced reactor engineering with 3D printing for the continuous-flow synthesis of silver nanoparticles. *Reaction Chemistry & Engineering*, **2**, 129–136.

30 Peris, E., Okafor, O., Kulcinskaja, E., Goodridge, R., Luis, S.V., Garcia-Verdugo, E., O'Reilly, E., and Sans, V. (2017) Tuneable 3D printed bioreactors for transaminations under continuous-flow. *Green Chemistry*. doi: 10.1039/C7GC02421E

31 Jähnisch, K., Hessel, V., Löwe, H., and Baerns, M. (2004) Chemistry in microstructured reactors. *Angewandte Chemie, International Edition*, **43**, 406–446.

32 Cath, T.Y., Childress, A.E., and Elimelech, M. (2006) Forward osmosis: principles, applications, and recent developments. *Journal of Membrane Science*, **281**, 70–87.

33 Powell, C.E. and Qiao, G.G. (2006) Polymeric CO_2/N_2 gas separation membranes for the capture of carbon dioxide from power plant flue gases. *Journal of Membrane Science*, **279**, 1–49.

34 Dittmeyer, R., Hollein, V., and Daub, K. (2001) Membrane reactors for hydrogenation and dehydrogenation processes based on supported palladium. *Journal of Molecular Catalysis A, Chemical*, **173**, 135–184.

35 Femmer, T., Kuehne, A.J.C., and Wessling, M. (2014) Print your own membrane: direct rapid prototyping of polydimethylsiloxane. *Lab on a Chip*, **14**, 2610–2613.

36 Jani, J.M., Wessling, M., and Lammertink, R.G.H. (2011) Geometrical influence on mixing in helical porous membrane microcontactors. *Journal of Membrane Science*, **378**, 351–358.

37 Blume, I., Schwering, P.J.F., Mulder, M.H.V., and Smolders, C.A. (1991) Vapour sorption and permeation properties of poly (dimethylsiloxane) films. *Journal of Membrane Science*, **61**, 85–97.

38 Au, A.K., Bhattacharjee, N., Horowitz, L.F., Chang, T.C., and Folch, A. (2015) 3D-printed microfluidic automation. *Lab on a Chip*, **15**, 1934–1941.

39 Tsuda, S., Jaffery, H., Doran, D., Hezwani, M., Robbins, P.J., Yoshida, M., and Cronin, L. (2015) Customizable 3D printed 'plug and play' millifluidic devices for programmable fluidics. *PLoS One*, **10**, e0141640.

40 Bhargava, K.C., Thompson, B., and Malmstadt, N. (2014) Discrete elements for 3D microfluidics. *Proceedings of the National Academy of Sciences of the United States of America*, **111**, 15013–15018.

41 Lee, K.G., Park, K.J., Seok, S., Shin, S., Kim, D.H., Park, J.Y., Heo, Y.S., Lee, S.J., and Lee, T.J. (2014) 3D printed modules for integrated microfluidic devices. *RSC Advances*, **4**, 32876–32880.

42 Lee, K.G., Wi, R., Park, T.J., Yoon, S.H., Lee, J., Lee, S.J., and Kim, D.H. (2010) Synthesis and characterization of gold-deposited red, green and blue fluorescent silica nanoparticles for biosensor application. *Chemical Communications*, **46**, 6374–6376.

43 Lee, M.P., Cooper, G.J.T., Hinkley, T., Gibson, G.M., Padgett, M.J., and Cronin, L. (2015) Development of a 3D printer using scanning projection

stereolithography. *Scientific Reports*, **5**, 9875.

44 Tumbleston, J.R., Shirvanyants, D., Ermoshkin, N., Janusziewicz, R., Johnson, A.R., Kelly, D., Chen, K., Pinschmidt, R., Rolland, J.P., Ermoshkin, A., Samulskiand, E.T., and DeSimone, J.M. (2015) Continuous liquid interface production of 3D objects. *Science*, **347**, 1349–1352.

45 (a) Khare, V., Sonkaria, S., Lee, G.Y., Ahnand, S.N., and Chu, W.S. (2017) From 3D to 4D printing – design, material and fabrication for multi-functional multi-materials. *International Journal of Precision Engineering and Manufacturing-Green Technology*, **4**, 291–299; (b) Tibbits, S. (2012) Design to self-assembly. *Architectural Design*, **82**, 68–73; (c) Tibbits, S. and Cheung, K. (2012) Programmable materials for architectural assembly and automation. *Assembly Automation*, **32**, 216–225.

46 Ge, Q., Qi, H.J., and Dunn, M.L. (2013) Active materials by four-dimension printing. *Applied Physics Letters*, **103**, 131901.

**Part Three
Enabling Technologies**

used to analyze this product analyze other products. The off-line laboratory is often justified in terms of human resources as it allows chemists to contribute to the analysis of several manufacturing processes, and it facilitates the contributions from the more experienced chemists. Unfortunately, bench chemists in the centralized off-line laboratory are frequently focused only on performing the analyses, and are unfamiliar with the manufacturing process, which limits their possible contributions. Furthermore, the manufacturing process is usually completed by the time when the results of the analyses are reported. The results are then used to determine whether products meet the established specifications. Thus, a number of industries have implemented at-line process laboratories, where dedicated instruments are installed in close proximity to a process line. The advantages of the at-line laboratory include a reduction in analysis time as the instruments are already set for the incoming samples, and chemists more closely connected with the manufacturing process. The at-line laboratory is often part of "a manufacturing cell" with personnel focused on a specific process. The at-line laboratory could be cost effective for high volume products with constant demand, but is not cost effective when manufacturing demands vary significantly. Even though the at-line laboratory is much faster, the results provided will not describe the current state of the process. The off-line and at-line instrument setups have been challenged by a field first known as PAC that conceived the idea of using analytical instruments to monitor a process, providing this information to engineers, and developing multidisciplinary teams to control and optimize a process [2].

Kowalski and collaborators at the Center for Process Analytical Chemistry provided the vision that process analytical chemistry should be considered a "worthy sub-discipline of analytical chemistry, and it requires an interdisciplinary systems approach so that progress can be made" [2]. The goals of PAC are described in the left side of Table 10.1. This description emphasizes the role of the analytical chemist providing quantitative and qualitative information about chemical processes. The second sentence regarding PAC emphasizes the use of information to monitor and control a process and to optimize its efficient use of energy, time, and raw materials. The third sentence visualizes environmental benefits (simultaneously minimize plant effluent release) at the same time that product quality and consistency was improved. The vision of a new industrial scheme with measurement systems at the process to enhance productivity, quality, and environmental impact instead of waiting for samples to reach the laboratory was an innovation [3]. This vision developed more than 30 years ago and defined a new subdiscipline of analytical chemistry to achieve environmental and product quality benefits.

The information to control a process requires measurements performed through online or in-line analysis. The measurement system should be five times faster than the fastest significant change that may occur in the process. Thus, if a significant change in concentration occurs every minute, the system should be capable of providing results every 12 s. This is a "rule of thumb" or guideline developed to assure that the analytical system is capable of detecting significant

Table 10.1 Comparison of definitions of process analytical chemistry (PAC) and process analytical technology (PAT).

Process analytical chemistry – PAC (1987)	Process analytical technology – PAT (2004)
"The goal of process analytical chemistry is to supply quantitative and qualitative information about a chemical process. Such information can be used not only to monitor and control a process, but also to optimize its efficient use of energy, time and raw materials. In addition, it is possible to simultaneously minimize plant effluent release and improve product quality and consistency."	"The Agency considers PAT to be a system for designing, analyzing, and controlling manufacturing through time measurements (i.e. during processing) of critical quality and performance attributes of raw and in-process materials and processes, with the goal of ensuring final product quality. It is important to note that the analytical in PAT is viewed broadly to include chemical, physical, microbiological, mathematical, and risk analysis conducted in an integrated manner. The goal of PAT is to enhance understanding and control the manufacturing process, which is consistent with our current drug quality system: *quality cannot be tested into products; it should **be** built-in or should be by design.*"

process changes. *Online* analysis uses an automated sampling system to remove a sample from the process and present it to an analytical instrument for measurement. The analytical instrument could be located in an enclosure necessary for process safety, to avoid vibrations or high temperatures that may affect instrument performance. The analytical instrument may also require some sample conditioning, before the measurement is performed [4].

In-line analyses consist of *in situ* chemical analysis performed within the process, using a probe that is chemically sensitive. The *in-line* system is able to monitor the process as it progresses. This may involve *noninvasive* analyses where the probe does not physically contact the sample, or the probe may be immersed in the reactor or process line [5]. The *in-line* methods analyze the sample without changing the physical or chemical properties. Thus, reagents are not required for their analysis. In some cases, the methods are cross sensitive where an analytical method provides more than one type of information [6]. Some of the techniques, such as near-infrared spectroscopy (NIRS), provide information on both the chemical and physical properties of the material [7]. There are also applications where more than one sensor was used [8]. Many *in-line* instruments are designed without moving parts to facilitate its operation close to the manufacturing areas where conditions are not as controlled as in the analytical laboratory. *In-line* instruments may also be smaller than its laboratory designed counterparts to avoid cluttering the manufacturing area.

Most analytical instruments used in the *off-line* and *at-line* laboratories are not suitable for PAC as they require extensive sample preparation and analysis time. New methods have been developed with analytical techniques such as NIR and Raman spectroscopy that do not require sample preparation. These new nondestructive methods can now be performed in much shorter time. The nondestructive methods require multivariate methods to extract the desired information in methods that are called chemometrics [9]. The process may be monitored by several analyzers and the integration and interaction between these actuators and statistical models for a unit operation are the basis of supervisory control and data acquisition systems (SCADA) [10]. The analyzers generate considerable data that must be captured and stored in a specific manner to adequately record the context of the information [11]. PAC also requires significant training of plant personnel to adapt to the new analytical methods and to the generation of analytical measurements in the manufacturing site instead of the manufacturing floor [12]. There are also important sampling issues as the goals described in Table 10.1 can only be achieved if representative samples are obtained [13]. PAC moves chemists to the manufacturing area in contrast to the traditional *off-line* and *at-line* laboratories that require transportation of the sample to a remote laboratory. Thus, PAC is a subdiscipline that seeks to integrate the work of analytical chemists with process control engineers.

PAC has impacted industrial processes in a number of companies. Thus, PAC is not just a field of analytical chemistry but also a reality in the processing of petroleum derivatives, where gasoline and diesel oil are obtained by obtaining mixtures of different process streams. The information obtained from NIR spectrometers are used by control systems in the real-time optimization (RTO) in the production of petroleum derivatives and transportation fuels [14]. PAC is also important in the polymer industry where raw materials are identified using nondestructive NIR spectroscopy avoiding traditional wet chemistry analyses [15]. The PAC approach is also used in monitoring reactions in gas processing steps in polymer production [16].

The term process analytical technology (PAT) has become widely used since 2004. Table 10.1 also describes the definition of PAT [17]. The two definitions are very similar. However, when the PAT guidance was developed the intent was to encourage the pharmaceutical industry to design, monitor, and control processes to improve pharmaceutical processes. It was recognized that this required a systems approach, including the engineering control and supervising systems, and using the term "process analytical chemistry" could imply that only the chemistry was of interest [18]. The Food and Drug Administration (FDA) developed a PAT guidance to describe the Agency's current thinking, and to provide a regulatory framework for voluntary implementation of innovative pharmaceutical development, manufacturing, and quality assurance [17]. The guidance emphasized that it was written for a broad industrial audience from different disciplines and organizational units. Thus, the term PAT was used to emphasize a systems approach.

The PAT definition is focused on the goal of enhancing the understanding and control of the pharmaceutical manufacturing process, consistent with a quality system based on the principle that *quality cannot be tested into products; it should be built-in or should be by design*. The PAT approach is being applied to the manufacturing of small [19] and large molecule active pharmaceutical ingredients (APIs) [20]. This chapter retains the original PAC name even though the PAT term has been widely adopted in recent years [21]. The term PAT is reserved within this chapter for those applications that will be submitted to the FDA for evaluation.

In summary, PAC is an entire subdiscipline of analytical chemistry and not simply moving instruments into a manufacturing plant. PAC concepts challenge the status quo of industrial chemical processes and the way in which chemists and engineers interact. In PAC, the chemist is no longer waiting in the *off-line* analytical laboratory for the sample to arrive, which sounds different, maybe challenging, but it is not different to other disciplines that have already adopted this concept. For example, meteorologists do not rely on manual reading from weather stations to predict future weather conditions as it has been in the past. Nowadays, they rely on many automated *in situ* sensors and meteorologists do not wait for hurricane or tornado samples to be brought to an analytical laboratory. PAC has sparked many opportunities for process improvement to support green chemistry and quality improvement initiatives, but has also challenged companies and professionals to make many changes to obtain the desired benefits.

10.3
Vibrational Spectroscopy

Mid-infrared, NIR, and Raman spectroscopy are different in several aspects as shown in Table 10.2 [22]. However, they are all examples of vibrational spectroscopy and can used in-line, online, at-line, or off-line; in all the modes of analysis shown in Figure 10.1. Each technique has specific advantages and disadvantages;

Table 10.2 Comparison of most common vibrational spectroscopic methods.

	Mid-infrared	Near-infrared	Raman
Sample preparation	Required, except for ATR	Not required	Not required
Spectral interpretation (band assignment)	Frequently used for structure elucidation	Quite complex, rarely used for structure elucidation	Frequently used for structure elucidation
Band intensity	Change in dipole moment	Change in dipole moment and bond anharmonicity	Change in polarizability
Response to water	Strong	Strong	Weak
Deployment to production area	Possible with ATR, gas cells	Frequent (diffuse reflectance, liquid)	Frequent (immersion, and noncontact probes)

Source: Adapted from Ref. [22].

the best method depends entirely on the situation where analytical information is required. NIR and Raman can always provide spectra of samples without sample preparation – without the need to dissolve the sample, reduce their particle size, or extracting the analyte from the matrix of interest and then separating or purifying. NIR and Raman spectra can be obtained for materials in their native state eliminating the use of solvents and eliminating the time spent on sample preparation. Mid-IR spectra can be obtained without sample preparation when attenuated total reflectance (ATR) is used or with samples in the gas phase [23]. ATR methods have replaced the preparation of potassium bromide (KBr) pellets in many industries. KBr pellets require that the analyte be mixed with dry KBr and then compressed at approximately 20 000 pounds of pressure to make pellets [24]. The analyte of interest should be prepared at a concentration of 1–2% (w/w) within the KBr matrix. With the exception of ATR, most mid-IR methods will require some degree of sample handling or work in the laboratory.

The bands observed in mid-IR and Raman spectra may be assigned to specific functional groups and can be used to determine the structure of a known compound. Both are information-rich techniques and a number of books are available describing the information that may be obtained from spectra [25–27]. Mid-IR spectra result from the absorption of radiation that occurs when the frequency of light matches the frequency of vibration of a bond. Thus, mid-IR spectrometers include polychromatic light sources. Raman spectrometers include a monochromatic laser source that distorts (polarizes) the electron cloud surrounding the nuclei to form a short-lived state called a "virtual state" [26]. This virtual state is at a higher energy than the vibrational states and is not stable and thus the photon is quickly reradiated. Only one in approximately 10^6 photons undergoes Raman scattering; most of the photons are elastically scattered. Recent advances in optics (especially notch-filters) and instrumentation have made Raman spectroscopy practical, in spite of the fact that it is a weak effect. Raman spectra will show bands at the same frequency as the mid-IR spectra, but the intensity of the bands in the two spectra will be different as shown in the heptane spectra shown in Figure 10.2. Mid-IR spectra will vary according to the change in dipole moment that occurs in the molecular vibration, while the intensity of the Raman bands will vary according to the polarizability of the bond. Symmetrical groups such as —C=C— will show strong Raman bands, but weak mid-IR bands because of their low dipole moments. The —C=O group provides very strong bands in mid-IR spectra but these are considerably weaker in Raman spectra. The differences in intensities between the mid-IR and Raman spectra are not a disadvantage, they are very helpful to spectroscopists specialized in using this information to elucidate the structure of compounds.

Mid-IR and NIR spectra provide very strong bands for —O—H groups making it difficult to obtain spectra of compounds dissolved in water. The weak Raman scattering for O—H groups becomes an advantage as it is now possible to obtain spectra of compounds dissolved in water. Thus, Raman spectroscopy may be used in monitoring the crystallization of active pharmaceutical ingredients [28] and proteins in aqueous media [29]. Raman spectroscopy is helpful in detecting

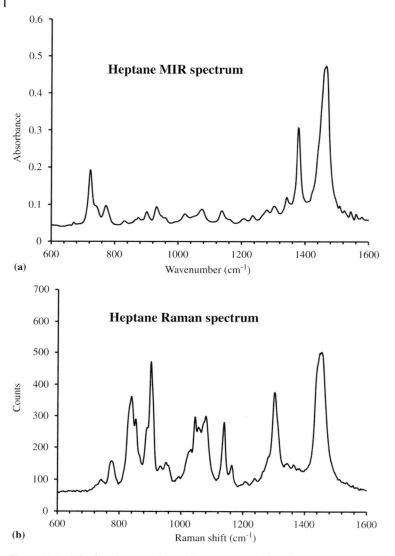

Figure 10.2 Mid-infrared spectra (a) and Raman spectra (b) in the spectral region from 600 to 1600 cm^{-1}.

the onset of crystallization and also monitoring the polymorphic form of the crystalline material obtained. The probe within a reactor, shown in Figure 10.1, could be a Raman immersion probe [5]. There are multiple applications where Raman has been used in this manner to monitor a chemical reaction [30] or monitor a crystallization process [8].

Figure 10.3 shows the Raman and NIR spectra for acetaminophen (APAP), displaying the differences between the two methods. The Raman spectrum

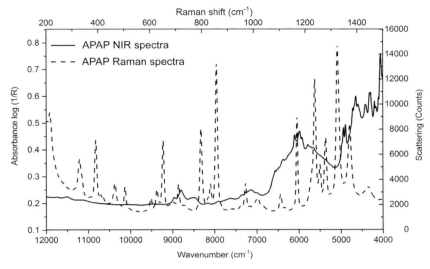

Figure 10.3 Near-infrared spectra of acetaminophen (wide bands) and Raman spectra (narrow) bands.

shows sharp bands, while the NIR shows broad overlapping bands. The Raman bands may be assigned to the vibrations of functional groups, while the NIR are not easily assigned to the vibrations of functional groups. NIR is not generally used to elucidate the structure of a compound, while Raman spectroscopy is widely applied for this purpose. NIR could also be used as an immersion probe as shown in Figure 10.1 [8]. The methods, NIR, and Raman spectroscopy, are widely used because they do not require sample preparation, spectra may be obtained without altering the physical properties of the material [22].

The first overtone of the O—H band (6600–6100 cm^{-1}) in the acetaminophen spectrum in Figure 10.3 is broad and overlapping. The broadening of this band is indicative of the hydrogen bonding in the molecule. In spite of the overlapping NIR bands and the difficulty in their assignment to functional groups, NIR spectroscopy is widely used in process analysis since it does not require sample preparation and is able to provide information on the chemical composition and physical properties of samples [31]. NIR spectra are the result of overlapping overtones and combination bands of fundamental vibrations observed in the mid-IR region. The spectrum of chloroform is an excellent example of the great challenge involved in the assignment of NIR bands. Chloroform, a molecule with only one C—H bond and six fundamental vibrational bands shows over 30 overtone and combination bands in the NIR region [32]. The wide overlapping bands provide spectra that show subtle differences for compounds with similar structures [31] and NIR spectra have been used to identify incoming raw material in pharmaceutical manufacturing facilities [33]. The intensity of NIR bands depends on the change in dipole moment that occurs in a vibration, and the anharmonicity of the transition. The combination of these two effects makes

NIR spectroscopy an excellent technique for detecting and investigating differences in hydrogen bonding [34].

O—H bands are among the most intense in the NIR spectrum. The intensity of the O—H bands decrease as drying progresses, and NIR has been shown to be highly effective in monitoring drying processes [35]. Drying processes are often based on fixed drying times, which could be unnecessarily long. The real-time monitoring of the process means that drying may be performed until the material reaches a desired level of moisture and not just to meet a specific drying time [36]. This approach may lead to considerable savings and adequate utilization of energy resources, and in some cases could avoid unnecessary product degradation. Real time monitoring is also useful to understand the stepwise loss of water throughout the drying process. Thus, process understanding is also increased.

10.4
Challenges to Overcome

The adequate sampling of the manufacturing process remains one of the principal challenges [37]. The first indication of this challenge stems from differences in the definition of sampling. Statisticians consider samples as group of objects selected from the population of all such objects, and are usually very concerned about the number of samples [38]. The theory of sampling (TOS) has evolved from scientists who have recognized the importance of how samples are collected. TOS defines samples as "mass reduction of lot L by selection of a certain subset of units, with the purpose – not always fulfilled – of obtaining a true, reliable sample S (when the conditions of sampling correctness are respected)" [39]. The conditions of sampling correctness are that all parts of the lot must have the same probability of being analyzed and the structure of the material analyzed should not be affected when the sample is obtained.

Sampling is the basis of all quality control. Brittain wrote three tutorial articles describing the practices for the determination of particle size of pharmaceutical materials and stated, "Samples are therefore defined as the units upon which a program of testing is conducted" [40]. The results of the most elaborate and costly instruments in a lab depend on the samples obtained [41]. However, most analytical chemistry and instrumental analysis textbooks hardly discuss sampling. The recent adoption of PAT methods in the pharma industry has sparked an important discussion on sampling issues [13]. Even though in-line methods do not remove a material from a reactor or industrial process, they do sample the material. The in-line spectrometer or instrument sample and analyze a reduced mass of Lot L at a given time. This is an "optical" sampling or non-destructive sampling step [42].

One of the tenets of TOS is that materials are always heterogeneous [39]. The spectroscopic methods used in process analytical chemistry analyze a very small fraction of a lot. For example, NIR radiation can penetrate up to 5 mm but the radiation that returns to a detector is mostly from the top 1 mm. A powder of liquid could be flowing and be 30 mm thick, but only the top 1 mm is being analyzed by the NIR spectrometer. The material next to the NIR spectrometer might not be representative of the material that is not interacting with the NIR radiation and is not being sampled.

Sampling issues also affect the development of calibration models. Calibration models consist of mathematical relationships based on the changes in spectra (NIR, Raman, or mid-IR) as the concentration or property of interest is varied [9]. Calibration models are widely used to determine drug concentration in pharmaceutical mixing processes [43, 44], and many other industrial processes. Once validated, the models are used to predict the desired quality attribute: drug concentration, moisture content, and so on. While developing calibration models, scientists often remove a sample from the process and send it for analysis to an off-line laboratory where an established reference method is used. The calibration model is then developed with the result received from the off-line laboratory and the spectra obtained with the spectrometer installed in-line. Unfortunately, the samples analyzed by the in-line and off-line methods are often not the same and this systematic error is usually the principal source of error in calibration models [45]. This systematic error was overcome in the determination of moisture in a pharmaceutical drying process by developing an interface that captured the granulated material prior to the acquisition of spectra [35]. The interface was later removed to send the granulated product to the reference laboratory for analysis. This approach assured that the in-line and off-line samples were similar.

The implementation of in-line or online sampling does not guarantee that representative is obtained [13]. The material adjacent to the probe shown in Figure 10.1 could be different from the material in other sections of the reactor, especially in process that is changing, such as a chemical reaction or drying process. Thus, composite sampling is necessary to obtain a representative sample. Representativeness requires minimizing the systematic error (bias) in sampling and also the random error (variance) of the sampling process. The random error is reduced by averaging the composition of a large number of increments, but not the systematic error. The reduction of systematic error in sampling requires that all potential increments of the lot have the same probability of being selected for analysis and that the composition of the lot is not altered by the sampling process.

Composite sampling is possible with spectroscopy. A number of spectra could be averaged and considered a composite sample. The reactor shown in Figure 10.1 could be stirred while NIR or Raman spectra are obtained [46]. In one application a better calibration model was obtained when the powder analyzed moved while the NIR spectra were collected [47]. Powder movement could be considered a factor adding complexity to the application. However, in this case it helped in providing a more representative composite sample. Process sampling has been called the "missing link" to understand the measurement errors observed when monitoring manufacturing process through PAC [13].

There are many opportunities for improving sampling in PAC and sampling is becoming an increasingly important area of research [41].

The successful implementation of PAC systems requires a network of collaborating units that often spans from centralized research labs to the manufacturing area [37]. Centralized research labs might propose a project based on process knowledge, and a vision of the important information that may be obtained from the PAC system [48]. A feasibility study might have indicated significant benefits from the proposed project. However, the technology and knowledge now has to be transferred to the manufacturing area. The project will require significant commitment from the manufacturing unit and project management to make it a reality. Engineering units will need to provide installation designs to guarantee safe operation and maintenance of the system. Plant personnel will need to understand the system, and training procedures are often developed. The data obtained must be properly collected and considered within the plants' quality system [49]. The information obtained could be very useful for continuous improvement [37], but resistance to change is frequent in many companies [50]. The financial value of continuous improvement must be clearly communicated by industrial leaders.

10.5
Applications of Process Analytical Chemistry and Nondestructive Analyses

10.5.1
Dairy Industry

The determination of moisture content in powdered infant milk is a good example of the need to understand the effects of moisture on product stability [51]. Several NIR instruments have been used to analyze powdered milk with measurements for samples brought to the laboratory and also during processing [52]. The development of nondestructive NIR methods for powdered milk started around 1990 and has been implemented by a number of companies [52]. NIR spectroscopy is now becoming a widely accepted method to monitor moisture in the dairy industry [51]. A new multiprobe NIR system was recently used to monitor the drying of powdered infant milk [51]. In its initial spray drying stage, a moisture content of 12–20% (w/w) is expected. The moisture content is then reduced to 8–10% (w/w) with fluid bed drying, and in a third stage to below 5% (w/w). Milk powders are usually controlled to a moisture content between 2 and 4.5% (w/w). NIR methods have been developed to control moisture in the food industry where a recent publication describes the use of NIR spectroscopy to control moisture in powdered infant milk [51]. In this case the control of moisture is critical for the storage and handling of this product. The excess of moisture would accelerate the conversion of anhydrous lactose to α-lactose monohydrate and cause sticking and caking problems. With NIR spectroscopy multiple spectra can be obtained providing an assessment of the variation of moisture throughout the lot. The use of conventional methods for moisture determination such as Karl Fischer titration or loss on drying would require a limited number of samples. This limited number of samples might not avoid sticking or caking in the product purchased by a customer.

NIR spectroscopy is not limited to powdered milk, it is widely used in the analysis of other dairy products [53]. These applications include measurements of the fat, protein, and moisture in cheese [53]. Milk is also analyzed for water content, fat, protein, lactose, and total solids. The prediction of sensory attributes such as creaminess, rancidity, pungency, hardness, chewiness, and buttery flavors is also possible. These sensory attributes cannot be attributed to a single band from a NIR spectrum and are determined with methods that relate changes in the NIR spectra to the evaluation provided by a panel of experts [53]. The development of a similar method based on destructive conventional laboratory analyses would be much more difficult, as these methods usually determine only one property of the sample. The NIR multivariate method captures these sensorial variations throughout the many spectral responses included in the model. The NIR method captures the changes in chemical composition, and also changes in the physical properties of the materials that are related to sensorial properties such as chewiness, creaminess, and hardness. If the cheese or dairy product were to be dissolved, the information related to the physical (sensorial) properties of the materials would be lost. Thus, the determination of these important sensorial properties is an advantage of the nondestructive methods discussed in this chapter.

10.5.2
Synthesis of Active Pharmaceutical Ingredients

Process analytical approaches are practically required in cases where the chemical species that must be analyzed is unstable. Thus, the material would be reactive during sample preparation for off-line chromatographic analysis and the composition would continue changing in the reactor. Off-line analysis is meaningless in these cases, and in-line/online analysis is the only suitable alternative. The unstable chemical species motivated the development of a process analytical method to monitor and control the formation of a key intermediate (lactam acetal) in the synthesis of an API [54]. Mid infrared spectroscopy (Mid-IR) with an attenuated total reflectance (ATR) accessory was used to monitor the synthesis.

The control of the lactam acetal synthesis is important to avoid formation of impurities and for environmental and economic reasons. Trifluoroacetic anhydride (TFAA) reacts with lactol in the first batch synthesis step to form the activated lactol. If the TFAA is overcharged, then it reacts with an expensive chiral alcohol used in the next step to produce unwanted process impurities. If the TFAA is undercharged then the residual lactol will react to form dimer impurities. However, thanks to mid-IR spectroscopy, the addition of TFAA (overcharged/undercharged) and the formation of activated lactol can be quantitatively monitored *in situ* without off-line sampling issues. The TFAA has strong carbonyl bands at 1875 and 1804 cm^{-1}. The carbonyl bands are observed when the TFAA is added, but rapidly disappear when it reacts with the lactol. The TFAA concentration was monitored throughout the process and the reaction end point was defined on the basis of a sustained slight excess of TFAA. TFAA was determined with a calibration model based on the response of the 1875 cm^{-1} peak area versus its concentration. The process analytical approach (calibration model based on the response of the 1875 cm^{-1} peak area versus its concentration) reduced the consumption of the expensive chiral alcohol and the need for additional processing steps to remove process impurities.

This method was not developed with equipment installed in an API plant. Instead an automated laboratory reaction monitoring system was used to develop the analytical method. This approach permitted method development without the interruption of plant operations. Several reaction monitoring systems with spectroscopic measurements are currently available. These systems may be used to develop PAC systems, and also play a major role in developing safe processes as some of the systems permit calorimetric measurements.

The manufacture of the anticancer drug Xeloda, for instance, is such a PAC application for a safer process control [55]. The API synthesis contains a reductive step at about 80 °C where a tosyloxy group is cleaved from a thermally labile tosylfuranoside in N-methylpyrrolidone (NMP) and triethylamine (TEA) by adding solid sodium borohydride. This reaction is accompanied with a heat-kick (temperature increase) and the intensive production of hydrogen gas, foaming, and precipitated tosylate salt. During the reaction unreacted sodium borohydride (up to 20%) can be accumulated, which causes a dangerous safety potential. Therefore, sodium borohydride is added in subsequent steps of small amounts and only when the reaction temperature has fallen below a predetermined security level.

Due to necessary improvements of the too hazardous and labor-intensive previous isolation step of tosylfuranoside, a new control procedure needed to be developed and tested. Since the improvement caused only small and unreliable, sometimes even nil, temperature increase, which were by then the main method for the safe process control, a mid-IR spectroscopic method was developed in the laboratory and later successfully implemented in the production. It could be proven that the reaction can be followed by watching the disappearance of unreacted borohydride peaks, the appearance of the borohydride − TEA complex, and the reduction of the tosylfuranoside, which led to a better process understanding and hence process control [55].

Many organic processes in pharmaceutical manufacturing remain problematic because of handling toxic, hazardous, or rapidly decaying molecules at large scales, which creates a background threat for the population and the environment due to their overly vicinity, as already described in the two examples. Therefore, recent developments in synthesis of APIs are toward faster, cleaner, and safer methods by flow chemistry, which influences the requirements of PAC probes significantly. While batch reactions, as the two examples above, are performed in relatively large-scale reactor tanks utilizing immersion probes for *in situ* measurements, flow chemistry processes are conducted in tubular microfluidic reactors requiring flow-based probes. The applications of such flow mid-IR spectroscopes have been successfully demonstrated by academic studies [56].

All organic compounds have functional groups providing characteristic absorption bands in the mid-IR region. Therefore, the technique is traditionally used for a wide range of applications ranging from the composition of gas or liquid mixtures in synthesis or suspension processes, such as crystallization. Moreover, mid-IR spectroscopy is much easier to calibrate compared to NIR spectroscopy because individual absorption bands can be monitored, often free from interferences, and does not require multivariate calibration [54]. Consequently, the application of mid-IR is associated with cost benefits, which might be a positive selling point compared to NIR when a decision for one or the other technique has to be made.

10.5.3
Preparation of Polymeric Strip Film Unit Dosage Forms

Raman spectroscopy, Raman imaging, and near-infrared chemical imaging (NIR-CI) have been used in support of a novel process for the preparation of polymeric strip films to improve the solubility of poorly soluble drugs [57]. These spectroscopic techniques were used to monitor the film preparation process and also to elucidate the structure of the polymeric strip films produced. Figure 10.4 describes the manufacturing process for the strip films starting with a particle size reduction process and steps to maintain a stable nanoparticle suspension. Raman spectroscopy was used to monitor the progress of particle size reduction during wet stirred media milling (WSMM) [58]. In this process an API suspension that contains a soluble adsorbing polymer and/or surfactant is formed. The WSMM method makes it possible to reduce particle size of the API to the nanometer size range as the particles are captured between the colliding beads [59]. Raman spectra were obtained in real time with an immersion probe inserted within the WSMM vessel, and each spectrum was the average of two 30 s long scans totaling 60 s of scanning time during milling operation. Raman spectra were obtained throughout the entire 64 min milling process. The intensity of the Raman bands decreased with milling time. The changes in Raman spectra were correlated with both the d_{50} (median) and d_{90} particle size that were measured off-line by a laser diffraction method. The correlation was developed for milled suspensions ranging from 0.1 to 6 μm for griseofulvin and 0.1 to 8 μm for

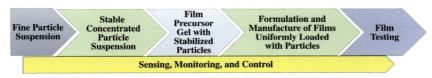

Figure 10.4 Processing steps involved in the preparation of novel polymeric strip films.

naproxen drug particles. The results obtained also suggest that particle breakage was faster initially (likely because coarser particles have higher breakage probability), but slowed down as the milling progressed [58]. Thus, the real-time method monitors particle size and also helps researchers to understand the mechanism of particle breakage.

The Raman spectra obtained during media milling were also used to evaluate whether the milling process affected drug crystallinity. Crystallinity changes were a concern due to the high impact energy of the milling beads. The Raman spectra acquired in real-time spectra were compared with spectra from a previous study where the effect of milling on the crystallinity of the griseofulvin drug particles was thoroughly studied [60]. Spectral comparison concluded that the crystallinity of the drug particles was not affected by the milling process. Thus, the Raman method served a dual purpose monitoring particle size reduction and drug particle crystallinity.

The nondestructive spectroscopic methods were also used to evaluate how different processing parameters affected the final product obtained: the polymeric strip films. While most analytical methods provide the average composition of a material within the area analyzed by the radiation, Raman and NIR imaging methods provide spatial information demonstrating the distribution of analytes of interest. The imaging methods made it possible to determine the size of the areas within the film where some drug aggregation occurred. These were important studies as the milled particles may irreversibly aggregate. The Raman and NIR-CI spatial information was then used to study the effect of process parameters on the size of the drug aggregates [61]. The nondestructive analytical methods played a key role in studying the effect of the stabilization process [62] and the effect of the drying process on the structure of the polymeric strip films [63].

Raman spectroscopy played a key role in the final film testing stage, where spectra were obtained during drying of the films. Raman spectroscopic methods were developed to determine the drug concentration in the films, and determine whether the crystallinity of the API was affected by the drying process [64].

This example shows the development of a novel polymeric strip film to improve the solubility of a poorly soluble API, and how nondestructive spectroscopic methods contributed at different stages of the process. It also shows research efforts to obtain a final product (films with a suspension of dispersed fine particles) and to use process information to obtain that final product. This goal is not achievable with methods of analysis that destroy or dissolve the sample matrix.

10.5.4
Polymer Industry

The xylene soluble content test measures the percent of amorphous fractions present in polypropylene (PP). Small amounts of undesirable amorphous mass is present in homopolypropylene (homo-PP) material that affects some properties of the polymer. However, 1–20% of amorphous material is incorporated in the production of impact PP and random PP copolymers to improve its strength performance and optical transparence. Thus, the amorphous material must be controlled to meet the desired specifications of the final product. Important properties in the manufacture of the polymer, such as rigidity, flexibility, tensile strength, stiffness, melting temperature, and optical transparency, are related with crystallinity. The amount of crystallinity and amorphous phases depends on stereoregularity of polymer structure that is called as tacticity. Tacticity is the spatial orientation of the methyl groups (CH_3) in the polypropylene that has three different stereochemical configurations: atactic (a-PP, methyl groups forming random arrangements), isotactic (i-PP, all methyl groups oriented at the same side of the polymer chain), and syndiotactic (s-PP, methyl groups are regularly alternated side to side on the polymer chain) [65].

Greener alternative analytical methods to determine xylene soluble content or tacticity can be used with spectroscopic techniques such as Fourier transform infrared (FTIR), low resolution nuclear magnetic resonance (^{1}H-NMR), and near-infrared (NIR) to replace the gravimetric test in quality control laboratories. Mid-infrared spectra obtained through attenuated total reflectance (ATR) could be used to obtain spectra of polymers completely eliminating solvent extraction [66]. The ATR spectra determine tacticity based on infrared absorbance bands at 841 and 998 cm^{-1} for crystalline and at 974 cm^{-1} band for amorphous structure [65, 66]. Variations in crystalline and amorphous composition also affect Raman spectra. These Raman spectral changes have been used to predict the mechanical properties (Young's traction modulus, tensile strength at yield, elongation at yield on traction, and flexural modulus at 1% secant) as homopolymer PP, random ethylene–propylene copolymer, and impact ethylene–propylene copolymer [67].

The polymer industry has also developed a method to monitor polymeric reactions through continuous sampling and dilution of approximately 0.050 ml min^{-1} of a stream from a reactor [4]. The system is called automatic continuous online monitoring of polymerization reactions (ACOMP) and is a method that relies on continuous extraction, dilution, and conditioning of a small stream of reactor contents such that light scattering, viscometric, spectroscopic, and other measurements are made on the diluted stream. The ACOMP's "front end" is an ensemble of pumps, mixing stages, and conditioning elements that ultimately produce the diluted, conditioned stream, which continuously feeds the detection train. The main advantage of ACOMP is that it permits the use of methods that are well understood and well adopted in the polymer industry, reducing the need to train personnel in near-infrared or Raman spectroscopic methods and chemometrics. ACOMP facilitates the use of gel permeation chromatography (SEC), light scattering methods, and viscosity measurements providing a flexible platform to which virtually any desired detector can be added. The disadvantage of the ACOMP is the complexity of the tubing and conditioning steps. The system has been used successfully to increase process understanding in polymerization reactions [68].

10.5.5
Process Analytical Chemistry for Biodiesel Production

The term biodiesel is defined by the American Society for Testing and Materials (ASTM) as the monoalkyl esters of fatty acids derived from vegetable oils or animal fats meeting the requirements of ASTM standard D 6751 [69]. Fatty acid monoalkyl esters are prepared using renewable lipid sources that react with an alcohol (mainly short chain) in the presence of a catalyst to form esters and glycerol. This reaction is known as transesterification or alcoholysis. Transesterification produces a mixture of alkyl esters of fatty acids that are biodiesel and glycerol [70, 71]. The scheme for the transesterification reaction is shown in Figure 10.5.

The renewable lipid sources are catalyzed by acid, alkali catalyst, or lipases, and are converted stepwise to diglyceride, monoglyceride, and finally glycerol. Major alkaline catalysts involve metallic oxides, NaOH, KOH, carbonates, and the corresponding sodium and potassium alkoxides, sulfuric acid, sulfonic acids, and hydrochloric acid that are usually used as acid catalyst [72]. The enzymatic synthesis of biodiesel includes lipases produced by microorganisms (fungi and bacteria), animals, and plants. Figure 10.5 shows that the transesterification reaction requires 3 moles of alcohol for each mole of renewable lipid source to produce 1 mole of glycerol and 3 moles of alkyl esters of fatty acids in presence of a catalyst. This is a reversible reaction and in practice, the transformation occurs when 6 moles of alcohol are mixed with 1 mole of renewable lipid source; this excess of alcohol over the stoichiometric ratio is used to increase the yields of the alkyl esters. Biodiesel can be produced from a variety of renewable sources, which can include some vegetable oils, fats, and microalgae-derived oils.

A number of chromatographic methods have been developed for monitoring the transesterification reaction. Thin layer chromatography (TLC) with flame ionization detection (FID) has been used to determine triglycerides, diglycerides, and monoglycerides [73]. Gas chromatographic (GC) methods have been developed but these require a derivatization – reaction with N,O-bis(trimethylsilyl) trifluoroacetamide (BSTFA) prior to analysis [74]. HPLC with density detection allows the determination of the overall content of triglycerides, diglycerides, and monoglycerides in biodiesel samples from methanolysis mixtures as well as the methyl esters detection [75]. The chromatographic methods require sample derivatization, and the analytical instruments are mainly used in the off-line and

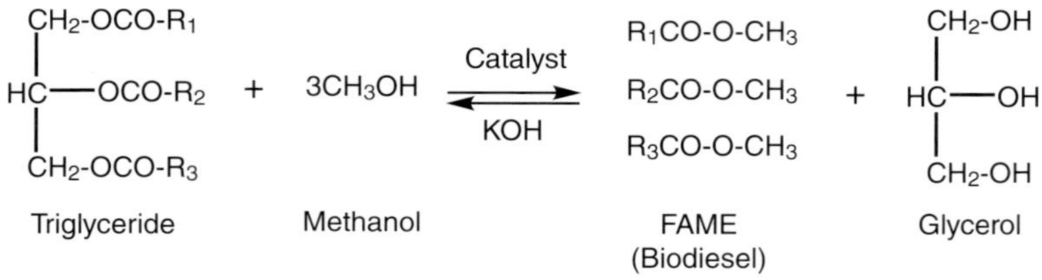

$$\begin{array}{c} CH_2\text{-}OCO\text{-}R_1 \\ | \\ HC\text{-}OCO\text{-}R_2 \\ | \\ CH_2\text{-}OCO\text{-}R_3 \end{array} + 3CH_3OH \underset{KOH}{\overset{Catalyst}{\rightleftarrows}} \begin{array}{c} R_1CO\text{-}O\text{-}CH_3 \\ R_2CO\text{-}O\text{-}CH_3 \\ R_3CO\text{-}O\text{-}CH_3 \end{array} + \begin{array}{c} CH_2\text{-}OH \\ | \\ HC\text{-}OH \\ | \\ CH_2\text{-}OH \end{array}$$

Triglyceride   Methanol   FAME   Glycerol
                                        (Biodiesel)

Figure 10.5 A general scheme of a transesterification reaction.

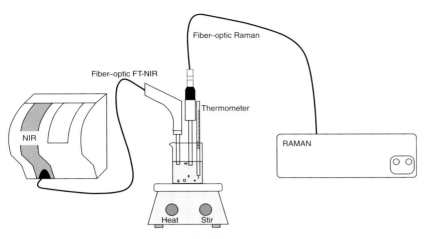

Figure 10.6 Schematic of the Raman and FT-NIR in-line instrument setup. (Reprinted with permission from Ref. [46]. Copyright 2013, SAGE Publications.)

at-line laboratories. Thus, the methodologies are not suitable for PAC and they do not follow the principles of green chemistry.

Vibrational spectroscopic methods have demonstrated to be greener methods than chromatographic methods for monitoring the transesterification reaction of oils/fats. In-line NIR and Raman spectroscopy were used to monitor the production of biodiesel [46]. The reaction is performed in a vessel that includes a thermometer, a hot plate with magnetic stirrer, and NIR and Raman optic probes. The instrument in-line setup for the simultaneous acquisition of NIR and Raman spectra during the transesterification reaction is shown in Figure 10.6. With this arrangement 180 Raman and NIR spectra are acquired every 30 s during the 90 min reaction time. Raman spectra showed the displacement of the carbonyl band from 1747 to 1744 cm^{-1} as the esterification progressed and the decrease of the 1000–1050 cm^{-1} band and the 1405 cm^{-1} band as methanol is consumed in the reaction. FT-NIR spectra also showed the decrease in methanol concentration with the band in the 5000–4700 cm^{-1} region; this signal is present in the spectra of the transesterification reaction but not in the neat oils. The variation in the intensity of the methanol band is an important factor to the in-line monitoring of the transesterification reaction using Raman and NIR spectroscopy. However, the multiple spectral changes make it difficult to follow the progress of the reaction based on the change of a single spectral band. Thus, principal component analysis (PCA) was performed with the Raman and NIR spectra to allow interpretation of the changes in the intensity of all the spectral bands simultaneously [76]. PCA allows a deeper understanding of the system than is possible when the change in only one band or spectral region is evaluated.

In PCA, each spectrum becomes a row vector in a matrix that holds all the spectra that will be subjected to analysis. The spectral wavelengths are the

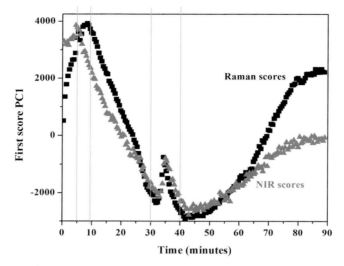

Figure 10.7 NIR and Raman scores for the first principal component showing the progress of the transesterification reaction. (Reprinted with permission from Ref. [46]. Copyright 2013, SAGE Publications.)

column vectors in this matrix. PCA performs a bilinear decomposition of the spectra that are transformed into a new set of orthogonal axes. The bilinear decomposition produces a set of scores or "map of samples" or map of spectra as shown in Figure 10.7 for the biodiesel reaction, and also a set of loadings that serve as a map of the variables or wavelengths [9, 77]. The loadings plot also provides valuable information for understanding the process. In the biodiesel reaction, the scores plotted along the first principal component (first orthogonal axis) showed the progress of the reaction. Thus, each score in Figure 10.7 represents a NIR or Raman spectrum. Figure 10.7 displays that during the first 5 min of the reaction the NIR score values increase, whereas the Raman scores increase during the first 10 min. The score values then decrease rapidly for up to 30 min for both Raman and NIR measurements. The scores are reflecting an acceleration of the reaction in this time. Then, the plots show a remarkable change between 30 and 40 min and finally the changes in the score plots are less marked after 40 min until completion (90 min). The results displayed in Figure 10.7 are consistent with the kinetics of the transesterification reaction, the reaction begins at a slow rate, proceeds at a faster rate, and then slows again as the reaction nears completion [71, 78]. This example clearly shows how PCA clearly contributes to the fundamental understanding of the reaction.

The transesterification reaction depends on the starting materials that are natural and complex materials [79]. The composition of the raw materials may vary significantly as these could include various vegetable frying oils, waste frying oils, and animal fats. Thus, the entire reaction will involve monitoring of a complex mixture. NIR or Raman spectra may be obtained during the reaction and evaluated through the use of PCA. PCA provides a means to track the progress of the reaction based on the analysis of the entire mixture. Thus, PCA is an alternative to traditional means of tracking the progress of the reaction based on the concentration of the expected product or the concentration of specific parameters or component concentration. The study of the reaction by PCA will provide a process signature or batch trajectory, such as that shown in Figure 10.7, that encompasses all the information contained in the spectra [79]. PCA then becomes a process supervision tool and the batch trajectory of a number of reactions may be studied. Batch trajectories and PCA may then be used to design space encompassing the multi-dimensional combination of input variables and process parameters needed to provide assurance of quality [80].

The final quality of biodiesel is evaluated by measurements of density, water content, acid value, methanol content, iodine value, glycerol content, and so on. These parameters are determined using analytical methods that are time consuming, require reagents, and are not suitable for analysis following the green chemistry approach. However, the water and methanol content in industrial and laboratory-scale biodiesel samples can be determined by NIR spectroscopy [81]. The O—H bonds in water and methanol provide strong bands that facilitate their determination by NIR spectroscopy. This methodology can also be used for quality control of the raw materials used in biodiesel production and to identify and quantify the type of oils in mixtures [82].

10.6
Future Trends in PAC

To date, modern spectroscopic *in situ*/online PAC with all its advantages and case studies in greener chemical engineering has not found the place it deserves in the manufacturing industry [83]. Particular in the pharmaceutical industry, PAT, as defined by the FDA [17], has not broadly passed the research and development (R&D) facilities to find its application into the real-world plant applications. One of the reasons is the historical restraint of the pharmaceutical industry to embrace new technologies, which would result in changes of validated manufacturing processes [84]. However, this is changing, in particular with the strong movement toward continuous manufacturing [85–91]. At the moment, the R&D departments within the companies envision the benefits of having the same PAT in their manufacturing sites, whereas the manufacturing personnel often do not see the need to have this level of analytical complexity [84]. This results in a different mindset between R&D and final manufacturing. The latter, oftentimes prefer measurements as simple as temperature, pressure, or time [84]. To tackle this entrenched mindset, companies are going to implement PAT earlier on in their drug development with an increasing extent [84] to bring their development of PAT methods closer to the manufacturing side and plant staff.

There is also a growing number of handheld devices being developed, which carry the laboratory to the sample to save time and effort compared to the need to take samples to bring them to the laboratory [92]. These new tools, including Raman, mid-IR, or NIR spectroscopy instruments, provide flexible application and have changed the way the companies are performing their raw material identity verification or cleaning validation of surfaces and reactors [93]. In recent years, the performance of these handheld instruments has significantly improved and their cost has been reduced [94]. A number of companies that were not considering NIR or Raman spectroscopy are now considering the low-cost handheld instruments [95].

The built-in algorithms for automatic data analysis of such handheld devices demand deeper understanding of the various algorithms available to choose the most suitable one for the desired application, which otherwise leads to the misuse of the technology. On this account, future trends are targeted not just to miniaturize PAC devices while keeping their performance quality at the level of laboratory bench devices but also to improve their ease of use [85]. The manufacturing staff will require expertise in chemometrics and the various spectroscopic methods to monitor their processes [12]. The search for miniaturization of lab-bench devices is an attainable goal considering the gained achievements in the telecommunication and information technology industry (e.g., smartphones and tablet computers). The ideal solution would be a multipurpose interface able to combine all spectroscopic instruments in one [85]. Here, GEA Pharma Systems has developed together with a German company, J&M Analytik, a fiber–optic interface that can be linked to several different spectroscopic techniques simultaneously [84]. Further research in this direction will contribute to deliver reliable and robust devices.

The ease of use of spectroscopic PAC devices to automatically monitor and control manufacturing processes needs further research efforts in the future. Thereby, it is not just analyzing the gathered data manually to determine the offset of the process compared to the set value. It is rather more complex by utilizing the extensive set of information that can be achieved with spectroscopic instruments and their analysis in real time for automatic decision-making via suitable control strategies [19]. These are actually the challenges and needs to be addressed in future industrial and academic research efforts [85, 96]. Here, multivariate chemometric modeling tools might help to close the gap [80]. Ultimately, the increasing trend toward continuous manufacturing in the pharmaceutical industry and the accompanied academic research efforts including flow and microreactors [56, 90, 91] will increasingly track the attention of the vendors of PAT instruments to provide solutions for the customers' needs [85, 97]. For instance, Mettler Toledo has developed their in-line ATR FT-IR system, known as FlowIR, to meet the demands of flow chemistry [98]. Such necessary innovative solution will even accelerate with the demands of tomorrow's small-scale pharmaceutical manufacturing platforms – the future of pharmaceutical manufacturing [91]. Here, a key challenge that needs to be overcome is the implementation of PAT suitable in size (miniaturization) while providing the same accuracy as laboratory bench devices [97, 99, 100]. In this context, handheld PATs, already implemented in the pharmaceutical industry for raw material validation and cleanliness checking [93], could be a potential solution. However, commercially available devices are currently lacking adaptors to be used as plug-and-play PAT in everyday reactor setups and thus require further development efforts in the future [97].

The ultimate objective is for a product to provide the performance and benefits that a customer expects and to obtain these benefits following the green chemistry approach. The implementation of process analytical chemistry contributes to green chemistry by monitoring and controlling the critical parameters of a process and reducing waste. The challenge is to advance from the traditional definition of quality as conformance to specifications to the delivery of the expected benefits [101]. This transition is already expected in the pharmaceutical industry where key leaders in regulatory agencies have defined a high-quality product as "a product free of contamination and reproducibly delivering the therapeutic benefit promised in the label to the consumer" [101, 102]. The use of nondestructive analytical methods will continue to contribute to the development of process understanding where the effect of processing conditions on the quality of the final product is understood.

Acknowledgments

This work is the result of many fruitful discussions with leading scientists. The support of the NSF ERC Structured Organic Particulate Systems EEC-0540855 grant is acknowledged. Authors are particularly thankful to graduate students Andres Roman Ospino and Eduardo Hernández Torres for the preparation of figures.

References

1 Anastas, P.T. and Warner, J.C. (1998) *Green Chemistry*, Oxford University Press, Oxford, UK.
2 Callis, J.B., Illman, D.L., and Kowalski, B.R. (1987) Process analytical chemistry. *Analytical Chemistry*, **59**, 624A–637A.
3 Pell, R.J., Seasholtz, M.B., Beebe, K.R., and Koch, M.V. (2014) Process analytical chemistry and chemometrics: Bruce Kowalski's legacy at The Dow Chemical Company. *Journal of Chemometrics*, **28**, 321–331.
4 Reed, W.F. (2013) Background and principles of automatic continuous online monitoring of polymerization reactions (ACOMP). *Monitoring Polymerization Reactions*, John Wiley & Sons, Inc., New Jersey, pp. 229–245.
5 Wang, F., Wachter, J.A., Antosz, F.J., and Berglund, K.A. (2000) An investigation of solvent mediated polymorphic transformation of progesterone using *in situ* Raman spectroscopy. *Organic Process Research & Development*, **4**, 391–395.
6 Pasikatan, M.C., Steele, J.L., Spillman, C.K., and Haque, E. (2001) Near infrared reflectance spectroscopy for online particle size analysis of powders and ground material. *Journal*

of *Near Infrared Spectroscopy*, **9**, 153–164.

7 Alcalà, M., Blanco, M., Menezes, J.C., Felizardo, P.M., Garrido, A., Pérez, D., Zamora, E., Pasquini, C., and Romañach, R.J. (2006) Near-infrared spectroscopy in laboratory and process analysis. *Encyclopedia of Analytical Chemistry*, John Wiley & Sons, Ltd, Chichester.

8 Simone, E., Saleemi, A.N., and Nagy, Z.K. (2015) *In situ* monitoring of polymorphic transformations using a composite sensor array of Raman, NIR, and ATR-UV/vis spectroscopy, FBRM, and PVM for an intelligent decision support system. *Organic Process Research & Development*, **19**, 167–177.

9 Miller, C.E. (2010) Chemometrics in process analytical technology (PAT). *Process Analytical Technology*, John Wiley & Sons, Ltd, Chichester, pp. 353–438.

10 Markl, D., Wahl, P.R., Menezes, J.C., Koller, D.M., Kavsek, B., Francois, K., Roblegg, E., and Khinast, J.G. (2013) Supervisory control system for monitoring a pharmaceutical hot melt extrusion process. *AAPS PharmSciTech*, **14**, 1034–1044.

11 Joglekar, G.S., Giridhar, A., and Reklaitis, G. (2014) A workflow modeling system for capturing data provenance. *Computers & Chemical Engineering*, **67**, 148–158.

12 Romañach, R.J., Hernández Torres, E., Roman Ospino, A., Pastrana, I., and Semidei, F. (2014) NIR and Raman spectroscopic measurements to train the next generation of PAT scientists. *American Pharmaceutical Review*, **17**, 82–87.

13 Esbensen, K.H. and Paasch-Mortensen, P. (2010) Process sampling: theory of sampling – the missing link in process analytical technologies (PAT). *Process Analytical Technology*, John Wiley & Sons, Ltd, Chichester, pp. 37–80.

14 Alves, J.C.L. and Poppi, R.J. (2015) Near-infrared spectroscopy in analysis of crudes and transportation fuels. *Encyclopedia of Analytical Chemistry*, John Wiley & Sons, Ltd, Chichester.

15 Seasholtz, M.B. (1999) Making money with chemometrics. *Chemometrics and Intelligent Laboratory Systems*, **45**, 55–63.

16 Le, L.D., Tate, J.D., Seasholtz, M.B., Gupta, M., Owano, T., Baer, D., Knittel, T., Cowie, A., and Zhu, J. (2008) Development of a rapid on-line acetylene sensor for industrial hydrogenation reactor optimization using off-axis integrated cavity output spectroscopy. *Applied Spectroscopy*, **62**, 59–65.

17 F.D.A. (2004) Guidance for Industry: PAT A Framework for Innovative Pharmaceutical Development, Manufacturing, and Quality Assurance. U.S. Department of Health and Human Services, pp. 1–19.

18 Hussain, A.S. (2015) Discussion on terms PAT vs. PAC.

19 Singh, R., Sahay, A., Muzzio, F., Ierapetritou, M., and Ramachandran, R. (2014) A systematic framework for onsite design and implementation of a control system in a continuous tablet manufacturing process. *Computers & Chemical Engineering*, **66**, 186–200.

20 Read, E.K., Park, J.T., Shah, R.B., Riley, B.S., Brorson, K.A., and Rathore, A.S. (2010) Process analytical technology (PAT) for biopharmaceutical products: part I. concepts and applications. *Biotechnology and Bioengineering*, **105**, 276–284.

21 Bakeev, K.A. (2010) *Process Analytical Technology: Spectroscopic Tools and Implementation Strategies for the Chemical and Pharmaceutical Industries*, 2nd edn, John Wiley & Sons Ltd., Chichester.

22 Heinz, W.S. (2007) Basic principles of near-infrared spectroscopy. *Handbook of Near-Infrared Analysis*, 3rd edn, CRC Press, pp. 7–19.

23 Clark, D. and Pysik, A. (2006) The analysis of pharmaceutical substances and formulated products by vibrational spectroscopy. *Handbook of Vibrational Spectroscopy*, John Wiley & Sons, Ltd., Chichester.

24 Chalmers, J.M. and Dent, G. (1997) Sampling techniques and accessories, in *Industrial Analysis with Vibrational Spectroscopy* (eds J.M. Chalmers and G.

Dent), The Royal Society of Chemistry, pp. 120–175.

25 Colthup, N.B., Daly, L.H., and Wiberley, S.E. (1990) *Introduction to Infrared and Raman Spectroscopy*, Academic Press.

26 Smith, E. and Dent, G. (2005) *Modern Raman Spectroscopy: A Practical Approach*, John Wiley & Sons, Ltd, Chichester.

27 Mayo, D.W., Miller, F.A., and Hannah, R.W. (2004) *Course Notes on the Interpretation of Infrared and Raman Spectra*, John Wiley & Sons, Inc., Hoboken, NJ.

28 Hu, Y., Liang, J.K., Myerson, A.S., and Taylor, L.S. (2005) Crystallization monitoring by Raman spectroscopy: simultaneous measurement of desupersaturation profile and polymorphic form in flufenamic acid systems. *Industrial & Engineering Chemistry Research*, **44**, 1233–1240.

29 Mercado, J., Alcalà, M., Karry, K., Ríos-Steiner, J., and Romañach, R. (2008) Design and in-line Raman spectroscopic monitoring of a protein batch crystallization process. *Journal of Pharmaceutical Innovation*, **3**, 271–279.

30 Clegg, I.M., Pearce, J., and Content, S. (2012) *In situ* Raman spectroscopy: a process analytical technology tool to monitor a de-protection reaction carried out in aqueous solution. *Applied Spectroscopy*, **66**, 151–156.

31 Alcalà, M., Blanco, M., Menezes, J.C., Felizardo, P.M., Garrido, A., Pérez, D., Zamora, E., Pasquini, C., and Romañach, R.J. (2012) Near-infrared spectroscopy in laboratory and process analysis. *Encyclopedia of Analytical Chemistry*, John Wiley & Sons, Ltd, Chichester.

32 Miller, C.E. (2001) Chemical principles of near-infrared technology. *Near-Infrared Technology in the Agricultural and Food Industry*, American Association of Cereal Chemists, St. Paul, pp. 19–37.

33 Blanco, M. and Romero, M.A. (2001) Near-infrared libraries in the pharmaceutical industry: a solution for identity confirmation. *Analyst*, **126**, 2212–2217.

34 Dziki, W., Bauer, J.F., Szpylman, J.J., Quick, J.E., and Nichols, B.C. (2000) The use of near-infrared spectroscopy to monitor the mobility of water within the sarafloxacin crystal lattice. *Journal of Pharmaceutical and Biomedical Analysis*, **22**, 829–848.

35 Green, R.L., Thurau, G., Pixley, N.C., Mateos, A., Reed, R.A., and Higgins, J.P. (2005) In-line monitoring of moisture content in fluid bed dryers using near-IR spectroscopy with consideration of sampling effects on method accuracy. *Analytical Chemistry*, **77**, 4515–4522.

36 Alcala, M., Blanco, M., Bautista, M., and Gonzalez, J.M. (2010) On-line monitoring of a granulation process by NIR spectroscopy. *Journal of Pharmaceutical Sciences*, **99**, 336–345.

37 Guenard, R. and Thurau, G. (2010) Implementation of process analytical technologies. *Process Analytical Technology*, John Wiley & Sons, Ltd, Chichester, pp. 17–36.

38 Miller, J.C. and Miller, J.N. (2010) *Statistics for Analytical Chemistry*, 6th edn, Ellis Horwood, Chichester, UK.

39 Gy, P. (1998) *Sampling for Analytical Purposes*, 1st edn, John Wiley & Sons, Inc., New York.

40 Brittain, H.G. (2002) Particle-size distribution II: the problem of sampling powdered solids. (pharmaceutical physics). *Pharmaceutical Technology*, **26**, 67.

41 Romañach, R.J. and Esbensen, K.H. (2015) Sampling in pharmaceutical manufacturing: many opportunities to improve today's practice through the theory of sampling (TOS). *TOS Forum*, **4**, 5–9.

42 Esbensen, K.H., Román-Ospino, A.D., Sanchez, A., and Romañach, R.J. (2016) Adequacy and verifiability of pharmaceutical mixtures and dose units by variographic analysis (theory of sampling) – a call for a regulatory paradigm shift. *International Journal of Pharmaceutics*, **499**, 156–174.

43 Vanarase, A.U., Alcalà, M., Jerez Rozo, J.I., Muzzio, F.J., and Romañach, R.J. (2010) Real-time monitoring of drug concentration in a continuous powder mixing process using NIR spectroscopy.

Chemical Engineering Science, **65**, 5728–5733.

44 Colón, Y., Florian, M., Acevedo, D., Méndez, R., and Romañach, R. (2014) Near infrared method development for a continuous manufacturing blending process. *Journal of Pharmaceutical Innovation*, **9**, 291–301.

45 Mark, H. (1991) *Principles and Practice of Spectroscopic Calibration*, John Wiley & Sons, Inc., New Jersey.

46 Fontalvo-Gomez, M., Colucci, J.A., Velez, N., and Romanach, R.J. (2013) In-line near-infrared (NIR) and Raman spectroscopy coupled with principal component analysis (PCA) for *in situ* evaluation of the transesterification reaction. *Applied Spectroscopy*, **67**, 1142–1149.

47 Mateo-Ortiz, D., Colon, Y., Romanach, R.J., and Mendez, R. (2014) Analysis of powder phenomena inside a Fette 3090 feed frame using in-line NIR spectroscopy. *Journal of Pharmaceutical and Biomedical Analysis*, **100**, 40–49.

48 Sistare, F., Berry, L.S.P., and Mojica, C.A. (2005) Process analytical technology: an investment in process knowledge. *Organic Process Research & Development*, **9**, 332–336.

49 Harbour, G.C. and Keiffer, R.G. (2006) Quality systems management. *Encyclopedia of Pharmaceutical Technology*, 3rd edn, Informa Healthcare, pp. 3075–3081.

50 Mojica, C.A., Sistare, F., and Pierre-Berry, L.S. (2007) Process analytical technology in the manufacture of bulk active pharmaceuticals? Promise, practice, and challenges. *Process Chemistry in the Pharmaceutical Industry*, vol. **2**, CRC Press, pp. 361–381.

51 Cama-Moncunill, R., Casado, M., Dixit, Y., Togashi, D., Alvarez-Jubete, L., Cullen, P., and Sullivan, C. (2015) Moisture determination of static and in-motion powdered infant formula utilising multiprobe near infrared spectroscopy. *Journal of Near Infrared Spectroscopy*, **23**, 245–253.

52 Holroyd, S., Prescott, B., and McClean, A. (2013) The use of in- and on-line near infrared spectroscopy for milk powder measurement. *Journal of Near Infrared Spectroscopy*, **21**, 441–443.

53 Holroyd, S. (2013) Review: the use of near infrared spectroscopy on milk and milk products. *Journal of Near Infrared Spectroscopy*, **21**, 311–322.

54 Chen, Y.D., Zhou, G.X., Brown, N., Wang, T., and Ge, Z.H. (2003) Study of lactol activation by trifluoroacetic anhydride via *in situ* Fourier transform infrared spectroscopy. *Analytica Chimica Acta*, **497**, 155–164.

55 Simon, L.L., Pataki, H., Marosi, G., Meemken, F., Hungerbuhler, K., Baiker, A., Tummala, S., Glennon, B., Kuentz, M., Steele, G., Kramer, H.J.M., Rydzak, J.W., Chen, Z.P., Morris, J., Kjell, F., Singh, R., Gani, R., Gernaey, K.V., Louhi-Kultanen, M., O'Reilly, J., Sandler, N., Antikainen, O., Yliruusi, J., Frohberg, P., Ulrich, J., Braatz, R.D., Leyssens, T., von Stosch, M., Oliveira, R., Tan, R.B.H., Wu, H.Q., Khan, M., O'Grady, D., Pandey, A., Westra, R., Delle-Case, E., Pape, D., Angelosante, D., Maret, Y., Steiger, O., Lenner, M., Abbou-Oucherif, K., Nagy, Z.K., Litster, J.D., Kamaraju, V.K., and Chiu, M.S. (2015) Assessment of recent process analytical technology (PAT) trends: a multiauthor review. *Organic Process Research & Development*, **19**, 3–62.

56 Ingham, R.J., Battilocchio, C., Fitzpatrick, D.E., Sliwinski, E., Hawkins, J.M., and Ley, S.V. (2015) A systems approach towards an intelligent and self-controlling platform for integrated continuous reaction sequences. *Angewandte Chemie, International Edition*, **54**, 144–148.

57 Jerez-Rozo, J.I., Zarow, A., Zhou, B., Pinal, R., Iqbal, Z., and Romañach, R.J. (2011) Complementary near-infrared and Raman chemical imaging of pharmaceutical thin films. *Journal of Pharmaceutical Sciences*, **100**, 4888–4895.

58 Ying, Y., Afolabi, A., Bilgili, E., and Iqbal, Z. (2014) Evaluation of in-line Raman spectroscopic monitoring of size reduction during wet media milling of biopharmaceutics classification system

class II drugs. *Applied Spectroscopy*, **68**, 1411–1417.
59 Bilgili, E. and Afolabi, A. (2012) A combined microhydrodynamics-polymer adsorption analysis for elucidation of the roles of stabilizers in wet stirred media milling. *International Journal of Pharmaceutics*, **439**, 193–206.
60 Zarow, A., Zhou, B., Wang, X.Q., Pinal, R., and Iqbal, Z. (2011) Spectroscopic and X-ray diffraction study of structural disorder in cryomilled and amorphous griseofulvin. *Applied Spectroscopy*, **65**, 135–143.
61 Sievens-Figueroa, L., Bhakay, A., Jerez-Rozo, J.I., Pandya, N., Romanach, R.J., Michniak-Kohn, B., Iqbal, Z., Bilgili, E., and Dave, R.N. (2012) Preparation and characterization of hydroxypropyl methyl cellulose films containing stable BCS Class II drug nanoparticles for pharmaceutical applications. *International Journal of Pharmaceutics*, **423**, 496–508.
62 Beck, C., Sievens-Figueroa, L., Gartner, K., Jerez-Rozo, J.I., Romanach, R.J., Bilgili, E., and Davee, R.N. (2013) Effects of stabilizers on particle redispersion and dissolution from polymer strip films containing liquid antisolvent precipitated griseofulvin particles. *Powder Technology*, **236**, 37–51.
63 Susarla, R., Sievens-Figueroa, L., Bhakay, A., Shen, Y.Y., Jerez-Rozo, J.I., Engen, W., Khusid, B., Bilgili, E., Romanach, R.J., Morris, K.R., Michniak-Kohn, B., and Dave, R.N. (2013) Fast drying of biocompatible polymer films loaded with poorly water-soluble drug nano-particles via low temperature forced convection. *International Journal of Pharmaceutics*, **455**, 93–103.
64 Zhang, J., Ying, Y., Pielecha-Safira, B., Bilgili, E., Ramachandran, R., Romanach, R., Dave, R.N., and Iqbal, Z. (2014) Raman spectroscopy for in-line and off-line quantification of poorly soluble drugs in strip films. *International Journal of Pharmaceutics*, **475**, 428–437.
65 Ozzetti, R.A., Oliveira, A.P.De., Schuchardt, U., and Mandelli, D. (2002) Determination of tacticity in polypropylene by FTIR with multivariate calibration. *Journal of Applied Polymer Science*, **85**, 734–745.
66 Karacan, I. and Benli, H. (2011) The influence of annealing treatment on the molecular structure and the mechanical properties of isotactic polypropylene fibers. *Journal of Applied Polymer Science*, **122**, 3322–3338.
67 Banquet-Terán, J., Johnson-Restrepo, B., Hernández-Morelo, A., Ropero, J., Fontalvo-Gomez, M., and Romañach, R.J. (2016) Linear and nonlinear calibration methods for predicting mechanical properties of polypropylene pellets using Raman spectroscopy. *Applied Spectroscopy*, **70**, 1118–1127.
68 McFaul, C.A., Drenski, M.F., and Reed, W.F. (2014) Online, continuous monitoring of the sensitivity of the LCST of NIPAM-Am copolymers to discrete and broad composition distributions. *Polymer*, **55**, 4899–4907.
69 ASTM-International (2005) Standard specification for biodiesel fuel blend stock (B100) for middle distillate fuels (D-6751-03). *Annual book of ASTM standards 2005*, ASTM International, Philadelphia, PA, pp. 609–614.
70 Ghesti, G.F., de Macedo, J.L., Braga, V.S., de Souza, A., Parente, V.C.I., Figueredo, E.S., Resck, I.S., Dias, J.A., and Dias, S.C.L. (2006) Application of Raman spectroscopy to monitor and quantify ethyl esters in soybean oil transesterification. *Journal of the American Oil Chemists Society*, **83**, 597–601.
71 Schuchardt, U., Sercheli, R., and Vargas, R.M. (1998) Transesterification of vegetable oils: a review. *Journal of the Brazilian Chemical Society*, **9**, 199–210.
72 Monteiro, M.R., Ambrozin, A.R.P., Lião, L.M., and Ferreira, A.G. (2008) Critical review on analytical methods for biodiesel characterization. *Talanta*, **77**, 593–605.
73 Freedman, B., Pryde, E.H., and Kwolek, W.F. (1984) Thin-layer chromatography flame ionization analysis of transesterified vegetable-oils. *Journal of the American Oil Chemists Society*, **61**, 1215–1220.
74 Freedman, B., Kwolek, W.F., and Pryde, E.H. (1986) Quantitation in the analysis

of transesterified soybean oil by capillary gas-chromatography. *Journal of the American Oil Chemists Society*, **63**, 1370–1375.

75 Trathnigg, B. and Mittelbach, M. (1990) Analysis of triglyceride methanolysis mixtures using isocratic HPLC with density detection. *Journal of Liquid Chromatography*, **13**, 95–105.

76 Esbensen, K.H. and Geladi, P. (2009) Principal component analysis: concept, geometrical interpretation, mathematical background, algorithms, history, practice, in *Comprehensive Chemometrics* (ed. S.D.B.T. Walczak), Elsevier, Oxford, pp. 211–226.

77 Esbensen, K.H. (2012) *Multivariate Data Analysis: In Practice*, 5th edn, CAMO Software.

78 Colucci, J., Borrero, E., and Alape, F. (2005) Biodiesel from an alkaline transesterification reaction of soybean oil using ultrasonic mixing. *Journal of the American Oil Chemists' Society*, **82**, 525–530.

79 Menezes, J.C., Felizardo, P., and Neiva Correia, M.J. (2008) The use of process analytical technology in biofuels production. *Spectroscopy*, **23**, 30–33.

80 MacGregor, J. and Bruwer, M.-J. (2008) A framework for the development of design and control spaces. *Journal of Pharmaceutical Innovation*, **3**, 15–22.

81 Felizardo, P., Baptista, P., Menezes, J.C., and Correia, M.J.N. (2007) Multivariate near infrared spectroscopy models for predicting methanol and water content in biodiesel. *Analytica Chimica Acta*, **595**, 107–113.

82 Baptista, P., Felizardo, P., Menezes, J., and Neiva Correia, M. (2008) Monitoring the quality of oils for biodiesel production using multivariate near infrared spectroscopy models. *Journal of Near Infrared Spectroscopy*, **16**, 445–454.

83 Chanda, A., Daly, A.M., Foley, D.A., LaPack, M.A., Mukherjee, S., Orr, J.D., Reid, G.L., Thompson, D.R., and Ward, H.W. (2015) Industry perspectives on process analytical technology: tools and applications in API development. *Organic Process Research & Development*, **19**, 63–83.

84 Thayer, A. (2014) Industry perspectives on process analytical technology: tools and applications in API development. *Chemical & Engineering News*, **92**, 8–12.

85 Page, T., Dubina, H., Fillipi, G., Guidat, R., Patnaik, S., Poechlauer, P., Shering, P., Guinn, M., McDonnell, P., and Johnston, C. (2015) Equipment and analytical companies meeting continuous challenges. May 20–21, 2014 continuous manufacturing symposium. *Journal of Pharmaceutical Sciences*, **104**, 821–831.

86 Rockoff, J. (2015) Drug making breaks away from its old ways. Wall Street Journal.

87 Stanton, D. (2015) Janssen and Rutgers expand R & D as continuous manufacturing picks up steam. Available at *in-pharmatechnologist.com*.

88 Brennan, Z. (2015) FDA calls on manufacturers to begin switch from batch to continuous production. Available at *in-pharmatechnologist.com*.

89 Alcala, M., Martinez, L., Esquerdo, R., Hausner, D., and Romañach, R.J. (2015) Continuous manufacturing, near infrared spectroscopy and process knowledge. *American Pharmaceutical Review*, **18**, 59–63.

90 Mascia, S., Heider, P.L., Zhang, H., Lakerveld, R., Benyahia, B., Barton, P.I., Braatz, R.D., Cooney, C.L., Evans, J.M.B., Jamison, T.F., Jensen, K.F., Myerson, A.S., and Trout, B.L. (2013) End-to-end continuous manufacturing of pharmaceuticals: integrated synthesis, purification, and final dosage formation. *Angewandte Chemie, International Edition*, **52**, 12359–12363.

91 Adamo, A., Beingessner, R.L., Behnam, M., Chen, J., Jamison, T.F., Jensen, K.F., Monbaliu, J.-C.M., Myerson, A.S., Revalor, E.M., Snead, D.R., Stelzer, T., Weeranoppanant, N., Wong, S.Y., and Zhang, P. (2016) On-demand continuous-flow production of pharmaceuticals in a compact, reconfigurable system. *Science*, **352**, 61.

92 Watt, R., Moffat, T., and Assi, S. (2011) Comparison of laboratory and handheld Raman instruments for the identification of counterfeit medicines, spectroscop. Spectroscopy.

93 Üstün, B. (2013) Raw material identity verification in the pharmaceutical industry. European Pharmaceutical Review, 13.

94 Alcala, M., Blanco, M., Moyano, D., Broad, N.W., O'Brien, N., Friedrich, D., Pfeifer, F., and Siesler, H.W. (2013) Qualitative and quantitative pharmaceutical analysis with a novel hand-held miniature near infrared spectrometer. *Journal of Near Infrared Spectroscopy*, **21**, 445–457.

95 Karry, K.M., Singh, R., and Muzzio, F.J. (2015) Fit-for-purpose miniature NIR spectroscopy for solid dosage continuous manufacturing. *American Pharmaceutical Review*, **18**, 64–67.

96 Myerson, A.S., Krumme, M., Nasr, M., Thomas, H., and Braatz, R.D. (2015) Control systems engineering in continuous pharmaceutical manufacturing. May 20–21, 2014 continuous manufacturing symposium. *Journal of Pharmaceutical Sciences*, **104**, 832–839.

97 Stelzer, T., Wong, S.Y., Chen, J., and Myerson, A.S. (2016) Evaluation of PAT methods for potential application in small-scale, multipurpose pharmaceutical manufacturing platforms. *Organic Process Research & Development*, **20**, 1431–1438.

98 Carter, C.F., Lange, H., Ley, S.V., Baxendale, I.R., Wittkamp, B., Goode, J.G., and Gaunt, N.L. (2010) ReactIR flow cell: a new analytical tool for continuous flow chemical processing. *Organic Process Research & Development*, **14**, 393–404.

99 Stelzer, T., Wong, S.Y., Chen, J., and Myerson, A.S. (2014) Ultrasound a versatile process analytical tool for quantitative in-line monitoring of pharmaceutical oral/injection formulations, Proceedings of 19th International Symposium on Industrial Crystallization, Toulouse, France.

100 Stelzer, T., Wong, S.Y., Chen, J., and Myerson, A.S. (2014) Quantitative in-line monitoring of pharmaceutical formulations by ultrasound, Ultrasonics-2014, Lisbon, Portugal.

101 Woodcock, J. (2004) The concept of pharmaceutical quality. *American Pharmaceutical Review*, **17**, 1–3.

102 Yu, L. (2008) Pharmaceutical quality by design: product and process development, understanding, and control. *Pharmaceutical Research*, **25**, 781–791.

11
NMR Spectroscopy and Microscopy in Reaction Engineering and Catalysis

Carmine D'Agostino, Mick D. Mantle, and Andrew J. Sederman

11.1
Introduction

The application of Nuclear magnetic resonance (NMR) methods to problems relevant to the chemical and process industry has been constantly growing during the past decades. Chemical and reaction engineering represent some of the most promising fields in terms of exploiting NMR techniques to address important questions such as mass transport, hydrodynamics, and molecular interactions in catalytic systems, both homogeneous and heterogeneous, as well as real-time reaction monitoring and spatially resolved measurements of reactor performances, such as conversion and selectivity. Numerous methodologies have been developed and it can now be said that an "NMR toolkit" is available to study a broad range of physical phenomena in these areas. Various NMR spectroscopic methods, including one- and two-dimensional NMR spectroscopy as well as polarization transfer techniques, can be used to monitor the progress of chemical reactions in industrially relevant systems [1]. Pulsed field gradient (PFG) NMR is a powerful tool to noninvasively and selectively probe translational dynamics in a variety of systems, including fluids confined in porous materials, multicomponent liquid mixtures, emulsions as well as other systems where conventional diffusion measurements are particularly challenging to perform [2]. The method allows not only the probation of mass transport properties but it may also be used to yield important information on intermolecular interactions as well as structural features of porous materials under investigation. Surface properties of fluids confined in porous systems may also be probed by performing measurements of NMR relaxation times [3, 4]. Such parameters can be considered as "fingerprints" of molecular dynamics of molecules and are very sensitive to the surrounding environment experienced by fluids confined in the pore space. These measurements, previously used to study wettability and surface affinity of oil/water systems in rocks [5], can be extended to heterogeneous catalysts and can yield important information on adsorbate/adsorbent surface interactions for these materials. Such knowledge can, in turn, be used to optimize and understand performances of catalytic systems and adsorbent media.

It is the aim of this chapter to introduce the reader to the basic principles or such NMR methodologies and introduce some of the latest developments and applications in chemical engineering, with a focus on reaction engineering and catalytic systems.

11.2
Basic Principles of NMR

In this section a brief introduction to the basic principles of NMR will be discussed. More detailed treatments of the subject are available in the literature [2, 6–8].

11.2.1
Nuclear Spins and Bulk Magnetization

NMR is a consequence of the quantum mechanical property of atomic nuclei known as "nuclear spin" [9]. The nuclear spin angular momentum, **P**, is quantized and its magnitude is specified as follows:

$$P = \frac{h}{2\pi} \sqrt{I(I+1)} \tag{11.1}$$

where h is Planck's constant, $h = 6.626 \times 10^{-34}$ J s, and the quantity denoted by I describes the spin of an atomic nucleus and is called nuclear spin quantum number. Each nucleus has a characteristic value of I and its nuclear spin will have $2I+1$ possible orientations or levels, denoted by the magnetic quantum number, m. Only atomic nuclei with a non-zero value of the nuclear spin quantum number I will be NMR active. The value of I depends on the number of protons and neutrons that make up the atomic nucleus, or more precisely it depends on the number of unpaired protons and neutrons. The rules for determining the net spin of an atomic nucleus can be summarized as follows [6]:

- Number of protons *and* neutrons are both even ⇒ $I=0$.
- Number of protons *plus* number of neutrons is odd ⇒ I is a half-integer (i.e., 1/2, 3/2, 5/2).
- Number of protons *and* number of neutrons are both odd ⇒ I is an integer (i.e., 1, 2, 3).

Nuclei such as ^{4}He, ^{12}C, and ^{16}O have no net spin and they are, therefore, NMR inactive. Examples of nuclei with spin half-integer are ^{1}H, ^{13}C, and ^{19}F, with $I = ½$, whereas nuclei such as ^{2}H and ^{14}N possess an integer value of $I = 1$. For nuclei with a non-zero spin number, the nuclear spin angular momentum **P** has an associated magnetic moment **μ** given by the following relationship:

$$\boldsymbol{\mu} = \gamma \mathbf{P} \tag{11.2}$$

Equation (11.2) defines the parameter γ, which is a property of the type of nucleus and is known as the gyromagnetic ratio. In the absence of an external magnetic field, all the possible magnetic quantum states are degenerate. However, if an external magnetic field of strength B_0 is applied, the magnetic moment of the nucleus acquires an energy value E given by

$$E = -\boldsymbol{\mu} \cdot \mathbf{B}_0 = -\gamma \frac{h}{2\pi} m B_0 \quad (11.3)$$

The nucleus develops $(2I+1)$ energy levels corresponding to the different values taken by m. Transitions across different energy levels are quantized and the NMR selection rule governing such transitions is $\Delta m = \pm 1$. Commonly studied nuclei, such as ^{1}H and ^{13}C, with $I = \frac{1}{2}$, can therefore adopt two spin energy states, with $m = \pm \frac{1}{2}$. These states are described as being parallel and antiparallel and are often referred to as $|\alpha\rangle$ state $(+\frac{1}{2})$ and $|\beta\rangle$ state $(-\frac{1}{2})$, respectively, with the parallel orientation, α and B_0, being of lower energy. The energy required to induce a transition across the two energetic levels is given by

$$\Delta E = h\nu = \left| \gamma \frac{h}{2\pi} \Delta m B_0 \right| = \gamma \frac{h}{2\pi} B_0 \quad (11.4)$$

The last equation defines the precessional frequency of the nucleus, which is also termed as the Larmor frequency of precession and is written as

$$\nu_0 = \frac{\gamma B_0}{2\pi} \text{ (Hz)} \quad \text{or} \quad \omega_0 = \gamma B_0 \text{ (rads}^{-1}) \quad (11.5)$$

In a static strong magnetic field B_0 at thermal equilibrium, the population of the energy states α and β obeys the Boltzmann distribution:

$$\frac{N_\beta}{N_\alpha} = \exp\left(-\frac{\Delta E}{k_B T}\right) \quad (11.6)$$

where k_B is the Boltzmann constant and T is the temperature of the spin system. A surplus of nuclear spins will exist in the α configuration. The excess of nuclear spins in the α state can be visualized as a number of magnetic moments distributed randomly around a precessional cone. The ensemble of spins gives rise to a net magnetization parallel to the external magnetic field, along the z-direction precessing about the field $\mathbf{B}_0$ and the Larmor frequency $\omega_0 = \gamma B_0$. The random distribution of spins about the z-axis causes the net magnetization in the transverse plane to be zero. Hence, the whole spin system can be treated as a macroscopic *bulk magnetization* vector $\mathbf{M}$ aligned with the external magnetic field along the z-axis. This way of visualizing the spin magnetization is also known as the Bloch vector model and is a very useful approach to describe NMR experiments.

When the bulk magnetization is tilted away from the z-axis, a net transverse component on the x–y plane is generated. The precession of this transverse component of the bulk magnetization induces a current within a detection coil, which gives rise to the NMR signal, usually referred to as free induction decay

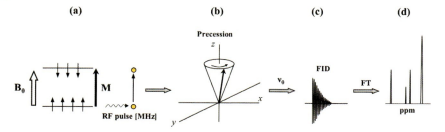

Figure 11.1 Schematic of a typical NMR experiment. (a) The spin system at equilibrium showing a net magnetization, **M**, aligned with B_0; a radio frequency pulse (RF) induces transition in the spin energy levels. (b) The magnetization undergoes a precessional motion generating a (c) free induction decay (FID) signal; (d) Fourier transform (FT) of this signal yields the NMR spectrum.

(FID). The Fourier transform of such time domain signal yields the frequency domain NMR spectrum. This is schematically shown in Figure 11.1.

The manipulation of the bulk magnetization **M** is achieved using so-called *NMR pulses*, which are essentially radio frequency pulses oscillating at the Larmor frequency, that is, the "on resonance condition." The rotation angle θ of **M** away from equilibrium induced by a pulse depends on the length of the pulse t_1 and its magnitude B_1 according to

$$\theta = \omega_1 t_1 = \gamma B_1 t_1 \text{ (rad)} \tag{11.7}$$

Most NMR "pulse sequences" make use of 90° ($\pi/2$) and 180° (π) pulses, shown schematically in Figure 11.2. A 90° pulse rotates the magnetization into the x–y plane, where it can be detected by a receiver coil. After such a pulse, often referred to as excitation pulse, the spin population of α and β states have been equalized and there is no net magnetization along the z-direction. Instead, a net polarization will exist on the x–y plane. A 180° pulse will invert the population of α and β states, causing a surplus of spins in the β states, hence yielding a negative net magnetization along the z-axis.

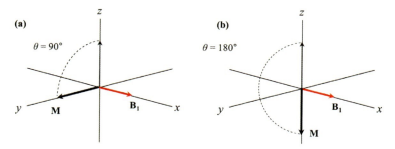

Figure 11.2 The effect of the RF pulse B_1 on the magnetization vector, **M** viewed in the rotating frame that rotate about the z-axis at the Larmor frequency [7]. The diagrams show: (a) 90° pulse and (b) 180° pulse. (From Ref. [7]. Copyright 2005, John Wiley & Sons, Inc.)

11.2.2
NMR Spectroscopy of Liquids

In the following section, a brief discussion on NMR spectroscopy in liquids will be presented. NMR spectroscopy of gases is based on similar principles, although several differences exist, which are partly related to large differences in atomic density. More details on this topic can be found elsewhere [10]. NMR spectroscopy is perhaps the most common set of magnetic resonance (MR) methods and is widely used to investigate molecular structures, hydrogen bonding, and intermolecular interactions in a variety of chemical and biological systems. A typical NMR spectrum is a plot of NMR resonance signals as a function of frequency. The Larmor frequency experienced by different nuclei in a molecule will differ slightly due to electronic shielding associated with the magnetic fields generated by the motion of the bonding electrons. The field experienced B by the nucleus may be written as

$$B = B_0(1 - \sigma) \tag{11.8}$$

where B_0 is the external magnetic field and σ is the shielding constant. The differences in resonant frequencies are characterized by the so-called chemical shift (δ) measured in parts per million (ppm). The chemical shift is usually defined relative to a reference molecule, typically, for ^{1}H NMR, tetramethyl silane (TMS), according to the following relationship

$$\delta = \frac{\nu - \nu_{TMS}}{\nu_{TMS}} \times 10^6 \tag{11.9}$$

where ν is the absolute frequency of the nucleus of interest and ν_{TMS} is the absolute frequency of TMS, which usually coincides with the operating frequency of the spectrometer, $\nu_{Reference}$.

Contributions to nuclear shielding include electronegativity and induced magnetic fields of neighboring groups, local diamagnetic and paramagnetic shifts, and hydrogen bonding. More details on shielding mechanisms are reported elsewhere [11].

Another important interaction, besides the chemical shift, is that of scalar or J-coupling. This is a through-bond interaction between two different nuclear spins, mediated by bonding electrons between these two spins. J-coupling further splits the fine structure of NMR spectra and gives rise to the hyperfine splitting, that is, the effective resonance frequency of the neighboring nucleus is split in $m = n + 1$ multiplets, where n is the number of the nearby coupled spins. Effects arising from J-coupling can be very useful for the determination of molecular structures. An example is given in Figure 11.3, which shows the ^{1}H NMR spectrum of chloromethane, where both the hyperfine structure for each of the NMR peaks, CH_3 and CH_2, can be observed.

J-coupling (always reported in Hz) is field independent (i.e., J is constant at different external magnetic field strength) and is mutual (i.e., $J_{AB} = J_{BA}$). Because

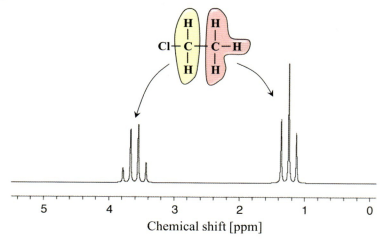

Figure 11.3 ^{1}H NMR spectrum of chloroethane. The two main peaks of the CH$_3$ and CH$_2$ groups are further split in multiplets, which arise from J-coupling. The multiplet corresponding to the CH$_3$ protons has a relative integration (peak area) of three (one for each proton) and is split by the two methylene protons ($n = 2$), which results in $n + 1$ peaks, that is, 3 which is a triplet. The multiplet corresponding to the CH$_2$ protons has an integration of two (one for each proton) and is split by the three methyl protons ($n = 3$). (Adapted from Ref. [12]. Copyright 1998, American Chemical Society.)

the effect is transmitted through the bonding electrons, the magnitude of J falls off rapidly as the number of intervening bonds increases.

The most exploited nucleus in NMR spectroscopy, especially in organic chemistry, is the ^{1}H nucleus, with a nuclear spin equal to $I = \frac{1}{2}$. This is due to the high natural abundance of the ^{1}H isotope (99.99%) and its inherent high NMR sensitivity. Both qualitative and quantitative information can be easily deduced by a proton NMR spectrum. Qualitative information can be deduced by the number of resonances, representative of different chemical groups or environments. Quantitative information can be obtained by the NMR signal intensity of different NMR resonances. The integration of such peaks is directly related to the number of protons making up a particular chemical group.

The power and usefulness of liquid ^{1}H NMR is in its ability to give quantitative composition and structure information. In some cases the use of quantitative ^{1}H NMR may become prohibitive when studying liquid phase reactions in porous catalysts as the relatively narrow chemical shift range of the ^{1}H nucleus together with the line broadening of ^{1}H resonances give rise to a large number of overlapping resonances, making the resulting NMR spectrum almost featureless and further analysis difficult, if not impossible. In such cases, ^{13}C NMR spectroscopy is undoubtedly the preferred approach. The use of ^{13}C as nucleus of choice gives a much wider chemical shift range than ^{1}H, reducing dramatically peak overlap typical of liquid species in porous materials. The major drawback of ^{13}C NMR is in the much lower signal-to-noise ratio compared to ^{1}H

NMR. The natural abundance of ^{13}C in a sample is very low (1.1%), hence enriched samples are often needed, which may have a significant economic impact on cost of materials. Moreover, the sensitivity of the ^{13}C nucleus is only 1.6×10^{-2} relatively to reference value of 1, which is given to the ^{1}H nucleus. However, these obstacles can be potentially overcome by using polarization transfer techniques.

11.2.3
NMR Relaxation

NMR relaxation is the process by which the bulk magnetization returns its equilibrium position. It has already been established that the nonequilibrium distribution of nuclear spins following an radio frequency (RF) pulse gives rise to components of both transverse and longitudinal magnetization.

11.2.3.1 Spin–Lattice Relaxation

At equilibrium, we have a net bulk magnetization **M**, while the transverse magnetization is averaged to zero. The recovery of the longitudinal component of the magnetization, M_z, to its equilibrium position, is also known as *spin–lattice* or T_1 relaxation. The Bloch theory of NMR assumes that the equilibrium is restored according to

$$\frac{dM_z}{dt} = -\frac{(M_z - M_0)}{T_1} \tag{11.10}$$

where T_1 is the spin–lattice (or longitudinal) relaxation time constant. The solution of Eq. (11.10) gives the general result

$$M_z(t) = M_z(0)\exp\left(-\frac{t}{T_1}\right) + M_0\left[1 - \exp\left(-\frac{t}{T_1}\right)\right] \tag{11.11}$$

where $M_z(0)$ is the longitudinal magnetization at $t=0$, that is, immediately after the RF pulse. Thermal equilibrium is almost fully reestablished after a time $t = 5 \times T_1$. The spin–lattice relaxation time constant T_1 is typically measured with the inversion recovery pulse sequence [13], which is shown in Figure 11.4.

Figure 11.4 shows that the magnetization **M** is inverted by a 180° pulse. The magnetization recovers along the z-axis only during the time τ and has to be rotated into the x–y-plane to be measured; this is achieved by a 90° pulse. The recovery of the magnetization proceeds according to Eq. (11.11), with the initial condition $M_z(0) = -M_0$. Therefore, T_1 can be measured by performing multiple inversion recovery experiments with τ ranging from 0 to $t > 5 \times T_1$ and fitting the measured signal intensities to the following equation:

$$M_z(t) = M_0\left[1 - 2\exp\left(-\frac{t}{T_1}\right)\right] \tag{11.12}$$

Another technique used to measure the spin–lattice relaxation time is the saturation recovery pulse sequence. The details of which can be found elsewhere [13].

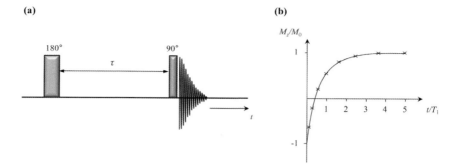

Figure 11.4 The inversion recovery experiment. (a) Inversion recovery pulse sequence. During the time delay τ the magnetization recovers. The experiment is repeated for a list of time delays. (b) Plot of M_z/M_0 against the time, t, normalized over T_1 in an inversion recovery experiment. The crosses represent experimental points acquired for a list of time delays τ.

11.2.3.2 Spin–Spin Relaxation

The process by which transverse magnetization decays away to its equilibrium position value of zero is called spin–spin (also transverse) or T_2 relaxation. An excitation pulse of a certain length may produce a transverse magnetization. As the magnetization evolves in the transverse plane, the nuclear spins will start to lose phase coherence. This is because the local field varies from spin to spin, so the precession frequency is slightly different for each spin. Consequently, over time, the precession of individual spins becomes incoherent leading to a reduction of the NMR signal. The evolution of the magnetization in the transverse plane is governed by the following equation:

$$\frac{dM_{x,y}(t)}{dt} = -\frac{M_{x,y}(t)}{T_2} \Rightarrow M_{x,y}(t) = M_{x,y}(0)\exp\left(-\frac{t}{T_2}\right) \tag{11.13}$$

It is important to note that for most physically relevant systems $T_2 \leq T_1$. Besides the loss of coherence due to differences in the local magnetic field, inhomogeneities in the static magnetic field B_0 will also contribute to the dephasing of transverse magnetization. Therefore, the loss of signal in the transverse plane is determined by the contribution of those two distinct processes.

It is appropriate to highlight that while the loss of signal due to variations of the local magnetic field is an *irreversible* process (due to the nature of the system), the loss of signal caused by the inhomogeneities of the static magnetic field B_0 is a *reversible* process. Therefore, an apparent transverse relaxation constant T_2^* is often quoted and is related to the irreversible, T_2, and reversible, $T_2(\Delta B_0)$, components of spin–spin relaxation by the expression:

$$\frac{1}{T_2^*} = \frac{1}{T_2} + \frac{1}{T_2(\Delta B_0)} \tag{11.14}$$

The decay time constant T_2^* manifests itself not only in the decay of the FID, but also in the width of the Fourier transform of the FID, that is, the resonance line shape. For liquids associated with a single resonance frequency, the spectrum is Lorentzian in shape and has a width $\Delta\nu$ at half its maximum height (FWHM) given by:

$$\Delta\nu = \frac{1}{\pi T_2^*} \tag{11.15}$$

which is usually on the order of a few hertz. In solids or liquids interacting with a solid surface, the relaxation rate is much faster giving rise to line widths ranging from 10^2–10^4 Hz. Two methods are commonly used to determine the relaxation constant T_2 and they are based on what is known as spin echo [2]. The Hahn echo sequence is based on the application of a 90° pulse and a subsequent application of the 180° pulse, after a time τ, which refocuses the reversible part of the transverse magnetization giving rise to an *echo* signal. The experiment is repeated for a list of τ time delays and the data are then fitted to Eq. (11.13). If significant molecular diffusion takes place, spin refocusing will be incomplete and there will be a further loss of signal, leading to an under estimation of the relaxation constant. The loss of signal due to diffusion may be particularly significant when dealing with liquids imbibed in porous media, where changes in magnetic susceptibility within the porous material may cause local magnetic field gradients, which in turn are responsible for the signal loss due to diffusion. To overcome this problem, the Carr–Purcell–Meiboom–Gill [14] or CPMG sequence is used, which consists of a series of refocusing 180° pulses and short τ delays. The use of short values of the delay time τ reduces significantly the contribution of molecular diffusion to the signal loss. Both the Hahn echo and the CPMG pulse sequences are depicted in Figure 11.5. In terms of data acquisition, either the Fourier transform of the echo signal or a single point at the echo-maximum can be used. The latter strategy is called single-shot CPMG and reduces significantly the acquisition time, although does not provide any NMR spectral resolutions as the acquired NMR signal does not undergo Fourier transform in this case.

11.2.4
Pulsed Field Gradient NMR

NMR can be used to measure both coherent (i.e., advective flow) and incoherent (i.e., diffusion) motion by exploiting phase-encoding methods. This is achieved by applying a magnetic field gradient **g**, which causes the Larmor frequencies to vary with position **r** according to

$$\omega(\mathbf{r}) = \gamma(B_0 + \mathbf{g} \cdot \mathbf{r}) \tag{11.16}$$

Removing the dependence on the base Larmor frequency (as it is appropriate when the signal is heterodyned at the Larmor frequency), the phase of the

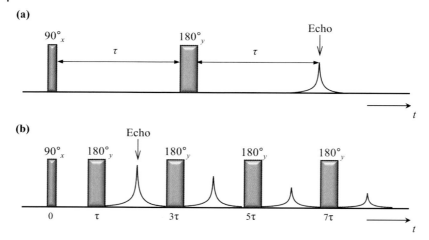

Figure 11.5 (a) Hahn spin echo pulse sequence; the sequence is repeated for a list of time delays τ and the data are fitted to Eq. (11.13). (b) CPMG spin echo pulse sequence; increasing values of time delay are achieved by a series of increasing number of 180° pulse. The application of frequent 180° pulses in the CPMG sequence minimizes the influence of molecular diffusion on signal decay. For the data acquisition, either the Fourier transform of the echo signal or a single point at the echo-maximum can be used.

precessing spins varies with position according to

$$\phi(\mathbf{r}) = \gamma \delta \mathbf{g} \cdot \mathbf{r} \tag{11.17}$$

where δ is the duration of the gradient pulse, $\mathbf{r}$ is the position of the nuclear spins, and $\mathbf{g}$ is the pulsed field gradient. At this point, the nuclear spins are effectively labeled with regard to their position. If, at a time Δ after the first gradient pulse, a second pulse is applied in such a way as to reverse the effect of the first, it becomes possible to detect a change in position of the spins (i.e., motion during the observation time, Δ). Nuclear spins that have not experienced any motion should have their phase fully rewound. Spins that have moved will possess a net phase given by

$$\phi(\mathbf{r} - \mathbf{r}') = \gamma \delta \mathbf{g} \cdot (\mathbf{r} - \mathbf{r}') \tag{11.18}$$

where $\mathbf{r}$ and $\mathbf{r}'$ denote the initial and final spin position, respectively. This net phase shift gives access to the displacement of nuclear spins as a consequence of coherent or incoherent motion.

Motion is typically probed using pulsed field gradient (PFG) methods. The basic method was originally developed by Stejskal and Tanner and consists in a Hanh spin echo pulse sequence with a gradient pulse applied on either side of the 180° [15]. The sequence is referred to as pulsed gradient spin echo (PGSE) depicted in Figure 11.6.

Other PFG NMR methods have been developed including the pulsed gradient stimulated echo (PGSTE) [16] and the pulsed gradient stimulated echo bipolar

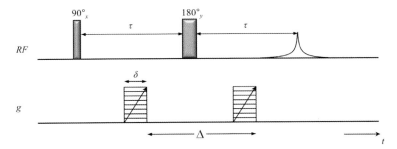

Figure 11.6 NMR pulsed gradient spin echo (PGSE) pulse sequence. (From Ref. [15]. Copyright 1965, AIP Publishing.)

(PGSTEBP) [17], which offers advantages over the PGSE sequence, such as longer observation times for diffusion and minimization of background gradients due to differences in magnetic susceptibility, which may affect diffusion measurements of molecules confined within porous materials.

In the case of molecular self-diffusion along a certain spatial direction (i.e., incoherent motion), summation of the individual phase shifts following PGSE encoding gives rise to a signal attenuation, defined as the ratio of the echo amplitude for a particular value of g, $E(g)$, and the echo amplitude in absence of a gradient, E_0, and has the following expression [15]:

$$\frac{E(g)}{E_0} = \exp\left[-\gamma^2 g^2 \delta^2 D\left(\Delta - \frac{\delta}{3}\right)\right] \quad (11.19)$$

The plot of $E(g)/E_0$ on logarithmic scale versus $\gamma^2 g^2 \delta^2 (\Delta - \delta/3)$, also called the "b-factor," is called a Stejskal–Tanner plot [15] and is used to calculate the molecular self-diffusion coefficient D from PFG-NMR experiments.

In the case of coherent motion (i.e., flow), the displacement in Eq. (11.18), $(\mathbf{r} - \mathbf{r}')$ can be expressed as a function of velocity $\mathbf{v}$ and observation time Δ which gives

$$\phi = \gamma \delta \mathbf{g} \cdot \mathbf{v} \Delta \quad (11.20)$$

In this case, the net phase shift is proportional to the velocity field. Hence, the PGSE, PGSTE, or PGSTEBP techniques previously described may also be used to measure velocity as well as diffusion with a combined effect on NMR signal given by equation [2]

$$\frac{E(g)}{E_0} = \exp\left[i\gamma \delta \mathbf{g} \cdot \mathbf{v} \Delta - \gamma^2 g^2 \delta^2 D\left(\Delta - \frac{\delta}{3}\right)\right] \quad (11.21)$$

11.3
The NMR Toolkit in Reaction Engineering and Catalysis

The operation of chemical reactors and understanding of their performances is a central issue in reaction engineering. This is particularly the case for heterogeneous catalytic reactions, where a series of phenomena may arise during the operation, including mass transport limitations, presence of hot spots along the reactor, nonuniform distribution of flow, and spatial heterogeneities in conversion and selectivity within the reactor. Industrial processes are mostly carried out in either batch or continuous reactors, such as fixed beds or fluidized beds. The catalyst is often used in the form of pellets for fixed bed or powder in the case of fluidized beds. Conventional reactor analysis usually requires the analysis of the inlet and outlet streams from the chemical reactor in order to obtain valuable information on reactor performance, such as reaction kinetics, conversion, and selectivity. However, this approach does not give a comprehensive picture of the process as it gives an average indication of the various reactor performances obtained by analyzing the outlet bulk fluid samples; hence, it does not yield any information about what is actually happening within the reactor and within the catalyst particles, in particular within the pore space, where the reaction is actually taking place.

It is clear that in this way a series of crucial phenomena are missing from the overall picture, which includes the following:

- Extent of fluid/solid contacting
- Adsorption properties of the fluids confined within the catalytic pore space
- Hydrodynamic regime and possible flow heterogeneities
- Extent of axial dispersion within the reactor
- Mass transfer between different phases
- Concentration gradients between the inter and intraparticle space of the reactor
- Catalyst wetting efficiency and liquid hold up
- Temperature variations within the bed

As a consequence, conversion and selectivity may vary greatly within the reactor. It is clear that a knowledge of such properties at a local level would certainly improve our perception of what is happening inside the reactor and ultimately help in the design and optimization of chemical reactors. In this context, NMR is a technique that can certainly provide an answer to many of these different challenges. The methodology is quantitative, noninvasive, and chemically selective. It offers a variety of techniques to address key questions in reaction engineering and catalysis, including studies of reaction kinetics, mapping of temperature, conversion and selectivity within chemical reactors, and mass transport and adsorption. In the following sections, various applications of NMR in catalysis and reaction engineering are discussed.

11.3.1
NMR Spectroscopy in Catalysis and Reaction Engineering

NMR spectroscopy can be a very useful tool to study heterogeneously catalyzed reactions and obtain information on reaction kinetics and product distribution. In a recent work, 2D NMR correlation spectroscopy (COSY) has been used in conjunction with compressed sensing in order to elucidate kinetics and identification of reactive species in the homogeneously catalyzed Meerwein–Ponndorf–Verley (MPV) reduction of propionaldehyde to 1-propanol [18]. The approach allowed sufficient temporal resolution for *in situ* kinetic studies without the loss of structural information and the quantitative nature of the measurements. The method showed the ability to retain all the intrinsic advantages of NMR with regard to quantitation and the ability, by exploiting the *in situ* nature of the measurement, to monitor changes in intermediate species as well as reactant and products.

Typical studies of heterogeneously catalyzed reactions rely on the analysis of the bulk fluid around the catalyst particles, for example, using gas chromatography. This approach, however, does not yield any information on the species confined within the catalyst pore; hence does not give a complete picture of the actual catalytic process. In many cases, for example, solvents, products and by-products may remain trapped inside the pore space; hence the analysis of the bulk fluid would not reveal their presence. In addition, competitive diffusion and adsorption processes of the different species involved in the reaction may lead to an intraparticle composition of the fluid that may differ significantly from the bulk, interparticle composition.

Conversely, the use of NMR methods is able to probe directly the behavior of the fluids inside the catalyst particles. Mantle *et al.* [19] have used ^{13}C distortionless enhancement polarization transfer (DEPT) NMR spectroscopy, at natural abundance, to study the isomerization and hydrogenation of pentenes over pure Al_2O_3 support and 1% Pd/Al_2O_3. The enhanced signal and unambiguous assignment of all ^{13}C peaks, resulting from the use of ^{13}C DEPT-45 NMR, allowed a clear distinction of all the reactants and products of the various reactions. The results showed that the pure Al_2O_3 support is inert to all single-component C_5 adsorption and to coadsorption of 1-pentene/hydrogen (Figure 11.7a) and *cis*-2-pentene/hydrogen (Figure 11.7b) at 298 K. However, coadsorption of *trans*-2-pentene/hydrogen (Figure 11.7c) was seen to cause hydrogenation to *n*-pentane. The adsorption of single-component pentene isomers on the Pd/Al_2O_3 catalyst showed rapid isomerization to predominantly *trans*-2-pentene. Additionally, the coadsorption of pentenes and hydrogen showed the formation of *n*-pentane in all cases. Initial variable temperature studies of coadsorbed 1-pentene/hydrogen on the Pd/Al_2O_3 catalyst also showed a rapid isomerization of 1-pentene to *trans*-2-pentene at 243 K, which subsequently reacted at 288 K to yield *n*-pentane.

In situ hydrogenation reactions using NMR have also been investigated by Koptyug *et al.* [20], who have used ^{1}H NMR to study the hydrogenation of α-methylstyrene (AMS) in a single Pt/γ-Al_2O_3 (15% Pt by weight) catalyst pellet. It was demonstrated that despite a substantial broadening of the ^{1}H NMR lines of liquids permeating the porous catalyst, the quantification of the relative amounts of α-methylstyrene and the major reaction product, cumene, was still feasible to some extent. By inspection of Figure 11.8 it can be observed that despite the much broader NMR line shapes of the liquid inside the catalyst pellets (Figure 11.8c) relative to the bulk liquid (Figure 11.8a), the contribution of

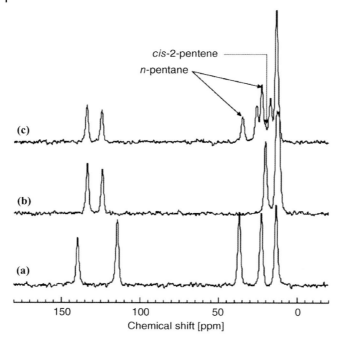

Figure 11.7 ^{13}C DEPT-45 spectra of coadsorption of C$_5$ hydrocarbon and hydrogen on Al$_2$O$_3$. Data were acquired at 298 K. (a) 1-pentene and hydrogen adsorbed in a molar ratio 2.6:1 at a hydrogen pressure of 2.4 bar. (b) cis-2-pentene and hydrogen adsorbed in a molar ratio of 2.0:1 at a hydrogen pressure of 3.0 bar. (c) trans-2-pentene and hydrogen adsorbed in a molar ratio of 2.7:1 at a hydrogen pressure of 2.8 bar; also shown the appearance of peaks associated with n-pentane and cis-2-pentene as indicated in the figure. (Adapted from Ref. [19]. Copyright 2006, Elsevier.)

the =CH$_2$ group of AMS can be recognized at 5.2 ppm. However, the presence of a broadened and distorted line of the phenyl ring at 7.3 ppm makes the quantification of the mixture composition hardly possible.

A recent work by Camp et al. [21] has demonstrated the use of combined NMR and Raman spectroscopy for *in situ* monitoring reaction progress in both homogeneous- and heterogeneous-catalyzed metathesis of alkenes. In particular, the progress of the homogeneous catalytic metathesis of 1-hexene using Grubbs' (I) catalysts was investigated using ^{13}C NMR and Raman spectroscopy. The metathesis of 2-pentene with excess ethene over Re$_2$O$_7$/Al$_2$O$_3$ catalysts was also investigated using ^{13}C MAS NMR and Raman spectroscopy with the results showing that the loss of catalytic activity as inferred by ^{13}C MAS NMR spectra can be attributed to a change in catalyst structure as revealed by Raman spectra. As an example, Figure 11.9 shows the evolution of ^{13}C NMR spectra as a function of reaction time for the homogeneous Grubbs'-catalyzed metathesis of 1-hexene, where it is possible to observe changes in composition over time of the various reactants and products.

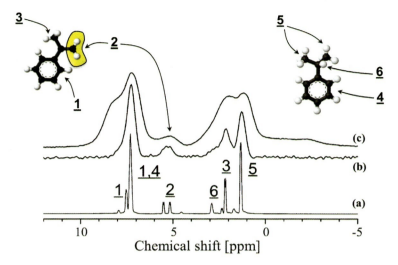

Figure 11.8 Proton NMR spectra of a 1 : 1 mixture of AMS and cumene detected for bulk liquid (a) and liquid permeating the catalyst pellet (b, c). Spectrum (b) was detected with spatial resolution. The number of acquisitions was eight (a, c) or two (b). The resonance at 5.2 ppm can be unambiguously assigned to the =CH$_2$ group of AMS (highlighted in yellow). (Adapted from Ref. [20]. Copyright 2002, American Chemical Society.)

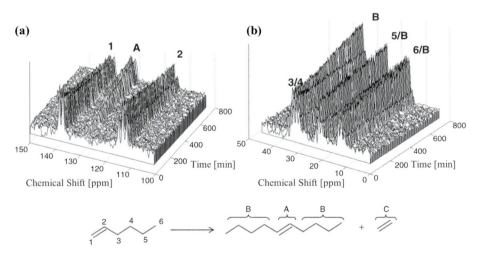

Figure 11.9 (a) ^{13}C NMR spectra as a stacked plot of the Grubbs-catalyzed metathesis of 1-hexene over time for 100–150 ppm range. (b) ^{13}C NMR spectra over the 0–50 ppm range. No peaks are observed outside these areas of the spectrum. No signal is observed from the ethene in the spectra due to its much lower spin density. Spectra are taken at 6 min intervals. (Adapted from Ref. [21]. Copyright 2014, AIP Publishing.)

In recent years, the use of hyperpolarization techniques to boost NMR signal intensity has also been growing. For example, hydrogenation reactions can be studied using *para*-hydrogen-polarized samples as molecular probe for NMR experiments. More details on the principles of *para*-hydrogen-induced polarization (PHIP) can be found elsewhere [22]. In brief, molecular hydrogen (H_2) is known to have two spin isomers, *ortho*-H_2 with the total nuclear spin $I = 1$ and *para*-H_2 with $I = 0$. The equilibrium of ortho/para ratio is approximately 3 : 1 at room temperature and it shifts toward almost pure *para*-H_2 at temperatures close to the hydrogen boiling point (20.25 K). Deviations of nuclear spin alignment from the statistical 3 : 1 ratio in *para*-hydrogen-enriched mixtures result in large NMR signal enhancement during hydrogenation reactions. When *para*-hydrogen is used in the hydrogenation reactions, pairwise addition of the two hydrogen atoms from the same H_2 molecule preserves their correlated nuclear spin state and this correlation can strongly enhance NMR signals of the reaction intermediates and products. This technique can, therefore, be used to elucidate reaction mechanisms and intermediates formed in hydrogenation reactions.

Koptyug *et al.* [23] used PHIP in the heterogeneous hydrogenation of styrene into ethylbenzene with homogeneous Wilkinson's catalyst immobilized on solid supports. The clear *para*-hydrogen spectral patterns confirmed that the addition of hydrogen to the C=C of styrene is pairwise. Indeed, the observation of PHIP effects is usually regarded as evidence that the addition of the two H atoms occurs pairwise. Unlike homogeneous-supported catalysts, the use of *para*-hydrogen in combination with supported metal catalysts has been postulated to be pointless as the reaction mechanism should not sustain pairwise addition of hydrogen to a substrate on multiatomic metal crystallites. However, in a recent work [24] it has been demonstrated that contrary to expectations, metal catalysts such as Pt/Al_2O_3 exhibit PHIP effects, suggesting that the techniques can be potentially used as a tool for mechanistic and kinetic studies on heterogeneous hydrogenation processes. Figure 11.10 shows an example of NMR signal enhancement due to pairwise addition of a *para*-H_2 molecule to a C=C bond during the hydrogenation of propylene over Pt/Al_2O_3 catalysts.

Unlike catalytic hydrogenation reactions, NMR spectroscopic studies of aerobic oxidation reactions in heterogeneous catalysts have been overlooked. The feasibility of NMR spectroscopy for the *in situ* high-pressure aerobic catalytic oxidations by studying the oxidation of 1,4-butanediol to γ-butyrolactone in a Pd/Al_2O_3 catalyst using ^{13}C distortionless enhancement polarization transfer (DEPT)-45 NMR was recently investigated [25]. The technique exploits the higher polarization of the ^1H nucleus relative to the ^{13}C nucleus, which is then transferred to ^{13}C nuclei, enhancing markedly the ^{13}C NMR signal. This helps overcome the low signal-to-noise ratio (SNR) of natural abundance ^{13}C samples while retaining the high-peak resolution of ^{13}C NMR, which makes it easier to differentiate species, particularly in catalysts. The reaction was carried out in a batch mode using a custom-made reactor cell of polymeric material, which was loaded with catalyst pellets saturated with the reacting mixture, heated to reaction temperature (45 °C), and pressurized with air (21 bar). The results are

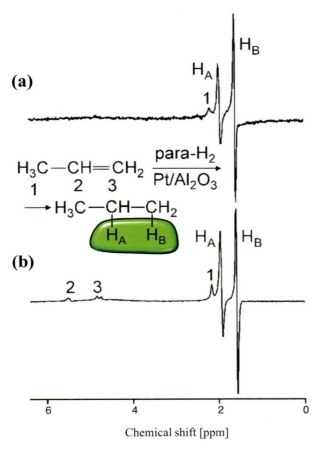

Figure 11.10 ¹H NMR spectra detected in the PASADENA experiments during *in situ* hydrogenation of propylene over Pt/Al₂O₃ catalysts with (a) 1.1 nm Pt particles and (b) 0.6 nm Pt particles (NA = 8). The two hydrogen atoms in the product that originate from the *para*-H₂ molecule are labeled A and B (highlighted in green). The residual NMR signals of the reactant (propylene) are labeled 1, 2, and 3. (Adapted from Ref. [24]. Copyright 2008, John Wiley & Sons, Ltd.)

shown in Figure 11.11. Despite the rather broad NMR line shapes and the presence of numerous NMR resonances, it was possible to unambiguously detect the γ-butyrolactone and other side products as they form within the catalytic particles. From this type of study, it is potentially possible to estimate reaction rates in the intraparticle space.

These preliminary studies suggest the feasibility of this methodology and can be applied to other catalytic oxidation reactions, which are recently attracting significant attention due to their potential for the production of fine chemicals from sustainable resources [26].

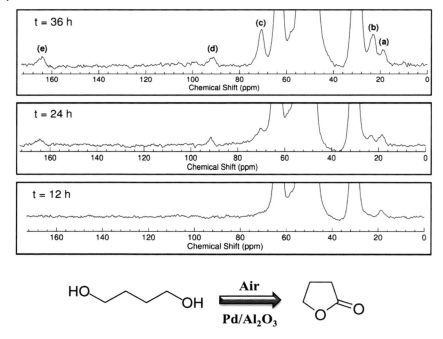

Figure 11.11 ^{13}C DEPT-45 NMR spectra of 1,4-butanediol imbibed in 0.5% Pd/Al$_2$O$_3$ catalyst in the presence of air at different reaction time. The new peaks are (a), (b), (c), (d), and (e). Peaks (b) and (c) represent the resonance of the main oxidation product, γ-butyrolactone. Peaks (d) and (e) are assigned to 4-methyl hydroxybutyrate. Peak (a) is likely to be an impurity in the feed mixture. The reaction was carried out at a temperature of 45 °C using air at 21 bar. (Adapted from Ref. [25].)

11.3.2
Diffusion of Fluids Confined in Porous Catalysts

Measurements of diffusion in porous media, particularly those used as catalysts or separation media, is perhaps one of the fields that fully exploits the potential and ability of PFG NMR spectroscopic techniques, which are able to probe self-diffusion of molecules confined in porous materials with a chemically selective and noninvasive approach.

The molecular displacement during a PFG NMR experiment can usually be described by the following Gaussian probability distribution, also called propagator [2]:

$$P(r, t) = (4\pi Dt)^{-d/2} \exp(-r^2/4Dt) \tag{11.22}$$

where d is the space dimension (1, 2, or 3), r is the position, and D is the diffusion coefficient observed over a certain period of time t, usually referred to as "observation time." The diffusion coefficient is related to the root mean square displacement (RMSD) according to

$$\text{RMSD} = \sqrt{\langle (r(t) - r(0))^2 \rangle} = \sqrt{2dDt} \tag{11.23}$$

Random motion obeying Eqs (11.22) and (11.23) is often termed as "ordinary diffusion" and describes the case of unrestricted (or free) diffusion. In a bulk liquid, except few cases, such as highly viscous ionic liquids [27], molecules undergo unrestricted diffusion, characterized by a diffusion coefficient D_0, hence this behavior is well described by a Gaussian propagator and the diffusion coefficient is independent of the observation time. When molecules are confined within pores, their diffusive behavior becomes very sensitive to physical and chemical characteristics of the porous medium such as pore size distribution, spatial heterogeneities, pore network connectivity, and surface properties. In addition, diffusive behavior may deviate significantly from that of a Gaussian propagator [2].

Porous structures of heterogeneous catalysts are usually formed by interconnected pores accessible from the outer surface of the catalyst particles. Reactants must diffuse and access the catalytic sites within the pore space and products must be able to diffuse out of the pore structure and back within the bulk of the fluid. The diffusive behavior within the interconnected pore space will depend on the interplay between the root mean square displacement (RMSD) experienced by molecules and the characteristic pore size, d. If the RMSD $\ll d$, the effect of the pore wall on molecular diffusion is negligible and the measured diffusion coefficient approaches that of the free unrestricted diffusion of the bulk liquid, D_0. However, when the RMSD $\gg d$, molecules will experience many collisions with the pore walls during their motion and the measured diffusion coefficient is that of the free unrestricted diffusivity reduced by the tortuosity factor, τ. Tortuosity may be measured performing PFG NMR experiments according to

$$\tau = \frac{D_0}{D_{\text{eff}}} \tag{11.24}$$

where D_0 is the diffusion coefficient of the free unrestricted bulk liquid and D_{eff} is the diffusion coefficient of the same liquid confined within the interconnected pore space measured during an observation time such that RMSD $\gg d$. Under this condition, which is usually the case for many low-viscosity liquids (and gases) diffusing within mesoporous materials, molecules will have sufficient time to explore a region of the pore space that is representative of the overall pore structure. In microporous materials, the characteristic pore dimension may become comparable with the typical molecular size of the probe molecule; hence, diffusion becomes strongly influenced by the interaction of the adsorbate molecules with the internal surface of the micropores. Under this condition, the obtained diffusivities reflect the adsorbate/adsorbent interaction rather than the pure geometrical restrictions described by the tortuosity of the pore structure [28]. Therefore, the tortuosity of microporous materials should be measured using guest molecules with a typical size that is smaller than the typical pore size; gases such as ethane have been used for this purpose [29]. It is worth noting that, depending on the typical pore size, diffusion of liquids and gases in porous media may have a different mechanism: diffusion of liquids is usually dominated by intermolecular interactions as molecules are tightly packed (that is, the mean free path is comparable to the typical molecular size), whereas for gases, due to the much lower density, molecule-solid collisions usually dominate. The latter is also called Knudsen diffusion mechanism and needs to be taken into account when measuring tortuosity in porous media using gases [29].

Tortuosity is a geometrical property of the porous medium and from a physical point of view it is defined as the average ratio of the winding pathway that a particle is forced to take due to the presence of the pore wall to the shortest, that is, straight, pathway the particle would take if there was no hindrance to its motion. This concept is clearly illustrated in Figure 11.12. Typical values usually range between 1 and 5 for many porous materials [30–34].

Recent diffusion studies [35] on mesoporous supports, such as TiO_2, Al_2O_3, and SiO_2, fully saturated with organic liquids, have demonstrated that in order for the ratio D_0/D_{eff} to be considered as an accurate estimate of tortuosity, τ, PFG NMR measurements should be conducted using weakly interacting guest molecules, such as alkanes. The τ-values were estimated using several hydrocarbons, which yielded essentially the same value of 1.6 for TiO_2, 1.7 for SiO_2, and 1.6 for Al_2O_3. Recent 3D electron tomography studies have reported values on tortuosity in TiO_2 that are very similar to the values probed by PFG NMR [31]. Conversely, the use of molecules with different functionalities often yields values that do not reflect the "true" tortuosity of the porous medium. This is because intermolecular interactions as well as interactions between molecules and the pore walls may affect the diffusion pathway of such molecules within the pore

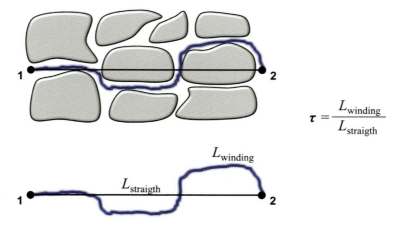

Figure 11.12 Geometric representation of the meaning of "tortuosity" of a porous material.

space, in addition to the effect of the physical structure. Therefore, generally speaking, the ratio D_0/D_{eff} is better defined as a PFG interaction parameter, ξ, which is rather an "apparent" tortuosity and may be very different from the "true" tortuosity of the porous materials. For many molecules confined in mesoporous materials $\xi > \tau$, which implies that the apparent tortuosity experienced by molecules is greater than the true tortuosity and this is attributed to the additional effect of physical interactions within the pore space in addition to geometric restriction of the porous structures. However, polyols systematically show values of $\xi < \tau$ [35, 36]. This implies that the apparent tortuosity experienced by polyols is lower than the true tortuosity of the porous medium, which also means that polyols diffuse faster than expected. It was observed that in the case of glycerol in mesoporous Al_2O_3, the diffusion within the pore space was even faster than that of the bulk liquid. This peculiar behavior of polyols is attributed to the ability of the porous medium to disrupt the extensive hydrogen bonding network within such molecules, which results in an enhanced rate of self-diffusion relative to compounds in which the hydrogen bonding network remained intact within the porous medium. The exact origin of this phenomenon is not yet fully elucidated but it is thought that such hydrogen bonding disruptions is cause by both geometrical features of the pore space as well as interaction of guest molecules with the pore surface. Indeed, local interactions with the pore wall may affect the diffusive motion of molecules. Enhanced self-diffusion has also been observed when studying diffusion of water [37, 38] and hydrocarbons [39] in partially filled porous materials and this has been attributed to molecular exchange between the liquid and vapor phase within the pore space.

Studies of cryo-NMR and diffusion of liquid hydrocarbons and their binary mixtures in mesoporous aluminosilicates under freezing conditions have also been reported by Krutyeva et al. [40]. Diffusion attenuation curves of several hydrocarbons and their mixtures inside the pores were close to monoexponential with a very slight nonlinearity; a similar finding was reported by Dvoyashkin et al. [41].

The linearity of the echo attenuation plot in mesoporous media is observed for materials that exhibit a homogeneous pore structure on the macroscopic length scale and is also defined as *quasihomogeneous* behavior. However, in many cases, porous materials may present *heterogeneities* in terms of diffusion environments, that is, the presence of different regions where guest molecules exhibit different diffusion coefficients. In general, exchange between molecules diffusing in different "diffusion regions" will often occur and two cases can be distinguished.

Fast Exchange
Fast exchange occurs when the typical observation time for diffusion is greater than the characteristic exchange time of the system, that is, the RMSD probed during a PFG NMR experiments is greater than the representative size of the single region. If fast exchange between n diffusion environments occurs, the

signal attenuation of a PFG NMR experiment is given by a mono-exponential decay [42]:

$$\frac{E(g)}{E_0} = \exp\left[-\gamma^2 g^2 \delta^2 D_{\text{average}}(\Delta - \delta/3)\right] \quad (11.25)$$

where $D_{\text{average}} = \sum_{i}^{n} D_i$ is the average diffusion coefficient of molecules present in the different n environments.

Slow Exchange

Slow exchange occurs when the typical observation time for diffusion is smaller than the characteristic exchange time of the system, that is, the RMSD probed during a PFG NMR experiments is smaller than the representative size of the single region. In the case of n diffusion regimes with slow exchange between different spin population p_i, the signal attenuation curve is given by the sum of individual exponential decays:

$$\frac{E(g)}{E_0} = \sum_{i}^{n} p_i \exp\left[-\gamma^2 g^2 \delta^2 D_i(\Delta - \delta/3)\right] \quad (11.26)$$

where D_i is the value of self-diffusivity in each regime.

Such behavior may be observed in liquid-phase catalytic systems, with catalyst pellets surrounded by the reacting fluid. In such a case, as depicted in Figure 11.13, two diffusion components may be observed, one representative of the bulk liquid surrounding the catalyst particles and another representing the intraparticle region (i.e., liquid inside the pore space).

It is noted that in many cases, the intraparticle region may give rise to a non-linear behavior due to the presence of different diffusion regimes within the catalyst particle itself. For example, in a single pore, several diffusion pathways may, in principle, be observed, as depicted in Figure 11.14.

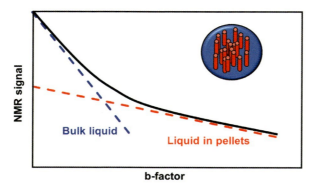

Figure 11.13 Schematic diagram of the NMR signal decay observable in a system of catalyst pellets surrounded by reacting fluid. Two diffusion regimes are visible, one associated with the bulk fluid and a second one associated with the fluid inside the catalyst pellets.

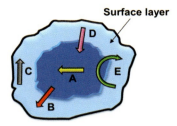

Figure 11.14 Schematic of a single irregular pore containing bulk pore (dark blue) and pore surface (light blue) liquid. Possible diffusion pathways observable during the observation time Δ in the PFG experiments are indicated by arrows. A: bulk diffusion; B: bulk to surface exchange; C: near-surface or surface diffusion; D: surface to bulk exchange; E: surface-to-surface diffusion *via* the bulk liquid. (Adapted from Ref. [43]. Copyright 2010, Royal Society of Chemistry.)

Kärger [44] developed a model to describe diffusion in a two-region system, which relates the NMR signal attenuation to the diffusion coefficients D_1 and D_2 and mean residence times τ_1 and τ_2 of molecules of population p_1 and p_2. This approach was used to model PFG NMR experimental data when probing surface diffusion of liquid alkenes in saturated hydrogenation porous catalysts [43]; diffusion coefficients and mean residence time of molecules in both bulk and surface regions were determined. A typical PFG NMR attenuation plot depicting this situation is reported in Figure 11.15.

The very steep signal decay in Figure 11.15 describes the diffusion across the pores whereas the much lower slope is assigned to molecules diffusing over the pore surface. The shift of this second slope with changing observation time is due to changes in population of molecules residing over the surface.

Knowledge of diffusion in porous catalysts has important practical implications as diffusion may affect significantly the performances of catalysts. Kortunov [30] studied diffusion in fluid catalytic cracking (FCC) catalyst particles and found that intraparticle diffusivity correlated well with the catalytic performance of FCC catalysts, highlighting the importance of mass transport limitations in such processes. The effect of mass transport limitations was also assessed for diol oxidation reactions within Au/TiO_2 oxidation catalysts in methanol and water/methanol solvents [45, 46]. The addition of water decreased significantly the diffusion rate of the diols and this behavior is attributed to the nonideal mixing of methanol/water mixtures, widely reported in the literature, where the hydrogen bonding between water and methanol molecules is responsible for the slower dynamics within the mixture.

11.3.2.1 Catalyst Deactivation Studies Using PFG NMR

PFG NMR diffusion measurements may also give very valuable insights in understanding catalyst deactivation processes. It is clear that self-diffusion measurements of guest molecules in porous materials are very sensitive to the structural features of the porous matrix; hence, changes in pore structure may be

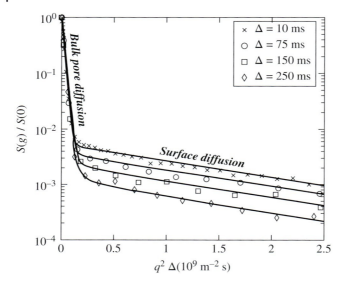

Figure 11.15 PGSTEBP data for 1-octene diffusing in porous 1 wt% Pd/γ-Al$_2$O$_3$ trilobes acquired with a range of Δ observation times. Two distinct diffusion coefficients are observed in each data set, one associated with the bulk pore diffusion and another associated with surface diffusion. The solid lines were determined by a global fit of the data to the model by Kärger for two-site exchange. (Adapted from Ref. [43]. Copyright 2010, Royal Society of Chemistry.)

probed *via* changes of diffusion coefficients of guest molecules. Changes in pore structures are very common while running catalytic reactions: sintering, thermal and mechanical degradation, and pore blockage by external species within the porous matrix may induce significant changes in pore structure, which can in turn affect the diffusion pathway of molecules, as shown in Figure 11.16.

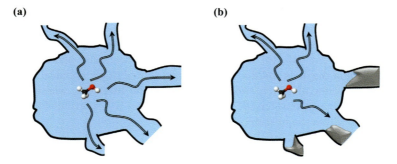

Figure 11.16 Schematic of (a) "fresh" catalyst pore and (b) "coked" catalyst pore. Blockage of pore channels results in less diffusion pathways available with a consequent decrease of the average molecular displacement (i.e., lower molecular self-diffusion within the pore structure).

Wood and Gladden [47] studied the effect of coke deposition upon pore structure and self-diffusion in deactivated industrial hydroprocessing catalysts, by probing self-diffusion of hydrocarbons within fresh and coked samples. The results revealed that as the coke content increases, the effective diffusivity of guest molecules within the catalysts decreases, which implies that coke deposition is changing the pore network connectivity, hence affecting the molecular displacement of guest molecules. Using the values of diffusion coefficients, it was estimated that reductions in effective diffusivity associated with coke deposition led to an increase in Thiele modulus of up to 40% and a decrease in effectiveness factor of up to 10%.

PFG NMR diffusion studies have also been used to understand deactivation of alumina-based catalysts used for the dehydrogenation of glycerol to acrolein [48], a process relevant for the production of value-added chemicals from sustainable resources. The PFG NMR signal attenuation decay for the stable catalyst, STA/CeO_2 on δ,θ-Al_2O_3, did not show significant changes between fresh and spent sample; conversely, for the catalyst that deactivated, STA/CeO_2 on α-Al_2O_3, a significant difference between the PFG NMR plots was observed, with diffusion within the spent sample being slower compared to the fresh sample. This can be clearly seen by analyzing the diffusion signal decay as shown in Figure 11.17. This suggests that the catalyst stability is related to the stability of the pore

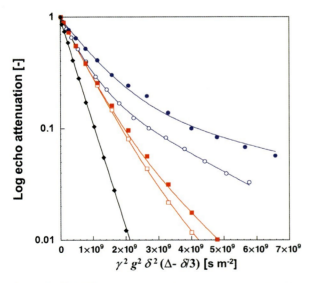

Figure 11.17 PFG-NMR log attenuation plots for n-octane in fresh (open symbols) and used (closed symbols) catalyst: ○/● = STA/CeO_2 on α-Al_2O_3, □/■ = STA/CeO_2 on δ,θ-Al_2O_3. The data indicated by (◆) represents unrestricted bulk n-octane blank run without catalyst sample. Solid lines are fittings to Eq. (11.26) using a two-component for the liquid inside the catalyst and a single component for the bulk liquid. (Adapted from Ref. [48]. Copyright 2014, John Wiley & Sons, Inc.)

structure, which can be inferred by performing diffusion measurements of guest molecules within the pore space.

In more recent work, the stability of AuPt/C was investigated by combining PFG NMR and batch reaction studies [49]. This catalyst is very active for the oxidation of 1,2-propanediol to lactic acid but it shows significant deactivation after each reuse. The deactivation is likely to derive from blockage of pores, particularly those in the microporous region, which reduces the diffusivity of reactant molecules and prevents them to access catalytic sites. In addition, mass transfer limitations within the micropores were also observed.

11.3.3
NMR Relaxation Time Analysis in Porous Catalytic Materials

The adsorption of molecules over solid surfaces is a topic of crucial importance in heterogeneous catalysis, particularly processes in which the catalyst has a large surface-to-volume ratio, which is the case of highly porous catalysts. In recent years, analysis of NMR relaxation times of guest molecules in porous media has become increasingly important in order to elucidate several aspects of molecular dynamics and adsorption of molecules in porous materials. NMR relaxation times have proven to be a useful tool to characterize wettability in oil bearing rocks [4, 50, 51] as well as hydration kinetics, microstructural evolution, and surface interactions in construction materials, such as concrete and gypsum [3, 4, 52, 53].

Analysis of the strength of surface interactions in porous materials is based on the measurement of the longitudinal, T_1, and transverse, T_2, NMR relaxation times. Briefly speaking, the T_1 and T_2 relaxation times can be considered as a fingerprint of molecular dynamics of molecular species, both rotational and translational motion [7, 8, 54]. T_1 relaxation processes drive the longitudinal magnetization to the equilibrium position, that is, aligned with the external magnetic field, while transverse T_2 relaxation processes determine the rate of loss of phase coherence of the magnetization in the transverse plane [7].

A molecular adsorption system, such as a porous catalyst containing a liquid phase, can be described as a dynamic equilibrium between molecules in the bulk phase of the pore and molecules within a layer close to the solid surface. This model, depicted in Figure 11.18, describes well the case of real adsorption

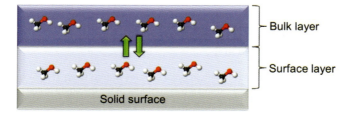

Figure 11.18 Schematic of molecules adsorbing over a solid catalyst surface. The model assumes an exchange between the bulk fluid and the fluid residing in the surface layer.

systems, where a dynamic exchange between molecules residing in the different layers takes place.

Assuming negligible internal magnetic field gradients, the observed relaxation times can be written as [53]

$$\frac{1}{T_{1,2}} = \frac{1}{T_{1,2,\text{bulk}}} + \frac{\varepsilon S}{V} \frac{1}{T_{1,2,\text{surface}}} \qquad (11.27)$$

In Eq. (11.27), $T_{1,2,\text{bulk}}$ represents the relaxation times of the molecules in the bulk layer, $T_{1,2,\text{surface}}$ represents the relaxation times of the molecules in the surface layer, ε the thickness of the surface layer, and S/V is the surface-to-volume ratio. In most porous materials, especially those with a high surface-to-volume ratio, the dominant contribution to the observed relaxation rate comes from the surface term [53].

The molecules residing in the surface layer, which separates the solid surface from the bulk liquid phase, are strongly affected by the force field of the solid catalyst surface. Because of the force field induced by the solid surface, the molecular dynamics within this layer may be significantly different from that of the bulk liquid phase (i.e., van der Waals interactions, hydrogen bonding with hydroxyl groups covering the surface, proton exchange with surface groups).

In bulk liquids, $T_1 \sim T_2$; reduced T_1 and T_2 relaxation times are observed when liquid molecules adsorb on a solid surface due to a change in the molecular mobility [55]. Both T_1 and T_2 are affected by changes in the rotational correlation time of the adsorbate molecules. However, T_2 is further influenced by a translational correlation time associated with surface diffusion [43, 56]. This can be understood by looking at their corresponding correlation functions, which describe how the random motion affects local magnetic fields experienced by spins, which, in turn, act as a source of relaxation. The NMR relaxation times can be written as [56]

$$\frac{1}{T_1} = \rho_1 \frac{S}{V} = C \frac{S}{V} \tau_m [2J(\omega_0) + 8J(2\omega_0)] \qquad (11.28)$$

$$\frac{1}{T_2} = \rho_2 \frac{S}{V} = C \frac{S}{V} \tau_m [3J(0) + 5J(\omega_0) + 2J(2\omega_0)] \qquad (11.29)$$

The spectral density $J(\omega_0)$ represents the contribution of the rotational motion, while $J(0)$ is the contribution of translational motion. The main point here is that T_1 and T_2 depend on rotational and translational motions to differing extents. Molecules adsorbed onto surfaces exhibit modified rotational dynamics [55] and slower translational diffusion [43]; hence, T_1/T_2 increases [53]. It has recently been shown, theoretically [56, 57] and experimentally [56] that T_1/T_2 can be considered to be analogous to the energy of adsorption on a surface.

Several techniques exist to probe NMR relaxation times, including two-dimensional T_1-T_2 experiments. A typical 2D T_1-T_2 NMR pulse sequence is illustrated in Figure 11.19.

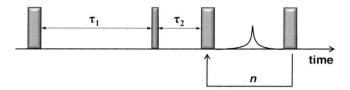

Figure 11.19 The T_1-T_2 pulse sequence showing the RF pulses a thin (90°) and thick (180°) vertical bars. The amplitude of each of the n spin echoes is recorded as a single data at the echo-maximum. The echo centers are separated in time by $t_e = 2\tau_2$.

A 180° radio frequency (RF) inversion pulse tips the spin ensemble, initially at equilibrium along the z-axis, onto the z-axis. Longitudinal T_1 relaxation occurs for a time τ_1; during this time the spins recover back to their equilibrium position in the static magnetic field of magnitude B_0. After this recovery time, a 90° RF excitation pulse is applied to tip the recovered spins into the x–y plane. A series of 180° RF refocusing pulses are then applied to generate a train of n spin echoes, each separated in time by $t_e = 2\tau_2$. The amplitude of each echo is recorded as a single point, and all echo amplitudes are recorded in a single scan (no chemical resolution). The amplitude of the initial echo is determined by the degree of recovery during τ_1; the envelope of the echo train is described by the transverse T_2 relaxation of the spin ensemble. By repeating the experiment for different τ_1 recovery times, a 2D data matrix is constructed. The 2D data are inverted numerically using a fast two-dimensional Laplace inversion [58] to form a 2D distribution of T_2 correlated against T_1. A typical data set is depicted in Figure 11.20, which shows the 2D T_1-T_2 distribution for water adsorbed in several porous oxides. The solid diagonal lines indicate the locus of points where $T_1 = T_2$, which is the case for bulk fluids. When fluids are confined within catalyst pores, their T_1/T_2 ratio is modified due to surface interactions. As a consequence, molecules exhibiting a stronger interaction with the surface, hence a higher T_1/T_2, will move away from the main diagonal line.

The ratio T_1/T_2 can be rearranged to define a dimensionless surface interaction parameter e_{surf} such that [56]

$$e_{\text{surf}} = -\frac{T_2}{T_1} \propto \Delta E \tag{11.30}$$

where ΔE is an activation energy for surface diffusion. Recent experiments have shown that indeed, such parameter correlates well with maximum activation energy of desorption observed in temperature-programmed desorption (TPD) experiments when studying water adsorption in a series of porous oxides. The data are shown in Figure 11.21.

NMR relaxation has been applied to study the strength of surface interactions of liquids within porous catalysts. T_1-T_2 two-dimensional experiments have been used to investigate the strength of surface interaction of water, 2-propanol, and 2-butanone within porous Ru/SiO$_2$ and Pd/Al$_2$O$_3$ catalysts. The results showed that water had a stronger interaction with both catalyst surfaces compared to

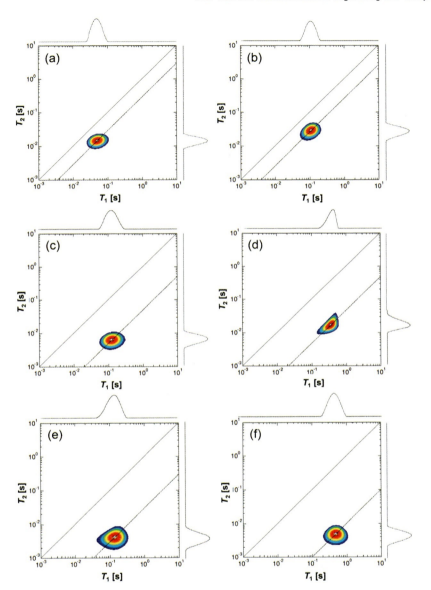

Figure 11.20 T_1-T_2 correlation plots for water in (a) TiO_2-anatase, (b) TiO_2-rutile, (c) γ-Al_2O_3, (d) SiO_2, (e) θ-Al_2O_3, and (f) ZrO_2. The solid diagonal line indicates $T_1 = T_2$; the dashed line indicates T_1/T_2 at the maximum of the peak. Projected $T_{1,2}$ distributions are shown for clarity. (Adapted from Ref. [56]. Copyright 2014, John Wiley & Sons, Inc.)

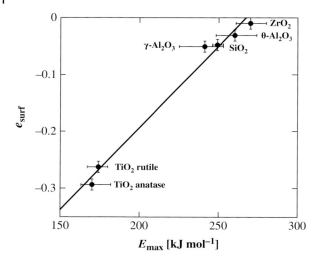

Figure 11.21 Comparison of e_{surf} (NMR) against E_{max} (TPD) for water on different porous oxide surfaces. A linear relationship (diagonal line) is observed. Error bars represent the uncertainty in determining the peak position in the T_1–T_2 distributions (e_{surf}) and maximum desorption energy from TPD analysis (E_{max}). (Adapted from Ref. [56]. Copyright 2014, John Wiley & Sons, Inc.)

2-propanol and 2-butanone, with the last showing the weakest interactions. This was also confirmed by displacement experiments, which showed that 2-butanone was not able to displace the water completely from the catalyst pores, whereas almost full displacement was observed when 2-propanol was the solvent adsorbed on the catalyst surface. From these measurements, information on the access of reactants to surface adsorption sites can be inferred.

In recent work, NMR relaxation measurements have been carried out to understand solvent effects and reactivity trends in the liquid-phase aerobic oxidation of diols in Au/TiO$_2$ catalysts [45, 46]. The results showed that addition of water in the diol/methanol system decreases significantly the T_1/T_2 ratio of the diol, hence its strength of surface interaction with the surface. This explained well the drop in catalytic activity observed by the addition of water, suggesting that the role of water molecules is that of inhibiting catalytic sites, hence preventing the access of reactant molecules to such sites. Studies of different diols also showed that the steric hindrance of the bulky groups affect the adsorption strength of the diol over the Au/TiO$_2$ surface, with more sterically hindered diols exhibiting lower strength of surface interaction.

NMR relaxation has also been used, in conjunction with temperature-programmed desorption (TPD) and diffusion reflectance Fourier transform spectroscopy (DRIFTS), to investigate the effect of mechanochemical treatments, that is, ball milling, on Ag/Al$_2$O$_3$ catalysts used in the selective catalytic reduction (SCR) of NO$_x$ in the presence of n-octane [59]. The outcome of this study suggests that the effect of the ball milling process is that of making the alumina surface more hydrophobic, which improves the conditions for the hydrocarbon to adsorb over the surface, which is in turn a fundamental step for the SCR reaction to proceed further.

NMR relaxation in catalysis is still at an early stage but the potential is significant. Its sensitivity to the chemical environment within a single molecule and to the different environments within a particular porous catalyst makes it suitable to study in details adsorption of guest molecules within porous environments, and may also be used to study simultaneously multicomponent adsorption and probe the effects of various reaction parameters such as competitive adsorption between reactants, solvents, products and side products of a particular reaction, effect of metal deposition over porous supports, changes in pore structure due to thermal degradation, and poisoning of catalytic surfaces. In addition, the NMR relaxation method can also be used as an analytical tool for mixtures in porous materials as it allows the separation of molecular species based on their relaxation times. This can be useful in case when the NMR spectra show significant overlap, which does not allow an unambiguous identification of the different molecular species. From a fundamental point of view, the investigation of further aspects of NMR relaxation times in porous media would certainly be of interest. This could be done combining the technique with other experimental methods as well as with density functional theory (DFT) calculation [60].

11.3.4
Combining NMR Spectroscopy with Magnetic Resonance Imaging

Typically, reactor performances and kinetic studies of a certain chemical process are carried out using an *ex situ* approach, whereby several samples are taken during the course of the reaction. These are then analyzed using a suitable analytic tool (i.e., gas chromatography, NMR, UV and visible spectrometry, and others). This approach is straightforward but is often not enough to characterize in detail chemical processes, particularly those occurring in the presence of porous catalysts, which is often the case in the chemical industry. Using the typical approach, whereby the bulk fluid around the catalyst particles is sampled and then analyzed, is enough to obtain basic information on reaction rate, conversion, and product distribution. However, this picture is missing several important aspects of the chemical process that need to be taken into account if one has to design a suitable catalytic process. For example, the *ex situ* analysis of the bulk fluid gives an average value of conversion and selectivity, whereas it is likely that these properties will vary significantly within the reactor. In addition, there might be species that are strongly adsorbed/constrained within the pores of the catalyst (i.e., poisoning species, by-products, and heavy molecular weight products) and cannot be detected in the bulk of the fluid.

In recent years, traditional NMR spectroscopy has attracted attention as a noninvasive tool for *in situ* studies of chemical processes occurring at industrially relevant conditions, particularly those processes that make use of heterogeneous catalysts. In addition, NMR spectroscopy may be combined with other magnetic resonance methods such as imaging, relaxation, and diffusion, which allow the investigation of mass transport and adsorption within the porous catalyst, as well as investigating hydrodynamics and residence time distribution of the fluid in continuous reactors; hence broadening the overall picture of the whole chemical process. Recent reviews and book chapters on the application of magnetic resonance techniques in chemical reaction engineering provide a good overview of the general ability of magnetic resonance to provide quantitative information and the reader is encouraged to consult Refs [61–63]. Here, we briefly review more recent research work that focuses on *in situ* and/or operando heterogeneous catalytic reactions studies that are in general a combination of NMR spectroscopy with other magnetic resonance techniques to probe spatial variation and mass transfer within packed bed reactors.

One of the main drawbacks with ^{1}H NMR spectroscopy in physically heterogeneous systems, that is, three-phase gas–liquid–solid, is that a ^{1}H spectrum can become difficult to quantify due to overlapping line shapes resulting from magnetic susceptibility-induced line broadening over a small chemical shift range. One way to alleviate this is to switch to a NMR active nucleus with a lower gyromagnetic ratio γ and greater chemical shift dispersion $\Delta\delta$, such as that of ^{13}C as shown in Section 11.3.1 [19]. A pioneering study by Akpa *et al.* [64] of a two-phase heterogeneously catalyzed organic reaction demonstrated the power of the combination of quantitative ^{13}C-DEPT spectroscopy and magnetic resonance imaging to be able to circumvent the problem of susceptibility broadening of line shapes and the lower sensitivity associated with the ^{13}C nucleus. Akpa *et al.* [64] studied the competing hydration and etherification reaction of 2-methyl-2-butene with water and methanol flowing over a packing of Amberlyst-15 catalyst spheres using ^{13}C-DEPT imaging techniques. The results from this study show significant in-plane heterogeneity in conversion within the reactor though selectivity remained reasonably constant throughout the reactor.

The first combined magnetic resonance ^{13}C spectroscopy and ^{1}H imaging study of an *in situ* three-phase heterogeneously catalyzed reaction in packed bed of catalyst particles was reported by Sederman *et al.* [65] who used spatially resolved ^{13}C-DEPT spectroscopy to investigate the hydrogenation of 1-octene in a 25 mm diameter glass reactor packed with 1.3 mm trilobe pellets of 1% Pd/Al$_2$O$_3$ catalyst. The spatially resolved ^{13}C-DEPT spectra were quantified at steady-state reaction conditions and showed that at low hydrogen (lean) flow rates, significant isomerization of the 1-octene to 2-octene occurred with increasing distance down the reactor from the inlet, but only small amounts of 3-, 4-octene and octane were produced. Increasing the hydrogen flow rate showed two major differences to that of lean hydrogen conditions: (1) hotspot formation was evident as shown by the disappearance of the ^{1}H signal associated with liquid octenes/octane in the associated two-dimensional ^{1}H images (see Figure 11.22), (2) the hydrogenation of 1-octene to 1-octane was more complete with a greater proportion of 3-and 4-octene isomers also being produced.

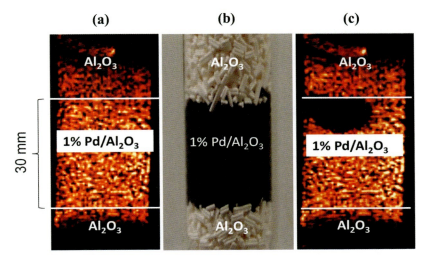

Figure 11.22 *In situ* 2D ^{1}H image of liquid distribution during the octene hydrogenation reaction for the bed operating at steady state for (a) gas and 1-octene flow rates of 32 and 1.0 ml min^{-1}, respectively and (c) gas and 1-octene flow rates of 64 and 1.0 ml min^{-1}, respectively. The loss of signal from the ^{1}H image depicted in (c) is due to vaporization of liquid octene into gas due to the formation of a hotspot induced by a localized exothermic reaction. (b) Optical image of the glass reactor showing the two zones of packing with inert alumina trilobes above and below the active catalysts shown as the black colored areas. (Adapted from Ref. [65]. Copyright 2005, Springer.)

In a related study, Gladden *et al.* [66] studied the hydrogenation of 1-octene over a 2 wt% Pd/Al$_2$O$_3$ catalyst in a fixed bed demonstrating the further ability of NMR to map temperature and chemical composition within the reactor at operating conditions. The temperature mapping was carried out using volume selective spectroscopy (VOSY) and measuring (wirelessly) the chemical shift difference between the two ^{1}H spectral peaks of ethylene glycol, contained in glass microspheres, placed at different locations within the reactor. Figure 11.23 shows a vertical 2D slice selective ^{1}H image from within the packed reactor illustrating the locations of the wireless temperature sensors and the catalyst pellets used.

The VOSY sequence allows localized ^{1}H spectra to be acquired only from the glass bulbs and the difference in chemical shift between the two ^{1}H peaks of ethylene glycol is known to be temperature dependent. Figure 11.23b shows the changes in temperatures of these glass bulbs shown in Figure 11.23a during the first 400 min of reaction. More recently, Jarenwattananon *et al.* [67] described a new *in situ* ^{1}H NMR thermometry method that exploits the inverse relationship between NMR linewidths and temperature caused by motional averaging in a weak magnetic field gradient. Gas temperature maps during the hydrogenation of propylene in model mini-reactors packed with metal nanoparticles and metal–organic framework catalysts were described and showed that this

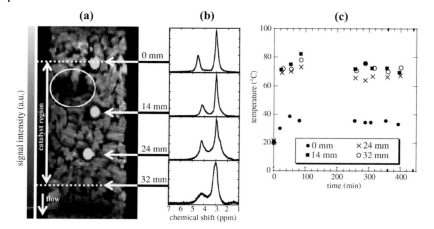

Figure 11.23 Temperature measurements during 1-octene hydrogenation over a 2 wt% Pd/Al$_2$O$_3$ catalyst. (a) 2D ^{1}H image showing the position of the sealed glass bulbs of ethylene glycol. The field of view was 50 mm (z) × 30 mm (x). The dotted lines indicate the beginning and end of the catalyst region and the solid circular line indicates the presence of a hotspot. (b) Localized ^{1}H spectra taken during reaction. (c) Evolution of the temperature for each of the glass bulbs shown in (a). (Adapted from Ref. [66]. Copyright 2010, Elsevier.)

technique could be useful for locating hot and cold spots in catalyst-packed gas–solid reactors.

Gladden et al. [66] also demonstrated the ability of multinuclear–multiphase magnetic resonance in the form of ^{1}H for liquid phase and ^{19}F for gas phase to image and quantify gas and liquid velocities during gas (SF$_6$) and liquid (H$_2$O) flow in a packed bed of glass spheres. Figure 11.24 summarizes the results from this study and shows combined gas–liquid velocity maps. In Figure 11.24a, there is no gas flow, and the liquid superficial velocity is 2.3 mm s^{-1}. For all images sown, the liquid superficial velocity remains constant and the gas superficial velocity is increased from (b) 11.2 mm s^{-1} to (c) 34.9 mm s^{-1}, through to (d) 61.3 mm s^{-1}.

The glass bead packing elements within the bed are shown as black (no ^{1}H signal). The location and velocity of the liquid phase is depicted by the green–blue color bar. The location and velocity of gas is shown by the red–yellow color bar. Figure 11.24 shows in general that the regions of highest gas velocity exist in the center of local elements of voidage between neighboring glass beads that are predominately gas filled. The glass beads forming the boundary of these pores tend to be wetted by near-stagnant liquid films. In general, the velocities across a gas–liquid interface are close to zero for both phases, although some regions of high relative (gas–liquid) velocity have been identified. This type of data is likely to be of great use to the numerical modeling community to help identify the appropriate closure law for use in numerical simulations; for example, computational fluid dynamics.

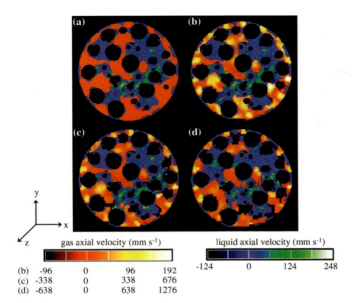

(b)	-96	0	96	192
(c)	-338	0	338	676
(d)	-638	0	638	1276

Figure 11.24 Magnetic resonance images of gas–liquid velocity fields within a model trickle bed reactor. Gas and liquid velocities are shown by the red-yellow color bar and blue-green color bar, respectively. The liquid superficial velocity for all images is 2.3 mm s^{-1}. The gas superficial velocity is (a) 0 mm s^{-1}, (b) 11.2 mm s^{-1}, (c) 34.9 mm s^{-1}, and (d) 61.3 mm s^{-1}. (Adapted from Ref. [66]. Copyright 2010, Elsevier.)

One of the major drawbacks with conducting magnetic resonance experiments in general is the limited space available for the samples under investigation. Today, most commercially available magnetic resonance systems, capable of performing both spectroscopy and imaging, are limited to incorporate cylindrical sample environments of around 60 mm in diameter. This is because of the cylindrical nature of the radio frequency coils and associated magnetic field gradient system. In addition, the materials used to contain the samples must not be metallic and certainly not ferromagnetic. Roberts *et al.* [68] gave the first report of an "MRI compatible" (40 mm outer diameter 20 mm internal diameter) heterogeneous reactor constructed from silicon nitride ceramic. This equipment was capable of operating up to 350 °C and 30 barg gas pressure. Roberts *et al.* [68] used ^{1}H NMR spectroscopy to follow the oligomerization of ethylene over a nickle-based silica alumina catalyst at 110 °C and 28 barg of ethylene pressure flowing at 14.4 Nl h^{-1}. The results showed that the ^{1}H spectra acquired over the time course of the reaction could be used to calculate the conversion and product distribution as a function of time on stream.

Heterogeneous catalytic microreactors have also been characterized using magnetic resonance microimaging. Bouchard *et al.* [69] showed that it was possible to use MRI and MRI velocimetry techniques combined with *para*-hydrogen (to enhance the ^{1}H signal) to visualize propane flow during the (*para*)-

hydrogenation of propene gas over a silica-gel-supported Wilkinson's catalyst. The authors concluded that heterogeneous flow patterns existed within the packing that were consistent with the nonuniform packing of the catalyst and this information could not have been deduced without the use of *para*-hydrogen as the thermal polarization of propene was 300 times lower than the hyperpolarized propane hydrogenation product.

11.4
Summary

With this chapter we hope to have given the reader a brief but useful insight into the use of NMR methods to study and understand a series of systems or relevance in reaction engineering and catalysis. These techniques are noninvasive, *in situ*, quantitative and chemically selective, and are able to provide a multifaceted approach to investigate the physical and chemical behavior of catalysts and chemical reactors, being able to provide data on kinetics, reactor performances, product distribution, mass transport, adsorption as well as hydrodynamics within chemical reactors and catalytic systems. In addition, NMR spectroscopy and microscopy combined with magnetic resonance imaging (MRI) are able to provide spatial resolution in chemical reactors, including spatial distribution of conversion, selectivity, temperature, velocity, adsorption, and diffusion properties within the reactor; hence obtain unique insights into reactor performances and process optimization.

References

1 Gladden, L.F., Alexander, P., Britton, M.M., Mantle, M.D., Sederman, A.J., and Yuena, E.H.L. (2003) *In situ* magnetic resonance measurement of conversion, hydrodynamics and mass transfer during single- and two-phase flow in fixed-bed reactors. *Magnetic Resonance Imaging*, **21**, 213–219.

2 Callaghan, P.T. (1993) *Principles of Nuclear Magnetic Resonance Microscopy*, Oxford University Press, Oxford, UK.

3 Gladden, L.F. and Mitchell, J. (2011) Measuring adsorption, diffusion and flow in chemical engineering: applications of magnetic resonance to porous media. *New Journal of Physics*, **13**, 46.

4 Korb, J.P. (2011) Nuclear magnetic relaxation of liquids in porous media. *New Journal of Physics*, **13**, 26.

5 Mitchell, J., Howe, A.M., and Clarke, A. (2015) Real-time oil-saturation monitoring in rock cores with low-field NMR. *Journal of Magnetic Resonance*, **256**, 34–42.

6 Levitt, M. (2001) *Spin Dynamics: Basics of Nuclear Magnetic Resonance*, John Wiley & Sons Inc., New Jersey.

7 Keeler, J. (2005) *Understanding NMR Spectroscopy*, John Wiley & Sons, Ltd, Chichester, UK.

8 Abragam, A. (1961) *The Pinciples of Nuclear Magnetism*, Oxford University Press, New York.

9 Rabi, I., Zacharias, J.R., Millman, S., and Kusch, P. (1938) A new method of measuring nuclear magnetic moment. *Physical Review*, **53**, 318

10 Jameson, C.J. (1991) Gas-phase NMR spectroscopy. *Chemical Reviews*, **91**, 1375–1395.

11 Hore, P.J., Davies, S.G., Compton, R.G., Evans, J., and Gladden, L.F. (1995) *Nuclear Magnetic Resonance*, Oxford Science Publications, Oxford, UK.

12 Ault, A. (1998) *Techniques and Experiments for Organic Chemistry*, University Science Books, California.

13 Fukushima, E. and Roeder, S.W. (1981) *Experimental Pulse NMR*, Addison-Wesley, Reading, USA.

14 Carr, H.Y. and Purcell, E.M. (1954) Effects of diffusion on free precession in nuclear magnetic resonance experiments. *Physical Review*, **94**, 630–638.

15 Stejskal, E.O. and Tanner, J.E. (1965) Spin diffusion measurements: spin echoes in the presence of a time-dependent field gradient. *Journal of Chemical Physics*, **42**, 288–292.

16 Tanner, J.E. (1970) Use of stimulated echo in NMR diffusion studies. *Journal of Chemical Physics*, **52**, 2523–2526.

17 Cotts, R.M., Hoch, M.J.R., Sun, T., and Markert, J.T. (1989) Pulsed field gradient stimulated echo methods for improved NMR diffusion measurements in heterogeneous systems. *Journal of Magnetic Resonance*, **83**, 252–266.

18 Wu, Y., D'Agostino, C., Holland, D.J., and Gladden, L.F. (2014) *In situ* study of reaction kinetics using compressed sensing NMR. *Chemical Communications*, **50**, 14137–14140.

19 Mantle, M.D., Steiner, P., and Gladden, L.F. (2006) Polarisation enhanced C-13 magnetic resonance studies of the hydrogenation of pentene over Pd/Al2O3 catalysts. *Catalysis Today*, **114**, 412–417.

20 Koptyug, I.V., Kulikov, A.V., Lysova, A.A., Kirillov, V.A., Parmon, V.N., and Sagdeev, R.Z. (2002) NMR imaging of the distribution of the liquid phase in a catalyst pellet during alpha-methylstyrene evaporation accompanied by its vapor-phase hydrogenation. *Journal of the American Chemical Society*, **124**, 9684–9685.

21 Camp, J.C.J., Mantle, M.D., York, A.P.E., and McGregor, J. (2014) A new combined nuclear magnetic resonance and Raman spectroscopic probe applied to *in situ* investigations of catalysts and catalytic processes. *Review of Scientific Instruments*, **85**, 063111.

22 Green, R.A., Adams, R.W., Duckett, S.B., Mewis, R.E., Williamson, D.C., and Green, G.G.R. (2012) The theory and practice of hyperpolarization in magnetic resonance using parahydrogen. *Progress in Nuclear Magnetic Resonance Spectroscopy*, **67**, 1–48.

23 Koptyug, I.V., Kovtunov, K.V., Burt, S.R., Anwar, M.S., Hilty, C., Han, S.-I., Pines, A., and Sagdeev, R.Z. (2007) para-Hydrogen-induced polarization in heterogeneous hydrogenation reactions. *Journal of the American Chemical Society*, **129**, 5580–5586.

24 Kovtunov, K.V., Beck, I.E., Bukhtiyarov, V.I., and Koptyug, I.V. (2008) Observation of parahydrogen-induced polarization in heterogeneous hydrogenation on supported metal catalysts. *Angewandte Chemie, International Edition*, **47**, 1492–1495.

25 D'Agostino, C. (2011) Advanced NMR techniques in sustainable chemistry, PhD thesis, University of Cambridge, Cambridge.

26 Kondrat, S.A., Miedziak, P.J., Douthwaite, M., Brett, G.L., Davies, T.E., Morgan, D.J., Edwards, J.K., Knight, D.W., Kiely, C.J., Taylor, S.H., and Hutchings, G.J. (2014) Base-free oxidation of glycerol using titania-supported trimetallic Au-Pd-Pt nanoparticles. *ChemSusChem*, **7**, 1326–1334.

27 Hayamizu, K., Tsuzuki, S., and Seki, S. (2008) Molecular motions and ion diffusions of the room-temperature ionic liquid 1,2-dimethyl-3-propylimidazolium bis(trifluoromethylsulfonyl)amide (DMPImTFSA) studied by ^{1}H, ^{13}C, and ^{19}F NMR. *Journal of Physical Chemistry A*, **112**, 12027–12036.

28 Stallmach, F. and Kärger, J. (1999) The potentials of pulsed field gradient NMR for investigation of porous media. *Adsorption*, **5**, 117–133.

29 Vasenkov, S., Geir, O., and Kärger, J. (2003) Gas diffusion in zeolite beds: PFG NMR evidence for different tortuosity factors in the Knudsen and bulk regimes. *The European Physical Journal E*, **12**, 35–38.

30 Kortunov, P., Vasenkov, S., Kärger, J., Elia, M.Fe., Perez, M., Stocker, M.,

Papadopoulos, G.K., Theodorou, D., Drescher, B., McElhiney, G., Bernauer, B., Krystl, V., Kocirik, M., Zikanova, A., Jirglova, H., Berger, C., Glaser, R., Weitkamp, J., and Hansen, E.W. (2005) Diffusion in fluid catalytic cracking catalysts on various displacement scales and its role in catalytic performance. *Chemistry of Materials*, **17**, 2466–2474.

31 Divitini, G., Stenzel, O., Ghadirzadeh, A., Guarnera, S., Russo, V., Casari, C.S., Bassi, A. Li., Petrozza, A., Di Fonzo, F., Schmidt, V., and Ducati, C. (2014) Nanoscale analysis of a hierarchical hybrid solar cell in 3D. *Advanced Functional Materials*, **24**, 3043–3050.

32 Vallabh, R., Banks-Lee, P., and Seyam, A.-F. (2010) New approach for determining tortuosity in fibrous porous media. *Journal of Engineered Fibers and Fabrics*, **5**, 7–15.

33 Salmas, C.E. and Androutsopoulos, G.P. (2001) A novel pore structure tortuosity concept based on nitrogen sorption hysteresis data. *Industrial and Engineering Chemistry Research*, **40**, 721–730.

34 Hollewand, M.P. and Gladden, L.F. (1995) Transport heterogeneity in porous pellets. 1. PGSE NMR studies. *Chemical Engineering Science*, **50**, 309–326.

35 D'Agostino, C., Mitchell, J., Gladden, L.F., and Mantle, M.D. (2012) Hydrogen bonding network disruption in mesoporous catalyst supports probed by PFG-NMR diffusometry and NMR relaxometry. *The Journal of Physical Chemistry C*, **116**, 8975–8982.

36 Mantle, M.D., Enache, D.I., Nowicka, E., Davies, S.P., Edwards, J.K., D'Agostino, C., Mascarenhas, D.P., Durham, L., Sankar, M., Knight, D.W., Gladden, L.F., Taylor, S.H., and Hutchings, G.J. (2011) Pulsed-field gradient NMR spectroscopic studies of alcohols in supported gold catalysts. *Journal of Physical Chemistry C*, **115**, 1073–1079.

37 Dorazio, F., Bhattacharja, S., Halperin, W.P., and Gerhardt, R. (1989) Enhanced self-diffusion of water in restricted geometry. *Physical Review Letters*, **63**, 43–46.

38 Dorazio, F., Bhattacharja, S., Halperin, W.P., and Gerhardt, R. (1990) Fluid transport in partially filled porous sol–gel silica glass. *Physical Review B: Condensed Matter*, **42**, 6503–6508.

39 Valiullin, R.R., Skirda, V.D., Stapf, S., and Kimmich, R. (1997) Molecular exchange processes in partially filled porous glass as seen with NMR diffusometry. *Physical Review E*, **55**, 2664–2671.

40 Krutyeva, M., Grinberg, F., Kärger, J., Chorro, C., Donzel, N., and Jones, D.J. (2009) Study of the diffusion of liquids and their binary mixtures in mesoporous aluminosilicates under freezing conditions. *Microporous and Mesoporous Materials*, **120**, 104–108.

41 Dvoyashkin, M., Valiullin, R., and Kärger, J. (2007) Temperature effects on phase equilibrium and diffusion in mesopores. *Physical Review*, **75**, 041202.

42 Schönhoff, M. and Söderman, O. (1997) PFG-NMR diffusion as a method to investigate the equilibrium adsorption dynamics of surfactants at the solid/liquid interface. *The Journal of Physical Chemistry B*, **101**, 8237–8242.

43 Weber, D., Sederman, A.J., Mantle, M.D., Mitchell, J., and Gladden, L.F. (2010) Surface diffusion in porous catalysts. *Physical Chemistry Chemical Physics*, **12**, 2619–2624.

44 Kärger, J. (1969) Zur Bestimmung der Diffusion in einem Zweibereichsystem mit Hilfe von gepulsten Feldgradienten. *Annals of Physics (Leipzig)*, **24**, 1–4.

45 D'Agostino, C., Kotionova, T., Mitchell, J., Miedziak, P.J., Knight, D.W., Taylor, S.H., Hutchings, G.J., Gladden, L.F., and Mantle, M.D. (2013) Solvent effect and reactivity trend in the aerobic oxidation of 1,3-propanediols over gold supported on titania: NMR diffusion and relaxation studies. *Chemistry - A European Journal*, **19**, 11725–11732.

46 D'Agostino, C., Brett, G.L., Miedziak, P.J., Knight, D.W., Hutchings, G.J., Gladden, L.F., and Mantle, M.D. (2012) Understanding the solvent effect on the catalytic oxidation of 1,4-butanediol in methanol over Au/TiO$_2$ catalyst: NMR diffusion and relaxation studies. *Chemistry - A European Journal*, **18**, 14426–14433.

47 Wood, J. and Gladden, L.F. (2003) Effect of coke deposition upon pore structure and self-diffusion in deactivated industrial hydroprocessing catalysts. *Applied Catalysis A - General*, **249**, 241–253.

48 Haider, M.H., D'Agostino, C., Dummer, N.F., Mantle, M.D., Gladden, L.F., Knight, D.W., Willock, D.J., Morgan, D.J., Taylor, S.H., and Hutchings, G.J. (2014) The effect of grafting zirconia and ceria onto alumina as a support for silicotungstic acid for the catalytic dehydration of glycerol to acrolein. *Chemistry – A European Journal*, **20**, 1743–1752.

49 D'Agostino, C., Ryabenkova, Y., Miedziak, P.J., Taylor, S.H., Hutchings, G.J., Gladden, L.F., and Mantle, M.D. (2014) Deactivation studies of a carbon supported AuPt nanoparticulate catalyst in the liquid-phase aerobic oxidation of 1,2-propanediol. *Catalysis Science & Technology*, **4**, 1313–1322.

50 Mitchell, J., Hurlimann, M.D., and Fordham, E.J. (2009) A rapid measurement of T_1/T_2: the DECPMG sequence. *Journal of Magnetic Resonance*, **200**, 198–206.

51 Godefroy, S., Korb, J.P., Fleury, M., and Bryant, R.G. (2001) Surface nuclear magnetic relaxation and dynamics of water and oil in macroporous media. *Physical Review E*, **64**, 13.

52 Song, K.M., Mitchell, J., Jaffel, H., and Gladden, L.F. (2010) Simultaneous monitoring of hydration kinetics, microstructural evolution, and surface interactions in hydrating gypsum plaster in the presence of additives. *Journal of Materials Science*, **45**, 5282–5290.

53 McDonald, P.J., Korb, J.P., Mitchell, J., and Monteilhet, L. (2005) Surface relaxation and chemical exchange in hydrating cement pastes: a two-dimensional NMR relaxation study. *Physical Review E*, **72**, 9.

54 Bloembergen, N., Purcell, E.M., and Pound, R.V. (1948) Relaxation effects in nuclear magnetic resonance absorption. *Physical Review*, **73**, 679–746.

55 Liu, G., Li, Y., and Jonas, J. (1991) Confined geometry-effects on reorientational dynamics of molecular liquids in porous silica glasses. *Journal of Chemical Physics*, **95**, 6892–6901.

56 D'Agostino, C., Mitchell, J., Mantle, M.D., and Gladden, L.F. (2014) Interpretation of NMR relaxation as a tool for characterising the adsorption strength of liquids inside porous materials. *Chemistry – A European Journal*, **20**, 13009–13015.

57 Godefroy, S., Fleury, M., Deflandre, F., and Korb, J.P. (2002) Temperature effect on NMR surface relaxation in rocks for well logging applications. *The Journal of Physical Chemistry B*, **106**, 11183–11190.

58 Song, Y.Q., Venkataramanan, L., Hurlimann, M.D., Flaum, M., Frulla, P., and Straley, C. (2002) T_1-T_2 correlation spectra obtained using a fast two-dimensional Laplace inversion. *Journal of Magnetic Resonance*, **154**, 261–268.

59 Ralphs, K., D'Agostino, C., Burch, R., Chansai, S., Gladden, L.F., Hardacre, C., James, S.L., Mitchell, J., and Taylor, S.F.R. (2014) Assessing the surface modifications following the mechanochemical preparation of a Ag/Al$_2$O$_3$ selective catalytic reduction catalyst. *Catalysis Science & Technology*, **4**, 531–539.

60 Delbecq, F. and Sautet, P. (2002) A density functional study of adsorption structures of unsaturated aldehydes on Pt(111): a key factor for hydrogenation selectivity. *Journal of Catalysis*, **211**, 398–406.

61 Gladden, L.F., Mantle, M.D., and Sederman, A.J. (2006) *Advances in Catalysis*, vol. **50** (eds C.G. Bruce and K. Helmut), Academic Press, pp. 1–75.

62 Stapf, S. and Han, S.-I. (2006) *NMR Imaging in Chemical Engineering*, Wiley-VCH Verlag GmbH, Weinheim, Germany.

63 Lysova, A.A. and Koptyug, I.V. (2010) Magnetic resonance imaging methods for in situ studies in heterogeneous catalysis. *Chemical Society Reviews*, **39**, 4585–4601.

64 Akpa, B.S., Mantle, M.D., Sederman, A.J., and Gladden, L.F. (2005) In situ C-13 DEPT-MRI as a tool to spatially resolve chemical conversion and selectivity of a heterogeneous catalytic reaction occurring in a fixed-bed reactor. *Chemical Communications*, 2741–2743.

65 Sederman, J.A., Mantle, D.M., Dunckley, P.C., Huang, Z., and Gladden, F.L. (2005) In situ MRI study of 1-octene isomerisation and hydrogenation within a trickle-bed reactor. *Catalysis Letters*, **103**, 1–8.

66 Gladden, L.F., Abegão, F.J.R., Dunckley, C.P., Holland, D.J., Sankey, M.H., and Sederman, A.J. (2010) MRI: operando measurements of temperature, hydrodynamics and local reaction rate in a

heterogeneous catalytic reactor. *Catalysis Today*, **155**, 157–163.

67 Jarenwattananon, N.N., Gloggler, S., Otto, T., Melkonian, A., Morris, W., Burt, S.R., Yaghi, O.M., and Bouchard, L.-S. (2013) Thermal maps of gases in heterogeneous reactions. *Nature*, **502**, 537–540.

68 Roberts, S.T., Renshaw, M.P., Lutecki, M., McGregor, J., Sederman, A.J., Mantle, M.D., and Gladden, L.F. (2013) Operando magnetic resonance: monitoring the evolution of conversion and product distribution during the heterogeneous catalytic ethene oligomerisation reaction. *Chemical Communications*, **49**, 10519–10521.

69 Bouchard, L.-S., Burt, S.R., Anwar, M.S., Kovtunov, K.V., Koptyug, I.V., and Pines, A. (2008) NMR imaging of catalytic hydrogenation in microreactors with the use of para-hydrogen. *Science*, **319**, 442–445.

12
An Introduction to Closed-Loop Process Optimization and Online Analysis

Christopher S. Horbaczewskyj, Charlotte E. Willans, Alexei A. Lapkin, and Richard A. Bourne

12.1
Introduction

Optimization of synthesis reactions is of utmost importance to many chemical manufacturing industries, such as pharmaceutical [1–3], fine chemical [4, 5], and agrochemical [6, 7], and has traditionally been carried out at laboratory scale using glass reactor vessels as batch processes and larger stirred tank reactors during scale-up and manufacturing. The objective of optimization studies is to obtain a set of operating conditions or an optimal trajectory of a reaction, a "recipe" that allows maximizing a specific objective function, typically a combination of reaction time, yield, selectivity, and cost. In addition, the objective of optimization may be to gain in-depth understanding of a process through exploration of a wide range of operating conditions. In both cases, there is a need for optimization strategy, or tools, that aid decisions about best experiments to perform to achieve the set objective.

Commonly, and especially in the academic setting, reactions are optimized using a one variable at a time (OVAT) approach, for example, varying temperature, while keeping all other parameters constant in the kinetic studies, and so on. However, this approach may not necessarily find the "true" optimum within the experimental parameter space, and neglects to gain a complete understanding of the chemical system [8]. It also suffers from inability to differentiate noise of the system from the calculated improvement, unless a significant number of reactions are carried out at each iteration [9].

An alternative approach of an increased mathematical complexity allows varying multiple process factors simultaneously and gaining insight by statistical treatment of the reaction outcomes. This method, known as design of experiments (DoE), may require a similar number of reactions to be completed as OVAT, but will result in a better profile of the reaction with much more detailed results, maximizing the chances that an optimum operation zone is found.

Increasing the complexity of the optimization process typically requires that the chemical process be automated[†] and, so far this has been, arguably, easily achieved for small-scale continuous-flow experiments, rather than stirred-tank-type experiments. This is mainly due to the technical realization of the feedback loop between experimental design and experimental outcome, being simpler in realization for continuous-flow experiments. However, it is possible for a batch process to be automated in this way, especially if the experiment is miniaturized from nanoliter scale to microliter scale, or if disposable reactionware is being used.

The automation of chemical syntheses using continuous-flow reactors has recently emerged as a highly promising generic methodology for process development and optimization. Significant benefits of the fully automated process design are the potential to significantly decrease the overall cost and energy use, while maximizing the output productivity [2]. This comes as an added benefit to the already widely recognized advantages of continuous-flow processes, such as ability to control hazardous reactions (e.g., with unstable reactive intermediates, toxic substances, large exotherms, or high pressures or temperatures) and tight control of the operating conditions [10, 11].

This chapter aims to outline the current approaches and potential future developments of the automated process optimization methodologies. Section 12.2.4 deals with the components of a self-optimization system, namely, experimental equipment, analytical methods, and algorithms. Section 12.5 provides examples of the applications of self-optimization in recent literature.

12.2
Principles of Self-Optimization and Requirements for Experimental Systems

An automated self-optimizing system consists of three main components: intelligent controller, process execution, and in-line/online analysis (Figure 12.1). The intelligent controller sets and monitors reaction parameters. The controller also receives the output data from analysis, processes the data by extracting all relevant information, and performs calculations such as standard deviations or percentage yields. The interaction between the controller and other pieces of equipment in the reactor allows the generation of a feedback loop. The feedback loop enables the reactor system to be in a constant mode of parameter updating and output monitoring to eventually allow the algorithm to move the system toward a global optimum within the specified design space. This method completely depends on the feedback loop to ensure the system is a closed loop for optimization. The intelligent controller is connected to all sections of hardware

---

†) Automation can refer to the use of robots to carry out specific jobs that require a high level of control or the use of computer programs that can take complete control of the chemical process system. Any reference to automation in the text refers to the latter definition unless specified otherwise.

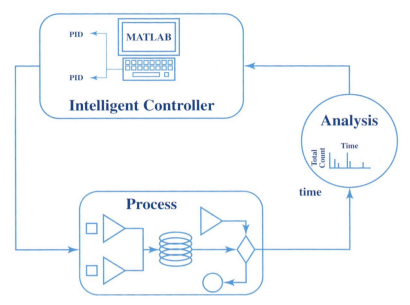

Figure 12.1 A schematic diagram of a generic closed-loop automated system containing an intelligent controller, a process, and online analysis. The intelligent controller has complete control of the system, monitors it, and induces changes when necessary. The intelligent controller is typically a design-of-experiments or optimization algorithm, and may make use of conventional control tools for process control, such as proportional–integral–derivative (PID) controllers. The process consists of reagent reservoirs (squares), pumps (triangles), a reactor (coil), a sample loop (diamond), sample collection (circle), and a pump to transport an aliquot to the chosen analytical technique.

in the system and can induce changes in all adjustable parameters upon demand, for example, by using lower level controllers, such as PID controllers.

Typically, the process, where the chemistry takes place, comprises reagent reservoirs, pumps, connecting tubes, reactors, sometimes a separation unit, and collection vessel(s). This section of the process is typically modular and can include any number of reagents and pumps, whether they are before the reactor or after to allow for different reaction modes (homogeneous, heterogeneous, recycling, controlled addition, reaction conditions gradients, and so on), quenching, and workup strategies. Over the last decade, there has been a significant advance in continuous-flow technology with several commercial providers offering large numbers of reaction and separation modules, including the more recent additions, such as cold reactors, photochemical and electrochemical reactors by Cambridge Reactor Design, Vapourtec, and Syrris.

A closed-loop system is created with the use of in-line/online analytical technique(s) that provide, ideally, real-time data on reaction outcomes and other process state variables. There are many analytical methods and sensors that can be used, discussed in more detail in Section 12.3. Online analysis is completed in exactly the same way as any offline analyses; however, the sensor or sampling is positioned in close proximity to the reaction and directly analyzes the system by either extracting a small portion into a parallel mobile phase [12] or by direct analysis of the reaction [13, 14].

Continuous-flow systems can allow reaction environments to be changed almost instantaneously, which is highly beneficial to allow the integration of an analytical method with a computer-based control/experimental design algorithm. This allows for a multitude of conditions, including, but not limited to, reactor temperature, flow rate of reagents, and system pressure to be automatically and rapidly changed within the automated experimental system. There are many algorithms that can be used in a closed-loop automated reactor system; algorithms are discussed in Section 12.4.

12.3
Analytical Techniques for Closed-Loop Optimization

The use of an algorithm in automating or optimizing a process necessitates the creation of a feedback loop. This loop relies on the output from an in-line/online analytical technique. The analysis of a continuous process can be completed with many different analytical techniques without the need to isolate product or to manually transfer it for offline analysis. However, there are specific aims for the analytical methods: (i) to minimize the delay between an experiment and its results being available in a form suitable for algorithmic evaluation against the set experiment objective, and (ii) to maximize the observability of the system states, which for chemical synthesis involve qualitative identification of molecules and analysis of their structure and quantification of composition. The first objective is necessary to ensure that the closed-loop experimental system is not idle, waiting for analytical results. The ability to perform large numbers of experiments with small quantities of substrates is the key differentiation of the automated self-optimization experiments over traditional process development/optimization approaches. The second objective is obviously critical for algorithmic discovery and optimization procedures.

Online analysis monitors the reaction by extracting a portion of the flow volume, while in-line analysis allows the monitoring of the entire flow stream whether it is destructive or nondestructive. The use of an analytical technique directly at the end of the chemical reaction can aid the optimization of the process by allowing reactants to be observed as well as products, intermediates, and impurities. Real-time reaction information is beneficial for any chemical process, yield calculations, kinetic data, and the search for optimum conditions. Each analytical technique has its own set of advantages and disadvantages, summarized in Table 12.1. It is possible to group them together in terms of acquisition speeds and level of detail the output provides. HPLC and GC both have slower analysis, but both tend to give higher quality data, which are easy to interpret and can be easily quantified. However, HPLC and GC are restricted by the

Table 12.1 An outline of analytical techniques and sensors that can be used in closed-loop optimization.

Technique	Information type	Sensitivity (mol%)	Acquisition speed (s)	Limitations
Mid-IR	Chemical identity, concentration. Solid, liquid, or gas samples	$\sim 10^{-1}$	~ 1	Short fibers Intolerant to water
Near-IR	Chemical identity, concentration	$\sim 10^{-1}$	~ 1	Less informative than mid-IR, tolerant to water
Raman	Chemical identity, crystal structure, concentration. Solid or liquid samples.	$\sim 10^{-1}$ Potentially to individual molecules in the case of SER(R)S	$\sim 1\text{--}100$	Fluorescence masking Raman signal
UV-vis	Chemical identity, concentration	$\sim 10^{-4}$	<1	Limited number of species
NMR	Molecular structure, I.D. of unknown compounds, concentration	$\sim 10^{-3}$	~ 10	At present, flow method is limited in sensitivity and resolution due to low field
GC	Concentration	$\sim 10^{-6}$	10–1500	Typically – slow. Cannot I.D. unknown compounds. Difficult to automate
HPLC	Concentration	$\sim 10^{-6}$	200–1500	Long method development times. Must be combined with MS for proof of molecular identity
MS or MS/MS	Chemical identity, concentration	$\sim 10^{-8}$	5–20	Requires cheminformatics expertise. MS/MS in more informative, but few process instruments on the market. Difficult method development on more advanced systems
Process Sensors	N/A	Pressure Temperature pH Conductivity Viscosity Dynamic light scattering Ultrasound		Sensor fouling Sensor fouling Requires specific inline cell Difficult for inline, requires dilution Presently used only for level Underdeveloped

Adapted with permission from Ref. [16]. Copyright 2015, Elsevier.

difficulty in determining molecular identity without the use of mass spectrometry [15, 16]. FT-IR, NMR, and MS all have fast acquisitions times, between 1 and 20 s, and can output data rapidly. It is possible to calibrate FT-IR to be quantitative, but this generally requires complex chemometric modeling with off-line quantitative chromatographic techniques or backed by NMR data. IR is also prone to spectral interference, which can lead to misinterpretation of results; spectra can be far more complex if impurities are present as they add a greater complexity to the system. NMR spectra can also be complex if large amounts of impurities are present. However, this method is far easier to calibrate, even with current bench-top instruments having limited resolution and sensitivity. NMR also has the benefits of being non-invasive and providing structural information on reaction components, like FT-IR. Quantification is possible through MS, but is difficult as ion suppression and mobile phase effects add greater complexity. Calibration would also have to take into account isotopic species from starting materials and products. A recent paper by Holmes *et al.* outlined how online mass spectrometry can be used quantitatively to optimize reactions. Quantitation was achieved by use of a 5:2 flow splitter and atmospheric pressure chemical ionization, which reduced the total concentration of the sample to be in the linear range of the mass spectrometer used [17]. However, ample structural data can be gained from MS analysis in minimal time and output spectra are relatively easy to interpret. The sample size needed to obtain MS spectra with good signal-to-noise ratio is very small, which allows this technique to be used even with very small-scale reactions. MS is a highly versatile and powerful technique to use as an online analytical tool coupled to a continuous-flow reactor and can provide detailed and rapid analysis in real time [15, 16].

12.4
Decision Algorithms in Closed-Loop Optimization

The core of a closed-loop optimization system is the intelligent controller, see Figure 12.1, which includes the instrument communication part and the decision/optimization algorithm. The communication part may have many possible practical solutions and largely reflect the preference of the end user. However, we should point to the increasing use of low-cost automation tools, such as Raspberry Pi-based instrument monitoring and control, which then requires the corresponding control coding in Python or similar platforms, as opposed to more established industrial solutions, such as LabView and others. Our main focus in this chapter will be on the decision algorithms.

We can identify three main tasks for automated closed-loop experiments, which require different decision algorithms:

1) Discovery of new materials or reactions.
2) Developing increased process understanding.
3) Identification of optimal operating conditions against a set objective function.

12.4.1
Algorithms for Discovery

Arguably, discovery is the hardest task for automation of chemistry. A very large number of current commercial successes in the chemical industry began as serendipitous discoveries, if not as failed experiments, being totally unplanned. Human ingenuity and curiosity leads to performing what may seem as unreasonable experiments. Observation of reality by a trained scientist allows spotting new phenomena. What we require from discovery algorithms is to do the same, but without human interference. This automatically rules out a large number of conventional approaches. For example, most knowledge-based approaches would not work, since our current level of knowledge likely does not contain the discovery we seek, or it would have already been found. Similarly, most data-driven approaches, including machine learning, are likely to fail: If the highly different data point is not present in the data set, the algorithms would not predict it.

An example of a discovery process, which we can use to identify the desired attributes of algorithms for discovery, is the work of the MacMillan group [18]. The "accelerated serendipity" approach was developed to promote discovery of new chemical transformations in the largely unexplored chemical space. The process is facilitated by robotic experiments to prepare reactants mixtures, dosing catalysts, running large numbers of reactions, and evaluating their outcomes using GC-MS. Based on the reactants and catalysts used in each reaction, a user would propose the likely reaction outcomes – obvious outcomes, which are then filtered out. The approach is specifically targeting unexpected reaction outcomes. For this, the evaluation of a reaction outcome focuses on significant GC-MS peaks and compound identification via the NIST MS spectra database. Thus, the first analysis of the reaction outcomes is with respect to whether the obtained result is predictable by expert scientists – in other words, by current knowledge. Based on structural identification of the compound, the next step is to confirm that the found transformation is "useful" or "important."

In the case of the pioneering work of MacMillan's group, the "importance" or "usefulness" is judged by the expert opinion or knowledge of frequency of appearance of a particular structural motif in end product, such as drug molecules. Thus, benzylic amines were present in 8 out of 100 top-selling pharmaceuticals in 2011. The algorithmic use of such criteria requires coding of heuristics and the ability to automate analysis of structures, identifying the main structural characteristics and comparing against a knowledge database. An alternative to this approach is to use the emerging Big Data approaches to chemical knowledge.

The recently published work on automated assembly of chemical routes [19] based on reported reactions suggests that it is possible to automatically evaluate which molecules are required for the most efficient transformations and what are the consequences of developing a new molecule, or a new reaction. The latter question would require a significant computational resource, as a reasonably large reaction network would be required to be reevaluated many times. However, it is a potential solution for algorithmic evaluation of usefulness of a new molecule or a new reaction, where the criteria will be "synthetic utility."

The issue of analysis of reaction outcomes is obvious for the application to "discovery": If we do not know what we are looking for, what analytical method to use in a "discovery" engine? If a problem is tightly bound and the range of likely outcomes is narrow, it is then possible to suggest what analytical method is appropriate. This is what has been done so far in the published studies on automation of discovery.

An example of a similar approach to accelerated serendipity developed by the group of MacMillan, but automated using active learning classification algorithm, has recently been published by the group of Cronin [20]. The problem was posed as follows: To find an "unknown," in this case formation of crystals yet not described in the literature, given a set of possible starting reactants and varying composition and synthesis conditions. The robotic experiment is driven by an active learner algorithm using the popular support vector machine classifier. The aim of the algorithm is to learn a fairly large experimental space and develop a classification model that would predict in which parts of the space there is an increased probability to find crystals. The problem is significantly simplified by removing all discrete variables – all reactants are considered to be present and their concentrations are continuous variables. The presence of discrete variables is a significant challenge to the application of machine learning algorithms in chemistry. This issue is discussed in a recent paper [21] and comes up again later in this section of the chapter.

Classification is a well-developed area of statistics with numerous methods developed for different problem settings with a large body of literature and available toolboxes. An example of the use of classification in chemistry is the problem of developing formulated products [22]. In formulations, the target functional specification, for example, a combination of sensory perception (olfaction, color, "feel" on touch), rheological behavior in use, main active function (antiperspirant, UV protection, food ingredient, paint, medicinal formulation), and shelf life, is attained by combining a number of ingredients such that their collective behavior under the conditions of use is close to the target specification. The combination of many ingredients with complex properties and behavior results in a complicated system that is largely not predictive by current mechanistic models. For this reason, statistical methods are useful, especially if a large amount of data are available. Dimensionality reduction and classification result in a statistical model of a formulation, valid within the range of tested input variables. The choice of methods used for dimensionality reduction and classification then depends on the complexity of the problem: number of dimensions, non-linearity, number of parameters in the product specification, and so on. As for any machine learning methods, the key requirement is access to data.

In the case when some data are available, there may be an alternative approach to discovery, based on fundamental understanding of the chemical system involved. If we are able to link the desired properties of a material, or a reaction, to be discovered with the structural properties of molecules via a structure–property model, we should be able to use that model in rational design of new, yet not discovered, materials and reactions. Arguably, this is not a "discovery" *per se*, but a facilitated development, a domain of computer-aided design. Certainly, the boundary between rational design and discovery is shrinking with the advent of automated methods discussed in this and other chapters of the book.

In this approach, the first step is to automate learning of the rules and connect the desired product's attributes or reaction outcome with molecular properties – molecular descriptors. The problem exists at several levels:

1) Discovery of fundamental laws governing an experimental system from data [23].
2) Development of a specific structure–property relation from data.
3) Development of complex structure–property relationships for multidimensional problems.

The problem of identification of a structure–property relationship in chemistry can be reformulated as identification of molecular descriptors – which function or property of a molecule (or a set of molecules) would translate into the desired property of the target material (or a reaction). This approach has been successfully demonstrated in a number of cases of chemical development, most notably in developing descriptors for solubility and the influence of solvents on reactivity [24–26]. A similar approach is now increasingly used to identify relevant descriptors and then use data mining and machine learning techniques to develop the required link between properties and descriptors, to be later used in the design [27, 28].

The more complex challenge is to identify molecular descriptors and rules that can be used in predicting chemical reactivity. This is an emerging field, combining data science, statistics, computational chemistry, and experimental synthetic chemistry. Some recent examples include different approaches to training of neural networks to chemical reactivity data and using the trained network to predict reaction outcomes [29–31]. There is, however, no guarantee that this approach could replace experiments and serendipity, as the quality of molecular descriptors today is insufficient to accurately describe reactivity [21].

12.4.2
Algorithms for Developing Process Understanding

The quality of process understanding can be represented by completeness of the process model. A list of what a process model would include can be represented as follows: (i) a set of thermodynamic properties over the full range of expected operating conditions of the system, (ii) detailed reaction kinetics, ideally based on the mechanism, validated and predictive, or at least an empirical kinetic expression, validated over the expected range of process conditions, (iii) a scale-up model, which includes behavior of pumps, valves, recycling streams, feeding schedule, and so on, at the process scale, and (iii) the purification/separation protocol. We can further classify everything in this list into two categories:

1) Rate processes, including both chemical and physical
2) Thermodynamic phenomena

Therefore, the problem of developing process understanding can be formulated as the problem of identification of relevant physical and chemical mechanisms that control the process within the range of experimental conditions of interest. Algorithmic identification of mechanisms is an emerging topic. One example of work in this field is the identification of physical laws from data [23]. A far simpler problem is that of model discrimination, when structure of a model, that is the nature of phenomena, is already known or a library of possible models exists, which then require to solve a simpler problem – selection of the suitable model and identification of its parameters. A number of approaches to this problem are discussed further in this chapter.

12.4.3
Algorithms for Automated Process Optimization

Within the framework of closed-loop process optimization, the set of algorithms targeting optimization of process conditions could be generally grouped together under the heading of design of experiments (DoE).

DoE is a statistical approach in which an optimum set of operating conditions is found by monitoring the relationship between variables from predefined experiments. A DoE approach needs an experimental design space to be constructed using as many numeric factors as possible, which can be controlled in the reaction system. This then dictates which reactions should be completed and, upon output analysis, outlines interactions and relationships between variables as well as creates a theoretical surface map containing the global optimum [32]. Relationships are determined by fitting the output process model to a mathematical function, typically to a polynomial with estimated coefficients that theoretically represents the entire design space. The book *A First Course in Design and Analysis of Experiments* by Gary W. Oehlert [33] outlines and explains the concept of experimental design in a coherent and concise manner.

When a system or process is optimized using a set of structured or predefined experiments, it is known as a designed experiment. DoE is a systematic method to determine the statistical relationship between process parameters and the output of that process. The optimized output is fully reliant on input variables, controllable factors, and noise factors. If the incorrect range of input variables is suggested, then the final output will provide an incorrect value. If factors are not controlled sufficiently, then the output value maybe distorted and not be representative of the initial conditions chosen [34].

There are several different forms of experimental design that are used in the development and optimization of processes. A factorial design uses a basic statistical model m^n where n is the number of parameters, qualitative or quantitative,

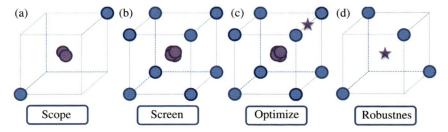

Figure 12.2 Four stages of an experimental design procedure. (a) Scope: exploration of the mildest to most forcing conditions. (b) Screen: narrow parameter range and find key parameters. (c) Optimization: use of statistical equations to derive an optimum set of conditions. (d) Robustness: test area around optimum to see if it provides a repeatable process.

in the specified design space and m is the number of values related to each parameter individually. As an example, in a three-parameter system (temperature, concentration, and flow rate) where each parameter has two values associated with it, there will be a total of 2^3 experiments needed, 8 in total to find an optimum output. The maximum number of reactions needed for an experimental design process is usually far lower than those carried out in an OVAT approach with far more information gained. This reduces the time spent on gained large amounts of useful data as well as improves the cost-efficiency as lower volumes of reagent and solvent will be used.

There are four stages in experimental design (Figure 12.2) that help understand the process or system in need of development. Initially, the system needs to be scoped, that is, explored from the mildest to the most forcing conditions, to check whether the proposed experimental space is appropriately wide. The design space is then subjected to a screening process where more combinations of parameters are tested to identify if any interactions are present. Parameters with a low statistical effect can be discounted at this stage. The optimum subsequently needs to be identified by performing a higher fidelity design of statistically significant parameters to identify interactions between parameters and curvature of the response surface, followed by identification of optimal operating conditions. Finally, a robustness design can be performed to verify conditions selected are robust and offer repeatable results.

The classical DoE approach requires predesign of a series of experiments and postanalysis of their results. An alternative is the sequential approach, where an optimization algorithm has access to each new experimental outcome, which may reduce the number of experiments required as well as allow for optimization toward a global optimal solution. Here we briefly describe a number of algorithms that have been used in the closed-loop optimization setting.

Spendley *et al.* first established the *simplex* method [35] for solving unconstrained problems of n dimensions, where n is a user-defined function equal to the total number of variables in the process. The number of experimental data points is then set, $n + 1$, to form the first simplex. A ranking system takes

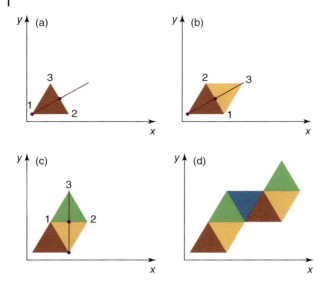

Figure 12.3 Simplex progression of a two-factor system: temperature (x) and pressure (y). (a) Formation of the first simplex with ranked points – 1 (highest), 2 (middle), and 3 (lowest) for the greatest output – with the midpoint set for reflection. (b) Reflection of the first simplex and formation of the second simplex with ranked points. (c and d) Reflection to form the third and subsequent simplexes through the vertices of the previous iterations.

place after each point has been trialed, ranking each point from lowest to highest output allowing the algorithm to proceed, between the two highest points, via a set midpoint. A perfect reflection of the simplex is performed in the opposite direction, away from the lowest ranking point, and a new set of conditions is selected to create the final point of the new simplex. Each new point is reranked and subsequent simplexes are formed after each repeated process until an optimum set of conditions is found.

The method could be extended and improved by varying the step size, such that the simplex converges to an optimum. However, the method is not guaranteed to find a global optimum (Figure 12.3).

Machine learning algorithms based on statistical surrogate models are particularly well suited to the problem of automated process optimization. This class of methods is based on the idea of developing an initial surrogate model, which is then iteratively improved with each consecutive experiment. The surrogate model is then used in combination with a suitable optimization algorithm to search for either a target or a global optimum solution. A popular, although not exclusive, surrogate model is Gaussian processes [36]. Then the construction of the overall algorithm depends on the objective (global versus target optimization), the complexity of the problem (dimensionality and nonlinearity), and available resources (number of experiments that can be performed within the constraints of time and budgets, computational resources available). One example of such complete algorithm, which has been successfully used in the optimization of chemical systems [37, 38], is the multiobjective active learner (MOAL) [39]. This algorithm was successfully used in target optimization problems, namely, when the optimization is terminated once a set target is reached, rather than when a global optimum is determined. This allows reducing significantly the number of sequential experiments needed, and also reducing the overall time for optimization. As the methods of surrogate functions and evolutionary optimization improve, there are likely to be significant advances in the algorithms for automated optimization of chemical systems in the near future.

12.5
Application Examples of Closed-Loop Discovery and Optimization

12.5.1
Discovery in Closed-Loop Self-Optimization

The most obvious application of automated experiments enabling self-optimization, to which most of this chapter will be dedicated, is the problem of optimization: When reaction chemistry is fixed and reaction conditions must be optimized to determine the optimal operating window against a set criteria of optimality, typically a complex cost function, incorporating space-time yield, product purity, and other relevant factors, for example, safety, environmental footprint of the process, and so on. However, one can envisage that a self-optimizer may be used for the purpose of discovery of new molecular structures.

In one of the first examples of using flow synthesis for molecular discovery, Cronin *et al.* reported discovery of new inorganic clusters, when they switched from conventional batch synthesis of such materials to a flow synthesis, which enabled them to exploit the reaction system states, not attainable within a batch reaction [40]. This was achieved in an open-loop approach, see Figure 12.2 for comparison of the open-loop versus closed-loop discovery systems. The power of such system was demonstrated in discovery of unknown inorganic clusters [41]. Other authors have also highlighted the possibility of discovery of novel chemical reactivity under unusual conditions of flow processes, but in the case of organic syntheses, for example, Ref. [42]. It is an obvious extension of the approach toward discovery of materials or molecules with predefined functional properties via a closed-loop approach.

This is shown schematically in Figure 12.2. The open-loop synthesis system allows generating novel materials, exploiting the convenience of automated dosing and control and the novelty of continuous-flow reactor, which allows the reactants to experience previously inaccessible states. If the resulting molecules or materials are subjected to analysis and discrimination against a fit to the desired properties target, then the synthesis system may be allowed to run until the desired properties are discovered. A simple application of the closed-loop control of target properties was demonstrated by the same group for the case of

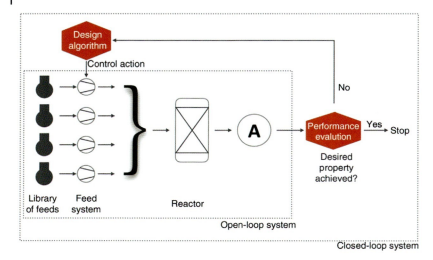

Figure 12.4 Schematic diagrams of open-loop and closed-loop chemical discovery systems. "A" denotes analytical method(s) used for identification of the results of the synthesis.

nanoparticle cluster size, where feed recipe was optimized to periodically obtain particles of the required cluster size [43]. The authors also utilize the feedback of the performance evaluation function, in this specific case a target UV response, to discover a new nanomaterial (Figure 12.4).

The same approach of closed-loop target search has been developed for discovery of novel bioactive compounds. The problem is well known within medicinal chemistry community, when multiple versions of a potential bioactive molecule are required to fully explore the bioactivity of the specific molecular motif. This is an extremely labor-intensive process, which, however, is the basis for a large number of small businesses, specializing in generating and evaluating such chemical libraries. A fully automated system using microfluidic synthesis, automated LC–MS analysis for identification of the synthesized compounds, a bioactivity assay, and an algorithm combining computational design of new compounds with synthesis design has recently been published [44]. The significant advance of this work, compared to most self-optimization studies, is the use of predictive modeling for design of targets, which makes use of the feedback data from the already performed syntheses and bioactivity assays. This approach of combining physical principle models with data-driven processes is highly promising and will be discussed further in this chapter.

12.5.2
High-Throughput Screening

The role of high-throughput screening is to synthesize libraries of chemical species, which have the potential to be used in drug discovery or catalyst discovery as well as to perform chemical, genetic, or pharmacological tests rapidly.

Processes performed in this way are completed by use of automated robots, data processing, and control software. This method forms the basis of drug/catalytic discovery and design by providing a large catalog of unique molecules with relevant data such as catalytic activity, interaction understanding, toxicological profiles, and so on.

High-throughput screening is typically carried out in microliter volumes on a plastic, gridded container (microplates) containing wells, anywhere from 96 to 3456 wells per plate, in which reagents are placed by robotic arms for automated processes. However, high-throughput methods were developed also for different scales of bio-processes (from few to tens of milliliters) and for inorganic syntheses, including hydrothermal synthesis of zeolites.

Catalytic discovery for organic reactions can be a time-consuming process when using conventional serendipitous means. Utilizing high-throughput screening can help screen large numbers of potential catalytic species for a wide variety of organic functional groups.

Robbins and Hartwig developed a laboratory-based method to rapidly screen multiple transition metal catalysts, ligands, and substrates where they discovered a copper-catalyzed alkyne hydroamination and two nickel-catalyzed hydroarylation reactions [45]. A set of 17 different reagents (Scheme 12.1) containing a single different functional group were explored where all 17 organic reagents were added to each well in a 384-well microplate. Twelve ligands were used, including phosphines, amines, phosphine oxides, phosphine sulfides, and amidinates, and added to each well of a column, while 15 catalysts were dispensed to each well in a row – species derived from Au, Co, Cr, Cu, Fe, Mn, Mo, Ni, Ru, W, and Yb. Analyses were completed by use of mass spectrometry (GC–MS and ESI–MS) where any catalytic processes that occur were observed by monitoring the mass range higher than any reactant mass, which corresponds to the coupling of two species. Due to the sheer number of experimental samples, all wells

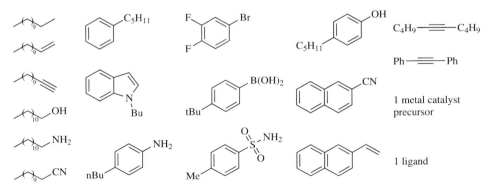

Scheme 12.1 The set of 17 organic functional groups, catalysts from 15 metal centers, and 23 ligands or absence of ligand used created more than 50 000 unique reactions during a high-throughput screening of a multitude of catalysts [45].

in each row were combined and added to a new well plate (eight reactions per well) and analyzed.

To determine what reactions, if any, had occurred, a reaction deconvolution procedure was developed, which splits reagents into four different categories and combines each in turn to find the previously observed product via GC–MS. Once detected, the groups could be whittled down eventually leading to the two reagents involved in the reaction. Detected species included a coupling reaction between 1-dodecyne and 4-butylaniline with multiple copper catalysts: CuCl, Cu(OAc)$_2$, and copper species with PBu$_3$, β-diketiminate, and tri-p-tolylphosphite as ligands. This reaction was shown to be a rare copper-catalyzed hydroamination where the highest yields were obtained with the CuCl catalyst. A side product species was also found from the coupling of diphenylacetylene and 4-*tert*-butylphenylboronic acid yielding a triarylalkene product – via a hydroarylation reaction – from several nickel catalysts: Ni(cod)$_2$, NiCl$_2$-dme (dme = 1,2-dimethoxyethane), and other nickel ligand catalysts (phosphines and an *N*-heterocyclic carbine). Other hydroarylation reactions were also observed with nickel catalysts to synthesize triarylalkenes from other reagents from the chemical system; hydroarylation reactions typically use expensive rhodium or palladium catalysts.

Although this method of synthesis and discovery is routinely used, the high-throughput process is still a brute force technique and rarely offers a fully optimized process for any species synthesized. Reactions discovered by high-throughput screening generally offer qualitative information even with integrated in-line/online analysis, which tends to be simpler instrumentation such as UV or fluorescence detectors. Any other analyses performed revolve more around offline techniques enabling a large catalog of data to be assembled quickly. This synthesis method is, however, very good at forming an initial starting point for reactions that may need to be fully optimized in the future.

12.5.3
Examples of One-Variable-at-a-Time Reaction Optimization

The one-variable-at-a-time approach (OVAT) is the most basic form of reaction optimization and sees that each parameter chosen be varied separately over a specified range giving the best response factor. This is then repeated while holding the previous parameter at the determined best value, so that the next parameter can be optimized (Figure 12.5).

A good example of this is the reaction between benzaldehyde and malonic acid to produce *trans*-cinnamic acid, which was first optimized for temperature (from a range of four temperatures) followed by optimizing the residence time, while keeping the optimum temperature constant (Scheme 12.2) [46]. This reaction is performed using pyridine as a solvent, which is notoriously difficult to remove and requires many steps to eliminate it from the reaction mixture. This creates the ideal scenario for the reaction to be monitored via in-line FTIR. A reaction

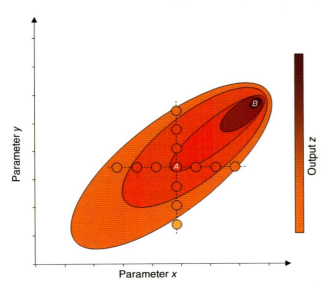

Figure 12.5 The progression of one variable at a time toward a local optimum. Parameter *x* is first optimized, where the highest output is found for that condition, before being held constant. Parameter y is then optimized, which finds the optimum output A. However, this method neglects to find the global optimum for the two-parameter system at point B.

screen was first established to map the positioning of individual peaks on FTIR spectra and determine where the desired product appears, allowing this peak to be monitored throughout the reaction. Initial runs optimized for temperature, selecting from the set of 80, 100, 120, and 150 °C, at a chosen residence time of 10 min. The best performing temperatures (100 and 120 °C) were run again at a residence time of 20 min before the best temperature (100 °C) was again run with a 30 min residence time. This final set of conditions was deemed to be the optimum operating set [46]. This form of optimization can add a great deal of time, resources, and cost to the overall development process.

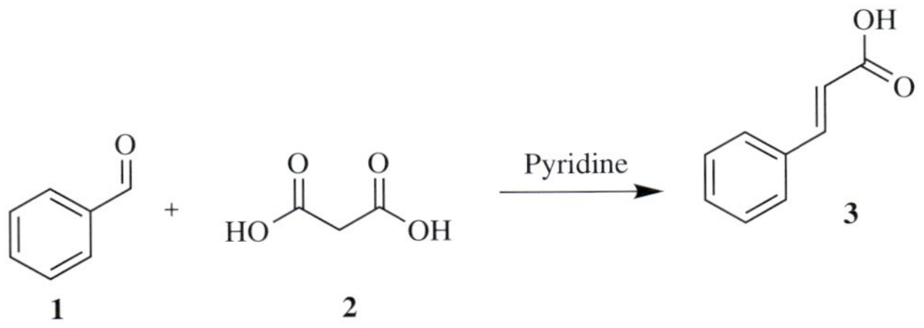

Scheme 12.2 The reaction between benzaldehyde (**1**) and malonic acid (**2**) to form *trans*-cinnamic acid (**3**) as the desired product. Pyridine is used as solvent, which is difficult to remove from the reaction that makes the system ideal for inline FTIR analysis [46].

12.5.4
Examples of Application of Design of Experiments

McWilliams *et al.* [47] used an experimental design procedure for the rapid optimization of metal-catalyzed processes for active pharmaceutical ingredients – the selective reduction of a nitro-containing compound to an aniline derivative (Scheme 12.3). Experimentation was completed with the use of an automated solid and liquid dispenser connected to a sealed reactor block, before assays were completed using a parallel pressure reactor (PPR) that contains six modules each containing eight reactors and a robotic arm to further automate the preparation of each reaction. Selected parameters included temperature ($-10 < T < 200\,°C$), pressure up to 1500 psi, and catalyst loading for each reaction.

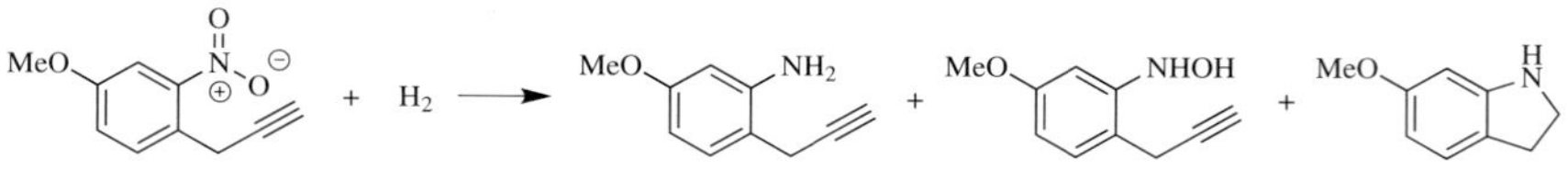

Scheme 12.3 Chemoselective catalytic reduction of a nitroaromatic nitrile, **4**, to an aniline derivative, **5**, as the desired product, a hydroxylamine, **6**, and indole, **7**, as other major products [47].

The reduction of a nitroaromatic nitrile (Scheme 12.3) used the three continuous variables stated above, ranging from temperature, pressure, and catalyst loading. A central composite (CCD) – face-centered (FCD) design with four replicate center points (totaling 18 experiments) – while focusing on evaluating and optimizing the yield of the aniline derivative **2** was used to optimize the output of the reaction. Subsequent data analysis showed that the yield of **2** was strongly dependent on the loading of catalyst but showed no relationship between temperature and pressure (Figure 12.6). As the temperature increases, the yield stays relatively constant; but as the catalytic loading increases, the yield also increases. The highest yields are achieved from 0.015 equiv of catalyst and above: 90–100% yield.

McWilliams *et al.* also explored the diastereoselective catalytic hydrogenation of an unsaturated ester as it was a key reaction step to a preclinical drug candidate (Scheme 12.4) [47]. Key aims were to ensure that the process created large amounts of candidate with minimum resources in the shortest time. Previous unoptimized conditions gave widely variable mixtures of **9** and **10** and formed an unknown impurity with quantities up to 10%. Initial screens of the catalytic process outlined that the reaction benefited from Wilkinson's catalyst using THF as solvent. Maximum conversion for Wilkinson's catalyst was 100% with 86% being the desired *cis*-product, **6**, and only 3% impurity. Key parameters in the process, temperature, pressure, and catalyst loading, were optimized using an experimental design method using the PPR system previously employed. The DoE procedure again was a central composite face-centered design with 4 replicate center points (18 experiments in total, Table 12.2).

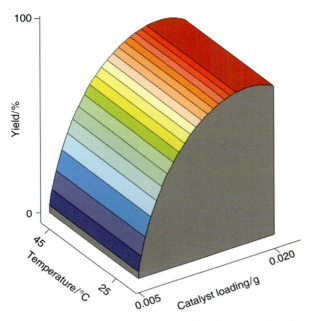

Figure 12.6 Surface profile for the yield of aniline, **2**, as a function of temperature and catalyst loading. The yield stays relatively constant with temperature, but reaches a maximum 100% when the catalyst loading is 0.02 equiv. (Redrawn from Ref. [47]. Copyright 2005, SAGE Publications.)

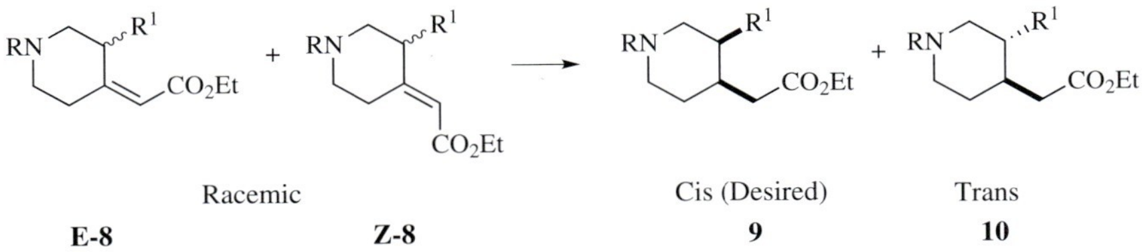

Scheme 12.4 Diastereoselective catalytic hydrogenation of a racemic mixture of unsaturated ester **8** to the desired *cis*-isomer, **9** [47].

Table 12.2 Experimental design parameter and conversion ranges and observed conversions to the desired product, **6**.

Variable	Range	Conversion (%)
$(Ph_3P)_3RhCl$	0.1–0.5 mol%	16–100
Temperature	25–75 °C	16–96
H_2 pressure	5–100 psig	16–100

Adapted from Ref. [47]. Copyright 2005, SAGE Publications.

Experimental design outlined that at 25 °C, in all cases even at highest catalytic loadings and pressures, the reaction had very low yields and suffered from incomplete conversion. Visual representation, via contour plots, of the data from DoE showed that the highest conversions were not observed at the convergence of maximum points of factors but in fact had a broader spanning range of conditions in the center of dual factor plots. Contour plots also showed that an increased catalyst loading allowed maximum conversions and yields to be maintained while allowing the operation temperature and pressure to be reduced. Optimum conditions were found to be between 70 and 80 psig of H_2, allowing a wide range of catalyst loadings to be used as well as minimizing the overall operating temperature (Figure 12.7).

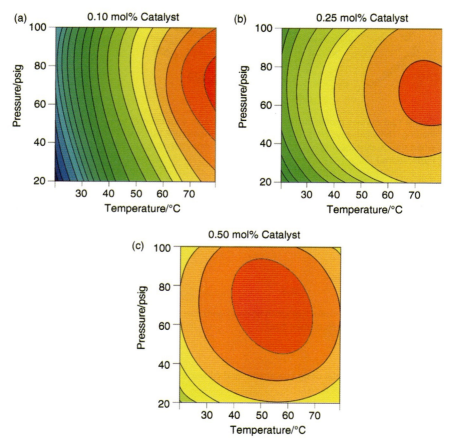

Figure 12.7 Contour plots for 0.1 mol% (a), 0.25 mol% (b), and 0.5 mol% (c) for parameter pressure and temperature. Optimum regions are not seen at convergence points of highest-valued parameters, but more toward the center of parameters. (Redrawn from Ref. [47]. Copyright 2005, SAGE Publications.)

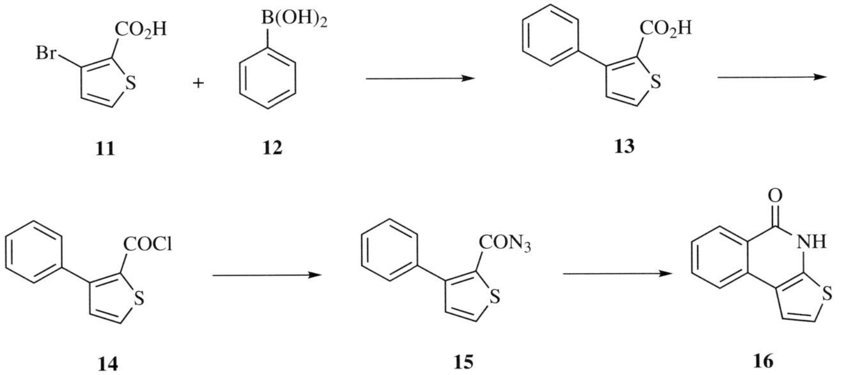

12.5 Application Examples of Closed-Loop Discovery and Optimization

Scheme 12.5 Current synthetic route of thieno[2,3-c]isoquinolin-5(4H)-one-A (TIQ-A) [34]. (Adapted with permission from Ref. [34]. Copyright 2014, American Chemical Society.)

Gioiello and coworkers [34] investigated the continuous-flow reaction for the synthesis of thieno[2,3-c]isoquinolin-5(4H)-one-A (TIQ-A) (Scheme 12.5, **16**) due to its potential therapeutic applications. A CCD was employed with the use of automated flow synthesizers for their set of chosen experiments. An appropriate model was initially defined for the reaction for the DoE to be the most effective for the system: 3-bromothiophene-2-carboxylic acid (1 equiv, **11**), phenylboronic acid (1.1 equiv, **12**), Pd (PPh$_3$)$_4$ catalyst (0.03 equiv), NaOH (2.0 equiv), a solvent system of H$_2$O–THF–PEG-400. Two variables, temperature and flow rate, were selected to be investigated by DoE analysis: a CCD comprising eight experiments plus four center point replicates. A summary can be found in Table 12.3. The maximum yield can be seen to be 98% at two very different

Table 12.3 CCD experimental matrix and measured responses for the synthesis of **13** [34].

Run	Type	A (°C)	B (ml min^{-1})	Yield (%)
1	Axial	140	2.00	89
2	Center	140	1.10	90
3	Factorial	182	0.46	92
4	Axial	200	1.10	68
5	Factorial	98	0.46	46
6	Factorial	98	1.74	28
7	Axial	80	1.10	9
8	Center	140	1.10	94
9	Center	140	1.10	91
10	Factorial	182	1.74	98
11	Center	140	1.10	93
12	Axial	140	0.20	98

Adapted with permission from Ref. [34]. Copyright 2014, American Chemical Society.

sets of conditions. A verification process was applied where two sets of targets were specified for their optimization for the model to be validated: (i) maximization of the yield, minimization of A, and keeping B within range; (ii) maximization of yield, A in range, maximization of B. The optimization variables A and B correspond to reaction temperature and feed flow, respectively. Predicted yields for both verification reactions 1 and 2 were 100% and experimental yields 91 and 92%, respectively. From both verification reactions, it was shown that the model from the experimental design had a high degree of correlation between predicted and experimental yields.

Automated open-loop optimization systems, like the examples above, allow processes to be optimized with human interaction as a major component to quickly run reactions with a wide range of conditions before falling upon an optimum operating region. This approach still allows a large amount of data to be found with minimal experiments, reagents, and solvents, but it is still a much more time-consuming method as it relies on people to prepare and run off-line samples and feed the data back into the computer programs to gain the desired information.

Holmes *et al.* performed a closed-loop optimization of a chemical reaction using an experimental design method. The reaction between methyl nicotinate and methylamine to form N'-methyl nicotinamide with niacin as an impurity was used for this purpose (Scheme 12.6). The DoE study employed a CCF design, which allowed any curvature of the response surface to be modeled. The final output model was constructed from reactions gained over a 5.5 h period. A maximum yield of 96% of the desired product, **18**, was found from conditions; flow rate of 0.1 ml min^{-1}, 9.7 equiv of MeNH$_2$, and a temperature of 7 °C (Figure 12.8). The statistics of the model show an excellent fit and predictability as R^2 was calculated to be 0.999 and Q^2 was 0.977.

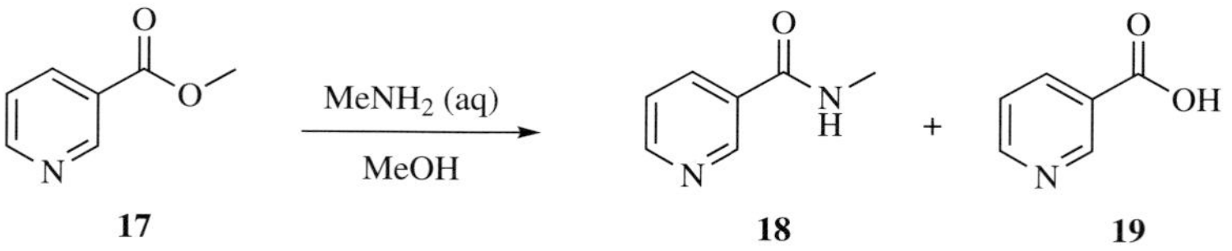

Scheme 12.6 The reaction of methylnicotinate **17** with aqueous methylamine to form the desired N'-methylnicotinamide, **18**, and niacin, **19**, as an impurity [17].

Closing the loop (by adding in-line/online analysis) will allow processes to be fully optimized without the need to prepare and run samples off-line while still gaining a large amount of data to validate reaction models and determine optimum operating conditions for the process in question.

12.5.5
Rate-Based/Physical Organic Approaches

Physical organic approaches and the use of kinetic parameters are also a possibility when it comes to the optimization of a chemical process. A model of the

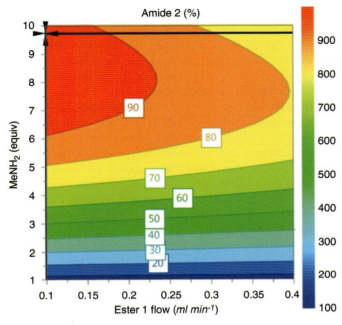

Figure 12.8 Contour plot, from the CCF DoE model, of the optimum conditions generated in MODDE. Temperature is fixed at 7 °C and the optimum point is emphasized by the crosshair. (From Ref. [17]. Copyright 2016, The Royal Society of Chemistry.)

system can be created and time series data can be compiled from calculable kinetic parameters, such as activation energies, rate constants, and equilibrium constants [48]. Availability of kinetic data allows this approach to be powerful as multiple factors can be optimized simultaneously, but prior knowledge must be sought to help understand the procedure before optimization, ultimately increasing the time and cost of this method. Automation is possible as a computer program can be used to run experiments – through the generation of a feedback loop – and automatically analyze the reaction, although the program will not "speak" directly to any online analytical techniques, which might be used. Data calculation and spectra interpretation will be completed manually by the chemical expert before the change in conditions is applied directly to the computer algorithm to proceed with the reaction optimization. This tactic is often an open-loop approach as each data point needs to be calculated by hand before the change in the reaction system can be made. Automation can occur this way but an initial experimental design needs to be inputted followed by best-fit literature data and a kinetic model.

To design a chemical reactor, several systems need to be considered: heat, mass, and component balances, heats of reaction, and kinetics for each reaction required to build a complete model. Each of the variables mentioned allows kinetic parameters, activation energies, and rate coefficients to be estimated using well-known theories: either transition state theory or kinetic theory. Wherever possible, use of direct experimental data is paramount to finding these parameters as accurately and as efficiently as possible. The kinetic models gained not only mean the possibility of scale-up to another larger flow reactor, but also other types of reactors that are still often used in industry, for example, large-scale batch CSTR (continuous-stirred tank reactor) [49, 50].

Finding kinetic data from continuous-flow reactions can be taken upon in several ways similar to reaction optimization via OVAT, DoE, and black box techniques. The simplest way of finding kinetic data is to carry out reactions manually at different temperatures and residence times to span a much larger kinetic area and to maximize the data output for a model. Efficiency can be improved by automation of the reaction conditions (using computer-controlled systems to automatically run a set of predefined reactions), although it may still need plenty of manual data mining, interpretation, and model fitting. New techniques are allowing chemists to fully adopt algorithms to automate the reaction, and use data provided from real-time analysis, to interpret and model fit without the need of extracting data manually [50].

Relevant and quantitative composition data from an analytical technique are essential to perform a kinetic study on a reaction, but quite often can be challenging to acquire. As with other types of reaction optimization, there are a few different types of analysis that can be used to get the data needed for a profile: extractive (at-line), in-line, and noninvasive. Each one offers a different angle for analysis of the chemical reaction and provides a unique data set, but anyone can be used to give the information required to outline a kinetic profile and other reaction parameters, such as activation energies and rate constants. Extractive techniques (GC, HPLC, and online MS) can be placed at any point on the microreactor, utilizing a sample injector, with the potential of providing information before, during, and after the reaction helping to maximize the data output. Reactor design can be a limiting factor when it comes to forming a kinetic profile or optimizing a reaction as certain analyses can only be completed at specific points on the reactor. Some reactors allow sampling to be made at multiple points before the reaction, on the reactor, or after the reactor, permitting a reaction profile to be formed where kinetic data can be derived [49, 50].

Kinetic data are widely used to determine mechanistic data from a reaction and gage an understanding of specific parameters of a system that help with understanding the formation of by-products and side products in a reaction. This is useful in laboratory-scale synthesis as it helps to expand current knowledge for new reactions to find details in reactions that may have previously been missed due to limited resources or technology. New techniques for determining these kinetic data allow parameters to be found and models fit much quicker and more efficiently than traditional means. Continuous flow allows complex reaction systems to be parametrized and optimized using less materials and more forcing conditions for a much larger modeling area increasing the accuracy and reducing the uncertainty of data. This is particularly useful in the scale-up from the laboratory (with volumes anywhere between 50 and 20 ml) to larger

12.5 Application Examples of Closed-Loop Discovery and Optimization

industrial processes (plant-scale reactors can be upward from 15 to 200 ml) when kinetic data can allow the inhibition of side reaction at specific conditions with a high degree of operation certainty. Large amounts of uncertainty may cause many problems or failure of large-scale processes, potentially leading to increased likelihood of side reactions being more dominant lowering yields of desired products. It is these data that are the most essential when scaling up a chemical process and new techniques, or the improvement of old techniques allow it to be sought much quicker and more accurately than in the past [49, 51, 52].

Mozharov et al. [50] monitored a Knoevenagel condensation reaction (Scheme 12.7) performing analysis after the reaction. Their use of noninvasive Raman spectroscopy allowed rapid and frequent sampling to generate a kinetic profile at selected temperatures (10 and 40 °C), only by varying the flow rate of reaction. Their reactor cell consisted of reagents that mixed together using a Y-piece, microfluidic pathway and detector at the end of the reaction route. Low flow rates (0.6 µl min^{-1}) were initially used (subsequent reactions being carried out at a higher flow rate, 14 µl min^{-1}) allowing longer residence times to be observed ensuring that the extent of reaction changed significantly from the beginning to the end of the reaction stream (Figure 12.9).

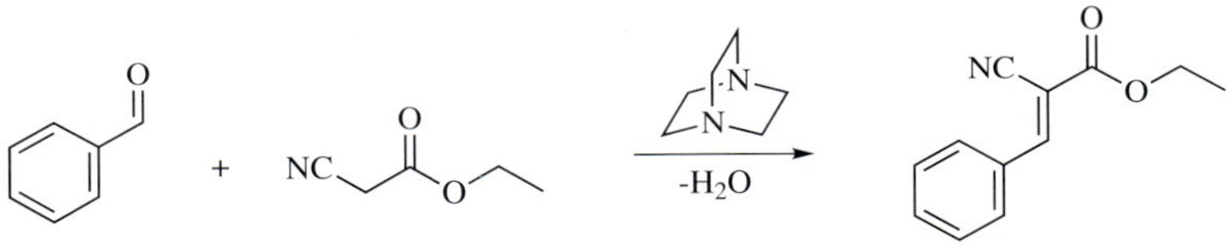

Scheme 12.7 Base-catalyzed Knoevengel condensation reaction between ethyl cyanoacetate and benzaldehyde.

It should be noted that the use of Raman spectrometry here allowed many data points to be performed as this spectroscopic technique has the potential to complete many analyses in a short space of time. Kinetic data can be calculated and creation of a reaction profile using known measurements, change in flow rate, time of the change, dimensions of the flow path. and times of analytical

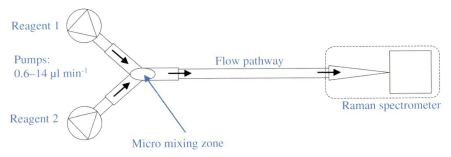

Figure 12.9 Schematic diagram of the flow system used by Mozharov et al. (Adapted with permission from Ref. [50]. Copyright 2017, American Chemical Society.)

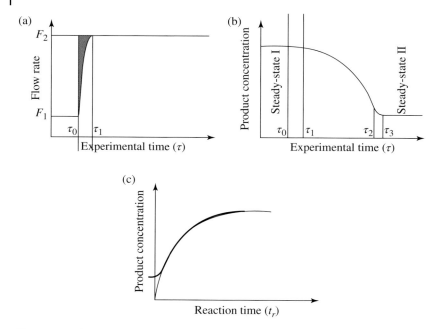

Figure 12.10 Curves outlining the changes in (a) flow rate and (b) concentration of the detected product producing (c) the kinetic curve calculated. (Reprinted with permission from Ref. [50]. Copyright 2017, American Chemical Society.)

measurements (Figure 12.10) shows how the change in flow rates affects the detected concentration of product in the reaction and the resultant kinetic curve that was calculated.

Two kinetic reactions and analysis methods were completed and compared to get the best kinetic motif possible in the shortest time and using the least material. The first method looked at different reaction times based on the flow rate and microreactors internal volume and the mean conversion at each flow rate. The generated model found reaction orders of 1.4 at 10 °C and 1.3 at 40 °C. Rate constants can be calculated using the reaction orders $3.35 \pm 0.32 \times 10^{-2}$ mol$^{-0.4}$ dm$^{1.2}$ s^{-1} and $24.4 \pm 3.2 \times 10^{-2}$ mol$^{-0.3}$ dm$^{0.9}$ s^{-1} for 10 and 40 °C, respectively.

The second method pumped the reaction at a low flow rate for a set amount of time before increasing to a higher flow rate; this was repeated for a range of different initial and final flow rates (Table 12.4). Raman measurements were made every 4 s and used to generate a more continuous reaction profile.

Both methods used by Mozharov *et al.* were shown to efficiently derive a kinetic model using the associated parameters for the Knoevenagel reaction studied. The second method surpasses the first in several unique ways, including its aptness at deriving a motif in a single experiment, using less reagents (about 10 times less) and having a much lower experimental time (about 5 times lower). However, for the technique as a whole, the analysis response times were

Table 12.4 Reaction orders (n) and rate constants derived using method 2 for different combinations of flow rates (F_1 and F_2).

	10 °C				40 °C		
F_1, F_2 (µl min^{-1})	n	k (10^{-2})	k (10^{-2}) (n = 1.4)	F_1, F_2 (µl min^{-1})	n	k (10^{-2})	k (10^{-2}) (n = 1.4)
0.4, 6	1.6	3.49	3.49	1.1, 20	1.1	23	30.3
0.4, 6	1.1	3.54	3.53	1.1, 20	1.2	27.5	31.7
0.4, 6	1.3	3.49	3.49	0.6, 14	1.1	23.3	30.8
0.4, 6	1.3	3.72	3.73	1.6, 20	1.3	24.8	24.8
0.2, 4	1.4	3.5	3.5	2.4, 10	1.1	22	25
0.8, 8	1.1	3.59	3.59	1.1, 6	1.1	23.1	31.4
0.6, 6	1.1	3.57	3.6	1.1, 14	1	24	26.2
average	1.27 ± 0.19	3.56 ± 0.08	3.56 ± 0.09	average	1.13 ± 0.10	24.0 ± 1.8	28.6 ± 3.1

Reprinted with permission from Ref. [50]. Copyright 2017, American Chemical Society.

convoluted by the "real-system" response time due to the step-change in conditions. These methods for kinetic modeling are also not transferable to alternative analytical systems that have longer acquisition times such as gas chromatography and liquid chromatography.

Hone *et al.* [53] explored the creation of kinetic motifs using automated flow reactions placing a greater emphasis on under- and overreaction products to extend the range of reaction conditions and give a much richer data set ultimately increasing the confidence of the fitted kinetic model. The reaction chosen was the nucleophilic aromatic substitution (S_NAr, Scheme 12.8) between 2,4-difluoronitrobenzene and pyrrolidine in ethanol to give a mixture of components: *ortho*-substituted (desired product), *para*-substituted, and bis-adduct (both side products) using HPLC as the online analytical technique.

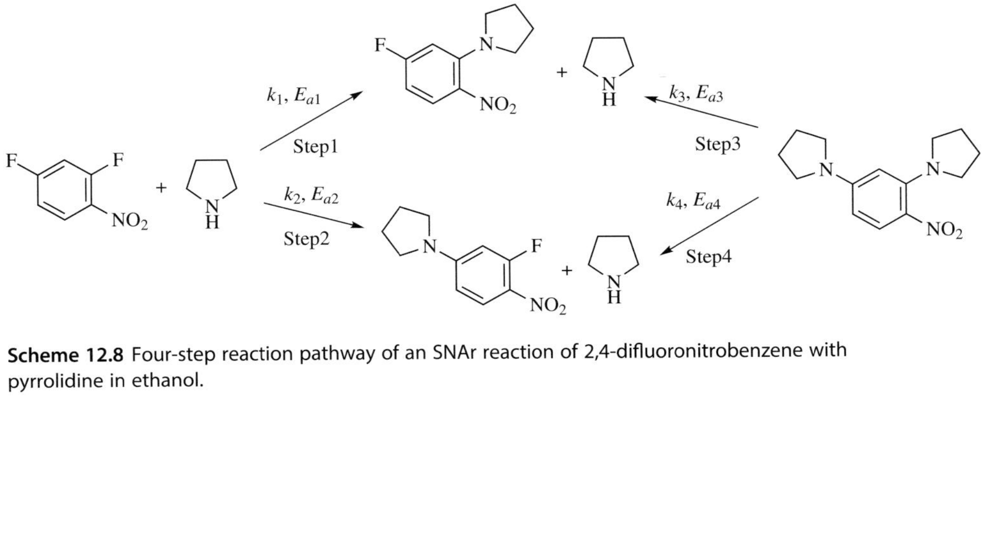

Scheme 12.8 Four-step reaction pathway of an SNAr reaction of 2,4-difluoronitrobenzene with pyrrolidine in ethanol.

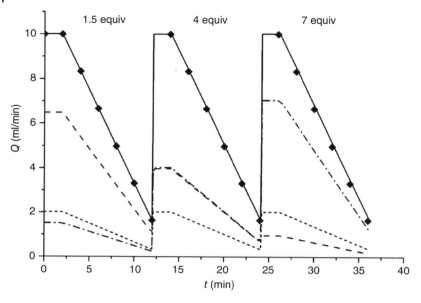

Figure 12.11 Changes in volumetric flow rate over time, where Q_{P1}, Q_{P2}, Q_{P3}, and Q_{total}; ···· pump 1 (Ar in EtOH), --- pump 2 (EtOH), -·-· pump 3 (pyrrolidine in EtOH), — total volumetric flow rate, and ♦ HPLC injection. The linear flow ramps correspond to pyrrolidine to 2,4-difluoronitrobenzene molar ratios: (i) 1.5: 1, (ii) 4: 1, and (iii) 7: 1 using Q_{total} from 10 to 1.5 ml min^{-1}. (From Ref. [53]. Copyright 2017, The Royal Society of Chemistry.)

Reactions were carried out using gradient flow ramps that allowed the exploration of a complete profile of the reaction in one transient experiment. The ramp began with maximum flow rates and a varied ratio of starting materials at a constant temperature. Steady state was reached before the flow rates were reduced at a constant rate (0.836 ml s^{-2}) to increase the total residence time while sustaining constant process concentrations. Another set of reactions developed (Figure 12.11), aimed at exploring conditions varying from mildest (1.5 mol equiv, 0.5 min, 30 °C) to harshest (7 mol equiv, 2 min, 120 °C) as based on the concept by Hessel *et al.* [54] of novel process windows that are rarely approached during organic synthesis. All process hardware was under the control of a MATLAB-based computer program that allowed all conditions to be changed instantaneously as well as all parameters being monitored constantly.

This method was applied to a variety of temperatures (30, 60, 90, and 120 °C) giving a total of 12 reaction profiles from 72 data points, which were collected over a short period of less than 3 h to offer a wider range of conversions for each individual reaction element (Figure 12.12). The best model for the data for the reaction scheme (Scheme 12.8) was found to be of second order and rate constants determined using the Levenburg–Marquardt algorithm [55] at isothermal conditions (90 °C). All data were simultaneously fit using DynoChem software (scale-up systems), with kinetic parameters matching closely to the experimental

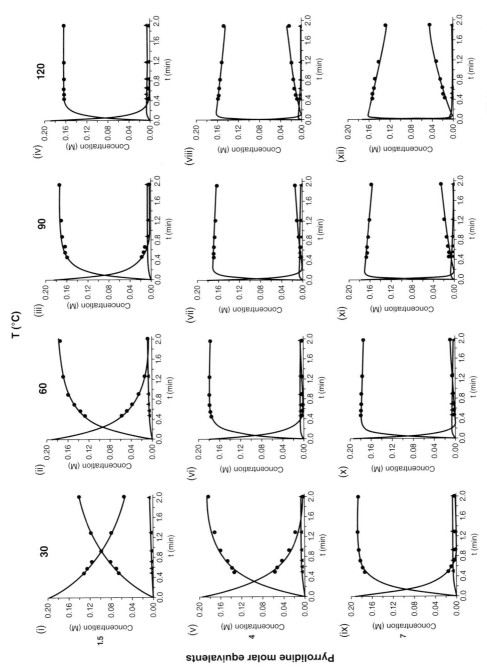

Figure 12.12 Concentration–time profiles from simultaneous parameter fitting; points = experiments ■ 2,4-DF, • ortho • para, ▼ bis, lines = model using kinetic parameter estimates. (From Ref. [53]. Copyright 2017, The Royal Society of Chemistry.)

data ($R^2 = 0.9995$), and determined rate constants for the *ortho*-side product being 20 times larger than the rate constant for the *para*-side product and activation energies being 33.3 ± 0.3 and 35.3 ± 0.5 kJ mol^{-1}, respectively. These parameters suggest that the overall temperature has an influence on the rate of reaction but does not have any effect on the positional selectivity in the molecule.

This work by Hone et al. [53] has shown that kinetic modeling can be completed using an automated continuous-flow platform where large amounts of data can be gained from many reaction profiles in a short period of time. This method was sufficient enough to fit a model to four reactions in one process, eight fitting parameters, and with a total uncertainty of 4%. This work allows kinetic profiles to be calculated without waiting for reactors to get to steady state helping to reduce the amount of material used and the time taken for analysis. The models can also be generated far earlier in process development to gage the sensitivity of product and important parameter changes that may be necessary. *In silico* model simulation is also a possibility as a form of alternative reaction optimization for specific scenarios, reactor, and equipment arrangements as well as minimizing costs and risks upon scale-up.

Being able to have a completely automated system that runs all required experiments, analyzes the data, and performs the necessary calculations to determine certain parameters and fit models is a much more complex approach and may have a large interest in industry for the optimization of parameters needed for process scale-up.

Reizman and Jensen [56] have explored the possibility of developing a system that is fully automated and calculates kinetic parameters and fits results to a model using computer algorithms. The chosen multistep reaction used (Scheme 12.9) was an S$_N$Ar reaction between 2,4-dichloropyrimidine and morpholine in ethanol to produce the desired product: 2-substituted aminopyrimidine and two side products (4-substituted aminopyrimidine and 2,4-substituted aminopyrimidine). This system was chosen because of its high

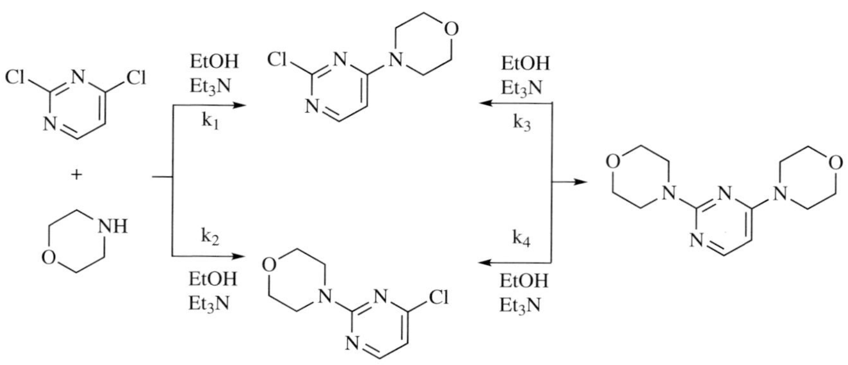

Scheme 12.9 Multistep reaction pathway for the S$_N$Ar reaction between 2,4-dichloropyrimidine and 4,4′-(2,4-pyrimidinediyl)bis-morpholine. (Reprinted with permission from Ref. [56]. Copyright 2017, American Chemical Society.)

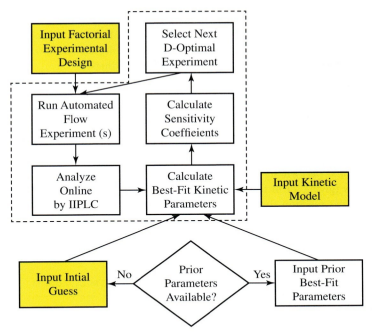

Figure 12.13 Logic flow diagram for the automated kinetic parameter estimation in continuous flow. (Reprinted with permission from Ref. [56]. Copyright 2017, American Chemical Society.)

impact in the pharmaceutical industry. These types of molecules have been shown to have a kinase inhibitory effect in cancerous cells.

Reizman and Jensen [56] explored the use of factorial experimental designs to run reactions with analysis via online HPLC to create a feedback loop and automatically calculate best-fit kinetic parameters and sensitivity coefficients. This stage of the cycle relies on a user-defined model to get the best agreement with experimental data. Progression of the algorithm to an optimum point (minimization of sensitivity coefficients) occurred by use of a D-optimal program that selects a latter experiment based on the prior experiment. However, this sort of program and optimization technique does not have an in-built cutoff code and so continues to iterate until the user terminates the design (Figure 12.13).

The experimental design procedure consisted of three individual stages, each with a specific desired outcome. The initial set of experiments completed were hoped to determine eight kinetic parameters based on the rate equations for each step in the reaction scheme and derive rate laws governing each species' generation and consumption. The second stage of the experimental design parametrized each step of the reaction in isolation, while the final stage calculated eight kinetic parameters concurrently using the values deduced from the previous experiments. The four factors explored were those essential to the reaction: residence time (t_{res}), temperature (T), initial concentration (C_{i0}), and the equivalents of morpholine used in each reaction. The user-defined model is based on assumptions and kinetic theory built into the automated program; in this case, it was assumed that the four reaction pathways followed a second-order kinetic system due to the reaction occurring in a bimolecular fashion, and that the entire flow system followed ideal plug flow reactor kinetics.

The initial factorial experimental design gave an output of 12 automated experiments, the data of which were plotted and a model fit. The D-optimal algorithm used ensured minimization of uncertainties in the parameter estimates and was employed after the initial set of experiments. One data point was collected at the harshest conditions and the parameter estimate calculated again with consideration of the upper and lower bounds of the parameters. This process of selecting and performing D-optimal experiments was continued for a further 12 reactions (giving 24 in total), which were again plotted as experimental conversions against reaction time with the model fitted to each (Figure 12.14) before manual termination. Although the results shown agree with both the final best-fit model parameters and the experimental data, the model underestimates the yields of the 2-substituted and 4-substituted species and overestimates the

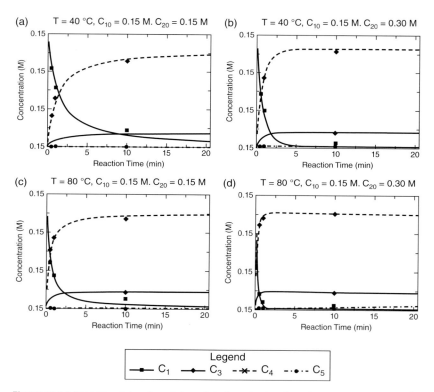

Figure 12.14 Experimental and model-predicted reactant and product concentration profiles after 24 automated experiments. Markers identify experimental data points. Lines indicate model prediction. (Reprinted with permission from Ref. [56]. Copyright 2017, American Chemical Society.)

Table 12.5 Optimal kinetic parameter estimates and uncertainties from final simultaneous experiments in isolated approach.

Number of experiment	$Log_{10}(A_1)$	E_{A1}	$Log_{10}(A_2)$	E_{A2}	$Log_{10}(A_3)$	E_{A3}	$Log_{10}(A_4)$	E_{A4}
Prior	3.4 ± 0.1	27.0 ± 0.7	3.5 ± 0.1	32.3 ± 0.7	4.8 ± 0.1	59.0 ± 1.7	3.0 ± 0.1	44.7 ± 1.7
12	3.5 ± 0.1	27.3 ± 0.7	3.5 ± 0.1	32.1 ± 0.7	5.0 ± 0.1	60.4 ± 1.7	3.2 ± 0.1	46.3 ± 1.8
13	3.5 ± 0.1	27.1 ± 0.7	3.5 ± 0.1	32.2 ± 0.6	4.8 ± 0.1	59.0 ± 1.7	3.0 ± 0.1	45.0 ± 1.7
14	3.4 ± 0.1	27.0 ± 0.6	3.5 ± 0.1	32.2 ± 0.6	4.8 ± 0.1	59.0 ± 1.7	3.0 ± 0.1	45.0 ± 1.8
15	3.4 ± 0.1	27.0 ± 0.6	3.5 ± 0.1	32.1 ± 0.6	4.8 ± 0.1	58.9 ± 1.7	3.0 ± 0.1	45.0 ± 1.8
16	3.4 ± 0.1	27.0 ± 0.6	3.5 ± 0.1	32.1 ± 0.6	4.8 ± 0.1	58.7 ± 1.7	3.0 ± 0.1	45.0 ± 1.8
17	3.4 ± 0.1	27.0 ± 0.6	3.5 ± 0.1	32.1 ± 0.6	4.9 ± 0.1	59.4 ± 1.6	3.0 ± 0.1	45.0 ± 1.8
18	3.4 ± 0.1	27.0 ± 0.6	3.5 ± 0.1	32.1 ± 0.6	4.9 ± 0.1	60.0 ± 1.6	3.0 ± 0.1	45.0 ± 1.7

Reprinted with permission from Ref. [56]. Copyright 2017, American Chemical Society.

conversion of the starting material, at long residence times. This forced looking at each reaction individually, finding isolated reaction kinetics for each using the same automated approach as described, but for the formation of both monosubstituted species and disubstituted species from each of the monosubstituted species. The results from kinetic parameter and uncertainty estimates from the simultaneous experiments via the isolated approach can be found in Table 12.5.

This work showed the possibility to automate parameter estimation experiments using design of experiments and get a plethora of kinetic data to form a model while optimizing eight parameters simultaneously. However, in terms of uncertainty for scale-up, the initial work for the simultaneous eight parameter optimization was not completely successful and although certain aspects of the approach were estimated well, other aspects contained too much uncertainty (as large as 20%) for the intended application. The isolation of individual experiments performed better as sensitivity parameters in the flow system were significantly reduced. This led to substantial improvements in the minimization of uncertainty, particularly with the additional maximum *a posteriori* estimations included in the calculations. Isolated reactions also proved important for determining the parameters for optimal yield for the formation of the 4-substituted species.

Reizman and Jensen [57] explored the use of droplet flow systems as an advancement of their previous kinetic parameter optimization in continuous flow. These types of systems allow a much wider range of variables to be manipulated and permit the exploration of discrete variables such as solvents or catalysts as well as the typical continuous variables already widely used, for example, temperature, reaction time, or concentration. A continuous-flow droplet system can provide profiles from mixing and heat transfer, due to the accuracy of controlling reagent compositions, mimicking those of microscale batch reactors. Sub-20 µl droplets were generated in their automated flow reactor coupled with online feedback, from LCMS, with the intention of changing and monitoring the effect of both continuous and discrete variables on the kinetics of the reaction (Figure 12.15).

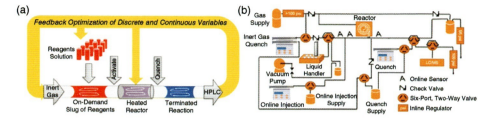

Figure 12.15 (a) Concept diagram for on-demand preparation, reaction, analysis, and feedback in automated flow screening. (b) Diagram of the droplet flow optimization system developed by Reizman and Jensen. (Reprinted with permission from Ref. [57]. Copyright 2017, American Chemical Society.)

The optimization platform consisted of manifold reagents, syringe pumps, automated liquid handler, reactor, LCMS sampling, and a downstream quench. Inert gas, nitrogen or argon, was used as the carrier phase to take the injected droplet to the reactor where the droplets could be viewed in Teflon FEP tubing. This was mainly used to inhibit complications in miscibility of organic solvents with fluorinated carrier phases and to make it easier to carry out oxygen-sensitive reactions, although use of inert gases cause other potential problems such as wetting, gas permeability, and compressibility.

The first application of the system employed an alkylation reaction (Scheme 12.10) where 10 different solvents were screened and three continuous variables: temperature (30–120 °C), residence time (1–10 min), and electrophile equivalents (0.7–2.0 equiv). The use of a D-optimal DoE approach allowed the progression of the algorithm to minimize uncertainty in kinetic variables and to find the best yield of the desired monosubstituted product from each of the three variables while also cycling through each of the 10 solvents, eliminating each if they were less likely to produce an optimum than another. This method found DMSO to be the best solvent to use and an optimum yield of 62% (Figure 12.16), which was verified using a gradient-based quasi-Newton search algorithm and a batch experiment that showed a yield of 61% by HPLC (59% isolated yield). Other feedback was also given from the algorithms response surface; polar aprotic of solvents gave faster conversion to the desired monosubstituted product and to side products. Information from the algorithm also suggested conducting these experiments

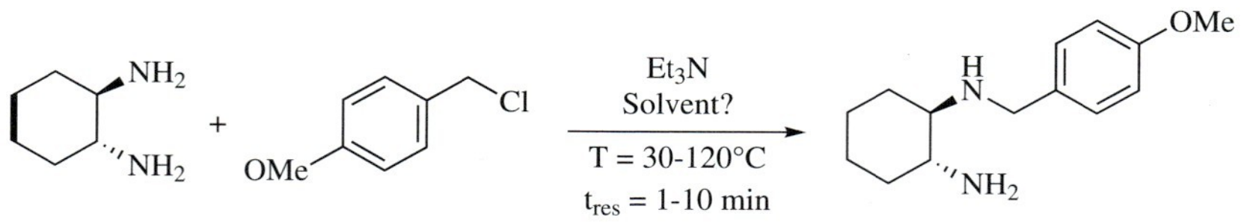

Scheme 12.10 Alkylation reaction studied for the simultaneous solvent screening and reaction optimization by Reizman and Jensen. (Reprinted with permission from Ref. [57]. Copyright 2017, American Chemical Society.)

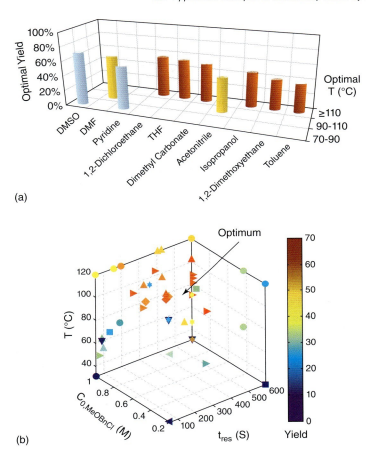

Figure 12.16 (a) Predicted maximum desired product yield as a function of solvent after **n** experiments. (b) Map of experiments conducted by algorithm while working toward the optimum operation conditions. (Reprinted with permission from Ref. [57]. Copyright 2017, American Chemical Society.)

at milder temperatures to limit the formation of side products. while it showed certain solvents would be better used at higher temperatures and for longer reaction times to produce a higher yield.

The advancement of this system has allowed a more complex reaction to be explored and optimized for multiple continuous variables and solvent that best performs (out of a total of 10). The use of experimental design techniques and D-optimal design saw uncertainty in kinetic values minimized to increase the confidence of the system while also finding an optimum for not only yield of the desired product but for minimization of side reactions.

Most syntheses completed in the laboratory are complex in nature, requiring a much more complex approach to help solve problems and create scalable results by providing as much data as possible. Being able to use algorithms for reaction optimization to allow the maximum yield of desired product, minimizing side reactions, finding kinetic data for model fitting, mechanistic detail and scale-up, and finding robust optimal operating conditions is becoming necessary as green chemistry is becoming more prominent in academia and industry. These types of approaches will allow menial tasks to be automated, giving chemists more time to focus on developing through models of chemical systems and designing novel, greener, faster, and cheaper routes for chemical synthesis.

12.5.6
Examples of Algorithm-Based Self-Optimization

Self-optimizing continuous-flow systems are able to sample a series of predefined reaction conditions generated from the optimization algorithm and perform evolutionary calculations to push the system toward optimization [58–60]. This method can improve the processes general speed and efficiency because the algorithm drives the system away from a minimum output toward a global maximum. The algorithm begins with a set of predefined reaction conditions and experimental constraints and performs automated analyses with the use of an online or in-line analytical technique. Subsequently, the program creates a new set of reaction conditions based on the previous set and the procedure is repeated until the maximum output is determined [61].

Coupling an online analytical technique with an automated reactor and adaptive feedback algorithms creates a simple and robust method to optimize a chemical reaction. This self-optimization process can progress through a multitude of variables and generate optimal conditions without human involvement [17, 62–64]. Continuous-flow reactors are highly apposite with integrated analytical methods because they are simplistic, offer quick measurement of the reactor stream, and all operating parameters such as flow rates, pressure, and temperature can be changed quickly and easily using the feedback loop between algorithm and equipment. The growth of the algorithm stems from the starting experimental conditions and the analyzed results. Based on these results, subsequent reaction conditions are set by the algorithm and the process repeated until optimum conditions are found (Figure 12.17). The use of optimization algorithms makes the majority of the chemical processes unmanned and involves no extra human interaction once the reaction is underway [63].

Skilton *et al.* used FT-IR to optimize the reaction of 1-pentanol with dimethyl carbonate to form pentylmethyl ether (Scheme 12.11), previous works used GLC [59]. FT-IR analysis significantly reduced the analysis time from 35 to 3.2 min due to the increased sampling rate of the FT-IR. This technique allowed them to measure chemical yields (>99%) but also when the reactor was at steady state. It was also necessary to calibrate the FT-IR instrument using online GLC that reduced the initial speed advantage of FT-IR. However, with the use of FT-IR, the volume of chemicals used was significantly reduced and a much wider image of the experimental space was permitted.

12.5 Application Examples of Closed-Loop Discovery and Optimization

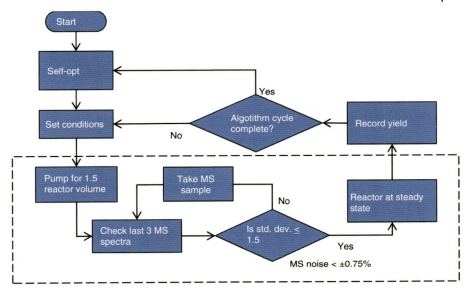

Figure 12.17 An outline of the procedure for a fully automated self-optimized reaction. Initial conditions are set – either manually or by the algorithm. Reagents are pumped for 1.5 reactor volumes until steady state is reached. Steady state is reached if the standard deviation is less than or equal to 1.5, this is calculated by the algorithm. Once at steady state, the output variable is recorded and the algorithm proceeds to the next set of conditions, based on the previous set, until an optimum is established.

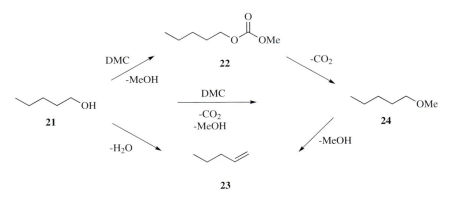

Scheme 12.11 Reaction scheme for the methylation reaction between pentan-1-ol (**21**) and dimethyl carbonate (DMC). At lower temperatures (150–200 °C), pentamethyl carbonate (**22**) is favored, while pentamethyl ether (**24**) is favored between 200 and 290 °C. At temperatures above 300 °C, pentamethyl ether decomposes and penta-1-ol is dehydrated to pentene (**23**) – much more significantly [59].

The SMSIM (supermodified simplex) was chosen to optimize the reaction that allowed >99% yield of the desired pentylmethyl ether to be synthesized in approximately 150 min. Overall analysis times were significantly reduced per data point acquisition, 3.2 min, from previous analyses using GLC (35 min). The main reason for this is FT-IR samples at a far greater rate than GLC while simultaneously being apt at calculating reaction yields and determining when the reactor had reached steady state. The reaction was repeated using a branch and fit algorithm, SNOBFIT, which selects random starting points in the design space of its own accord, increasing the overall confidence level of the determined global optimum. The drawbacks to using SNOBFIT are shown through longer sampling times, 8 min compared to 3.2 min, due to the fact that this algorithm covers a wider range of parameter space between iteration points, meaning the reactor takes longer to scope the novel conditions.

Further work into fully automated processes has been completed between teams from the University of Nottingham, Ningbo, China, Addis Ababa University, Ethiopia, and Jairton Dupont's Laboratory, Porto Alegre, Brazil [64]. Having a fully functioning flow setup, the Skilton *et al.* from Nottingham have setup Skype links and a cloud network (Figure 12.18) where teams across the world could automatically connect to the flow reactor and use it through an Internet connection. Initial tests of the system used the etherification of *n*-propanol in supercritical CO_2 over a γ-Al_2O_3 catalyst to form di-*n*-propyl ether (Scheme 12.12). Other reactions were tested with the setup of the flow reactor; however, analysis could not resolve different isomers of products.

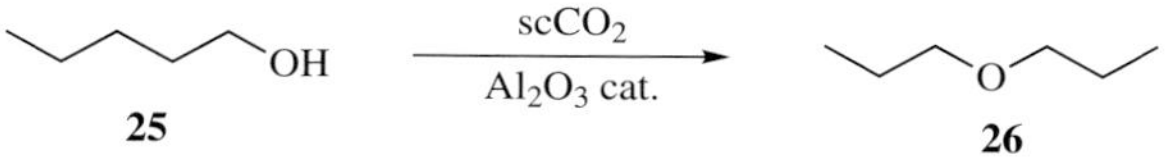

Scheme 12.12 An etherification reaction of *n*-propanol to form di-*n*-propyl ether over a γ-Al_2CO_3 catalyst in supercritical CO_2 for self-optimization studies [64].

All teams therefore used the initial reaction to scope the equipment and cloud technique even if they had no prior knowledge of self-optimization or automated reactors. It is possible to access the experiment mid-run at any time of day and can also be used as a teaching aid for research supervisors and their students. As automated reactors are expensive to buy, set up, and maintain, cloud chemistry may have a much broader application than just for teaching: completing larger industrial processes, for example. It may also help promote more international collaborations, industry–academic interactions, and increase the frequency and throughput of new industrial processes. With the use of modern robot technology, it may be possible to carry out more complex processes at a greater distance with a much greater reduction in human interaction.

More recent work to develop an Internet-controlled reactor has been conducted by a group from the University of Cambridge where their system can monitor and control different chemical syntheses via the Internet [65]. The limitations of current continuous-flow reactors impede discovery through their

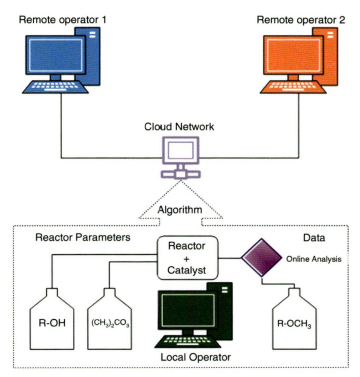

Figure 12.18 A schematic flow diagram outlining the principles of cloud chemistry. The remote operator can log in online and add information to the algorithm GUI, monitor the progress of the reaction, and view the outputted data. (Redrawn from Ref. [64]. Copyright 2015, Nature Publishing Group.)

rigidity and time-consuming nature to develop optimized processes. Continuous-flow reactors are also sparsely spread nationwide, and with current high costs of complete systems, usage is limited to the institute where they reside.

Fitzpatrick *et al.* have taken first steps into the development of a new system, LeyLab, which can be controlled by any researcher with only the use of a computer (Figure 12.19). LeyLab is a specially designed user–server and server–equipment communication program that can be accessed through the Internet from any Internet browser and converges with commonly found laboratory detectors (infrared instruments, mass spectrometers, etc.) and records all relevant data and parameters and also allows the modification of process parameters.

To test the system, an autonomous self-optimization was explored that utilized a modified simplex algorithm – Complex Method – which can manipulate any number of reaction variables that allows the selected output function to be maximized or minimized, for example, yield. The Complex Method of optimization progresses in the same way as previously explained by selecting $n + 1$ initial points, ranking each in turn, and forming the next experiments off previous results. This algorithm, however, begins with an even spread of iterative points

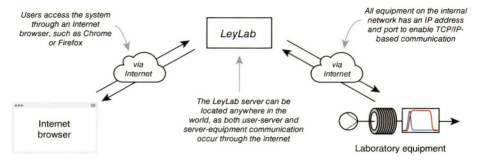

Figure 12.19 Communication array of LeyLab that allows control of laboratory equipment through the Internet (TCP/IP), creating a fully cloud-operated arrangement. The remote operator can log in online, via add information to the algorithm GUI, monitor the progress of the reaction, and view the outputted data [65].

throughout the design space and proceeds until an optimum has been determined. A simple reaction was chosen to be investigated first – the hydration of 3-cyanopyridine to its amide over MnO_2 (Scheme 12.13), where temperature, residence time, and inlet reaction concentration were the focus for optimization. An optimal set of conditions were found within 12 experiments that were carried out over the course of 17 h. The system automatically induced a safety shutdown sequence when pump fluctuations were detected, which shows how effective and intelligent an automated process can be.

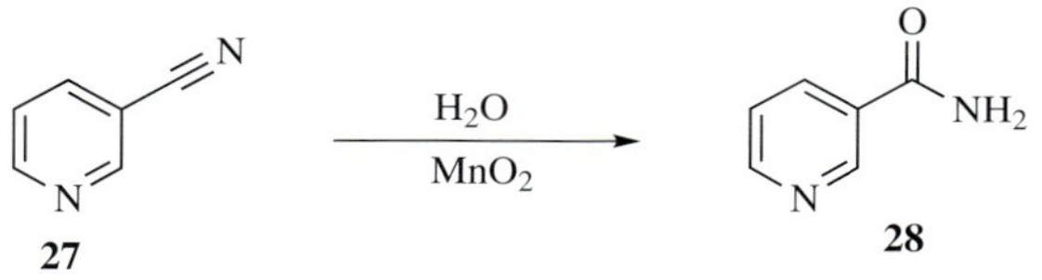

Scheme 12.13 The hydration of 3-cyanopyridine, **27**, to its amide, **28**, over MnO_2 was investigated by Fitzpatrick and used for preliminary self-optimization studies [65].

It is clear from current work with cloud-enabled automated continuous-flow reactors that this new technology can enable and facilitate reaction optimization without extensive or significant computer knowledge or without any prior training for chemists working in a different field. Although cloud-controlled chemical reactions have plenty of benefits in opening self-optimization reactions to a wide number of disciplines, this new method is still in development and may take considerable time to become well known and used in industry or between academics.

12.6

Conclusions and Future Directions

The use of continuous-flow systems with intelligent controllers enables a much greatly improved optimization process than traditional approaches (OVAT). The use of computer based algorithms for reaction optimization can be applied not only for optimization but also to run data-rich experiments, with significant amount of important and useful data allowing full models to be developed for the entire chemical process. Online analytics are an important tool employed for these systems, giving much greater control of experimental space, while requiring minimal human effort. Due to the nature of continuous flow, these systems are applicable to a large variety of reactions and may be applied to highly complex reaction systems and telescoped reaction pathways. The ultimate goal for these systems may not only be the optimization of reaction parameters but also ensuring chemists and chemical engineers put more focus on the development of more sustainable and cost-effective processes to enable a much greater focus on quality of results and on process understanding rather than a focus on running the experiments themselves.

Acknowledgments

Work on this chapter was enabled in part by the National Research Foundation, Prime Minister's Office, Singapore under its CREATE program. Funding from the EPSRC and AstraZeneca under the CASE studentship program also allowed this chapter to be developed.

References

1 Kockmann, N., Gottsponer, M., Zimmermann, B., and Roberge, D.M. (2008) Enabling continuous-flow chemistry in microstructured devices for pharmaceutical and fine-chemical production. *Chemistry: A European Journal*, **14**, 7470–7477.

2 Roberge, D.M., Zimmermann, B., Rainone, F., Gottsponer, M., Eyholzer, M., and Kockmann, N. (2008) Microreactor technology and continuous processes in the fine chemical and pharmaceutical industry: is the revolution underway? *Organic Process Research & Development*, **12**, 905–910.

3 Roberge, D.M., Ducry, L., Bieler, N., Cretton, P., and Zimmermann, B. (2005) Microreactor technology: a revolution for the fine chemical and pharmaceutical industries? *Chemical Engineering & Technology*, **28** 318–323.

4 Hartman., R.L. (2012) Managing solids in microreactors for the upstream continuous processing of fine chemicals. *Organic Process Research & Development*, **16**, 870–887.

5 Teoh, S.K., Rathi, C., and Sharratt, P. (2015) Practical assessment methodology for converting fine chemicals processes from batch to continuous. *Organic Process Research & Development*, **20**, 414–431.

6 Baumann, M. and Baxendale, I.R. (2013) An overview of the synthetic routes to the best selling drugs containing 6-membered heterocycles. *Beilstein Journal of Organic Chemistry*, **9**, 2265–2319.

7 Murray, P.M., Bellany, F., Benhamou, L., Bucar, D.-K., Tabor, A.B., and Sheppard, T.D. (2016) The application of design of experiments (DoE) reaction optimisation and solvent selection in the development of new synthetic chemistry. *Organic & Biomolecular Chemistry*, **14**, 2373–2384.

8 Friedman, M. and Savage, L.J. (1947) Planning experiments seeking maxima, in

Techniques of Statistical Analysis, McGraw-Hill, New York.

9 Weissman, S.A. and Anderson, N.G. (2015) Design of experiments (DoE) and process optimization. a review of recent publications. *Organic Process Research & Development*, **19**, 1605–1633.

10 Lapkin, A.A. and Plucinski, P.K. (2009) Engineering factors for efficient flow processes in chemical industries, in *Chemical Reactions and Processes under Flow Conditions*, The Royal Society of Chemistry, pp. 1–43.

11 Valera, F.E., Quaranta, M., Moran, A., Blacker, J., Armstrong, A., Cabral, J.T., and Blackmond, D.G. (2010) The Flow's the Thing . . . Or Is It? Assessing the merits of homogeneous reactions in flask and flow. *Angewandte Chemie, International Edition*, **49**, 2478–2485.

12 Bristow, T.W.T., Ray, A.D., O'Kearney-McMullan, A., Lim, L., McCullough, B., and Zammataro, A. (2014) On-line monitoring of continuous flow chemical synthesis using a portable, small footprint mass spectrometer. *Journal of the American Society for Mass Spectrometry*, **25**, 1794–1802.

13 Mettler-Toledo (2016) ReactIR 15. Available at http://us.mt.com/us/en/home/products/L1_AutochemProducts/ReactIR/ReactIR-15.html.

14 Mettler-Toledo (2016) FlowIR for Continuous Flow Chemistry. Available at http://us.mt.com/us/en/home/products/L1_AutochemProducts/ReactIR/flow-ir-chemis.html.

15 McMullen, J.P. and Jensen, K.F. (2010) Integrated microreactors for reaction automation: new approaches to reaction development. *Annual Review of Analytical Chemistry*, **3**, 19–42.

16 Houben, C. and Lapkin, A.A. (2015) Automatic discovery and optimization of chemical processes. *Current Opinion in Chemical Engineering*, **9**, 1–7.

17 Holmes, N., Akien, G.R., Savage, R.J.D., Stanetty, C., Baxendale, I.R., Blacker, A.J., Taylor, B.A., Woodward, R.L., Meadows, R.E., and Bourne, R.A. (2016) Online quantitative mass spectrometry for the rapid adaptive optimisation of automated flow reactors. *Reaction Chemistry & Engineering*, **1**, 96–100.

18 McNally, A., Prier, C.K., and MacMillan, D.W.C. (2011) Discovery of an alpha-amino C–H arylation reaction using the strategy of accelerated serendipity. *Science*, **334**, 1114–1117.

19 Lapkin, A.A., Heer, P.K., Jacob, P.-M., Hutchby, M., Cunningham, W., Bull, S., and Davidson, M.G. (2017) Automation of route identification and optimisation based on datamining and chemical intuition. *Faraday Discussions*, **202**, 483.

20 Duros, V., Grizou, J., Xuan, W., Hosni, Z., Long, D.-L., Miras, H.N., and Cronin., L. (2017) Human versus robots in the discovery and crystallization of gigantic polyoxometalates. *Angewandte Chemie, International Edition*, **56**, 10815–10820.

21 Skoraczyński, G., Dittwald, P., Miasojedow, B., Szymkuć, S., Gajewska, E.P., Grzybowski, B.A., and Gambin, A. (2017) Predicting the outcomes of organic reactions via machine learning: are current descriptors sufficient? *Scientific Reports*, **7**, 3582.

22 Peremezhney, N., Connaughton, C., Unali, G., Hines, E., and Lapkin, A.A. (2012) Application of dimensionality reduction to visualisation of high-throughput data and building of a classification model in formulated consumer product design. *Chemical Engineering Research and Design*, **90**, 2179–2185.

23 Schmidt, M. and Lipson., H. (2009) Distilling free-form natural laws from experimental data. *Science*, **324**, 81–85.

24 Famini, G.R. and Penski., C.A. (1992) Using theoretical descriptors in quantitative structure–activity relationships: some physicochemical properties. *Journal of Physical Organic Chemistry*, **5**, 395–408.

25 Folic, M., Adjiman, C.S., and Pistikopoulos, E.N. (2007) Design of solvents for optimal reaction rate constants. *AIChE J*, **53**, 1240–1256.

26 Adjiman, C.S., Galindo, A., and Jackson., G. (2014) Molecules matter: the expanding envelope of process design, in *Proceedings of the 8th International Conference on*

Foundations of Computer-Aided Process Design (eds M.R. Eden, J.D. Siirola, and G.P. Towler), Elsevier Science Bv, Amsterdam, pp. 55–64.

27 Oliynyk, A.O., Antono, E., Sparks, T.D., Ghadbeigi, L., Gaultois, M.W., Meredig, B., and Mar, A. (2016) High-throughput machine-learning-driven synthesis of full-Heusler compounds. *Chemistry of Materials*, **28**, 7324–7331.

28 Cole, J.M., Low, K.S., Ozoe, H., Stathi, P., Kitamura, C., Kurata, H., Rudolf, P., and Kawase, T. (2014) Data mining with molecular design rules identifies new class of dyes for dye-sensitised solar cells. *Physical Chemistry Chemical Physics*, **16**, 26684–26690.

29 Segler, M.H.S. and Waller, M.P. (2016) Modelling chemical reasoning to predict reactions. *Chem. Eur. J.*, **23**, 6118–6128.

30 Coley, C.W., Barzilay, R., Jaakkola, T.S., Green, W.H., and Jensen, K.F. (2017) Prediction of organic reaction outcomes using machine learning. *ACS Central Science*, **3**, 434–442.

31 Segler, M.H.S. and Waller, M.P. (2017) Neural-symbolic machine learning for retrosynthesis and reaction prediction. *Chemistry: A European Journal*, **23**, 5966–5971.

32 Goos, P. and Jones, B. (2015) *Exploring Best Practice in Design of Experiments*, RSC Chemistry World.

33 Jobson, N.K. (2008) The stereoselective synthesis of iodinated analogues of reboxetine: new imaging agents for the noradrenaline transporter. Thesis, Department of Chemistry, University of Glasgow.

34 Filipponi, P., Ostacolo, C., Novellino, E., Pellicciari, R., and Gioiello., A. (2014) Continuous flow synthesis of thieno[2,3-*c*] isoquinolin-5(4*H*)-one scaffold: a valuable source of PARP-1 inhibitors. *Organic Process Research & Development*, **18**, 1345–1353.

35 Spendley, W., Hext, G.R., and Himsworth, F.R. (1962) Sequential application of simplex designs in optimisation and evolutionary operation. *Technometrics*, **4**, 441–461.

36 Rasmussen, C.E. and Williams, C.K.I. (2006) *Gaussian Processes for Machine Learning*, The MIT Press, Cambridge.

37 Houben, C., Peremezhney, N., Zubov, A., Kosek, J., and Lapkin, A.A. (2015) Closed-loop multi-target optimisation for discovery of new emulsion polymerisation recipes. *Organic Process Research & Development*, **19**, 1049–1053.

38 Echtermeyer, A., Amar, Y., Zakrzewski, J., and Lapkin, A. (2017) Self-optimisation and model-based design of experiments for developing a C–H activation flow process. *Beilstein Journal of Organic Chemistry*, **13**, 150–163.

39 Peremezhney, N., Hines, E., Lapkin, A., and Connaughton, C. (2014) Combining Gaussian processes, mutual information and a genetic algorithm for multi-target optimization of expensive-to-evaluate functions. *Engineering Optimization*, **46**, 1593–1607.

40 Richmond, C.J., Miras, H.N., Oliva, A.R., Zang, H., Sans, V., Paramonov, L., Makatsoris, C., Inglis, R., Brechin, E.K., Long, D.-L., and Cronin, L. (2012) A flow-system array for the discovery and scale up of inorganic clusters. *Nature Chemistry*, **4**, 1037–1043.

41 Zang, H.Y., de la Oliva, A.R., Miras, H.N., Long, D.L., McBurney, R.T., and Cronin., L. (2014) Discovery of gigantic molecular nanostructures using a flow reaction array as a search engine. *Nature Communications*, **5**, 3715.

42 Hartwig, J., Metternich, J.B., Nikbin, N., Kirschning, A., and Ley, S.V. (2014) Continuous flow chemistry: a discovery tool for new chemical reactivity patterns. *Organic & Biomolecular Chemistry*, **12**, 3611–3615.

43 Sans, V., Glatzel, S., Douglas, F.J., Maclaren, D.A., Lapkin, A., and Cronin, L. (2014) Non-equilibrium dynamic control of gold nanoparticle and hyper-branched nanogold assemblies. *Journal of Chemical Sciences*, **5**, 1153.

44 Czechtizky, W., Dedio, J., Desai, B., Dixon, K., Farrant, E., Feng, Q.X., Morgan, T., Parry, D.M., Ramjee, M.K., Selway, C.N., Schmidt, T., Tarver, G.J., and Wright, A.G. (2013) Integrated synthesis and testing of

substituted xanthine based DPP4 inhibitors: application to drug discovery. *ACS Medicinal Chemistry Letters*, **4**, 768–772.

45 Robbins, D.W. and Hartwig, J.F. (2011) A simple, multidimensional approach to high-throughput discovery of catalytic reactions. *Science*, **333**, 1423–1427.

46 Goode, J.G., Mansfield, A., and Butters, C. Rapid Analysis and Optimization of Continuous Flow Reactions. White Paper, Mettler Toledo, http://uk.mt.com/gb/en/home/supportive_content/White_Papers.html, pp. 10. Available at http://uk.mt.com/gb/en/home/supportive_content/White_Papers.html.

47 McWilliams, J., Sidler, D., Sun, Y., and Mathre, D. (2005) Applying statistical design of experiments and automation to the rapid optimization of metal-catalyzed processes in process development. *Journal of the Association for Laboratory Automation*, **10**, 394–407.

48 Moore, J.S. and Jensen, K.F. (2014) "Batch" kinetics in flow: online IR analysis and continuous control. *Angewandte Chemie, International Edition*, **53**, 470–473.

49 Gargurevich, I.A. (2016) Foundations of chemical kinetic modelling, reaction models and reactor scale-up. *Journal of Chemical Engineering and Process Technology*, **7**, 6.

50 Mozharov, S., Nordon, A., Littlejohn, D., Wiles, C., Watts, P., Dallin, P., and Girkin, J.M. (2011) Improved method for kinetic studies in microreactors using flow manipulation and noninvasive Raman spectrometry. *Journal of the American Chemical Society*, **133**, 3601–3608.

51 Rostrup-Nielsen, J. (2000) Reaction kinetics and scale-up of catalytic processes. *Journal of Molecular Catalysis A: Chemical*, **163**, 6.

52 Donati, G. and Paludetto, R. (1997) Scale up of chemical reactors. *Catalysis Today*, **34**, 483–533.

53 Hone, C.A., Holmes, N., Akien, G.R., Bourne, R.A., and Muller, F.L. (2017) Rapid multistep kinetic model generation from transient flow data. *Reaction Chemistry & Engineering*, **2**, 103–108.

54 Hessel, V., Kralisch, D., and Kockmann, N. (2014) *Novel Process Windows: Innovative Gates to Intensified and Sustainable Chemical Processes*, Wiley VCH Verlag GmbH, Weinheim, Germany.

55 Levenberg, K. (1944) A method for the solution of certain non-linear problems in least squares. *Quarterly of Applied Mathematics*, **2**, 5.

56 Reizman, B.J. and Jensen, K.F. (2012) An automated continuous-flow platform for the estimation of multistep reaction kinetics. *Organic Process Research & Development*, **16**, 1770–1782.

57 Reizman, B.J. and Jensen, K.F. (2016) Feedback in flow for accelerated reaction development. *Accounts of Chemical Research*, **49**, 1786–1796.

58 Krishnadasan, S., Brown, R.J.C., deMello, A.J., and deMello, J.C. (2007) Intelligent routes to the controlled synthesis of nanoparticles. *Lab on a Chip*, **7**, 1434–1441.

59 Skilton, R.A., Parrott, A.J., George, M.W., Poliakoff, M., and Bourne, R.A. (2013) Real-time feedback control using online attenuated total reflection Fourier transform infrared (ATR FT-IR) spectroscopy for continuous flow optimization and process knowledge. *Applied Spectroscopy*, **67**, 1127–1131.

60 Sans, V., Porwol, L., Dragone, V., and Cronin, L. (2015) A self-optimizing synthetic organic reactor system using real-time in-line NMR spectroscopy. *Chemical Science*, **6**, 1258–1264.

61 Houben, C., Peremezhney, N., Zubov, A., Kosek, J., and Lapkin, A.A. (2015) Closed-loop multitarget optimization for discovery of new emulsion polymerization recipes. *Organic Process Research & Development*, **19**, 1049–1053.

62 Bourne, R.A., Skilton, R.A., Parrott, A.J., Irvine, D.J., and Poliakoff, M. (2011) Adaptive process optimization for continuous methylation of alcohols in supercritical carbon dioxide. *Organic Process Research & Development*, **15**, 932–938.

63 Holmes, N. and Bourne, R.A. (2015) *Chemical Processes for a Sustainable Future*, 1st edn, Royal Society of Chemistry, Cambridge.

64 Skilton, R.A., Bourne, R.A., Amara, Z., Horvath, R., Jin, J., Scully, M.J., Streng, E., Tang, S.L.Y., Summers, P.A., Wang, J., Perez, E., Asfaw, N., Aydos, G.L.P., Dupont, J., Comak, G., George, M.W., and Poliakoff, M. (2015) Remote-controlled experiments with cloud chemistry. *Nature Chemical Biology*, **7**, 1–5.

65 Fitzpatrick, D.E., Battilocchio, C., and Ley, S.V. (2015) A novel Internet-based reaction monitoring, control and autonomous self-optimization platform for chemical synthesis. *Organic Process Research & Development*, **20**, 386–394.

Index

a

ab initio methodology 32, 34, 35
Abraham descriptors 26
accelerated serendipity 335
accuracy 26, 38
acetaldehyde 28
acetaminophen 267
– spectrum 267
acetone 259
acetonitrile 33
ACOMP, see automatic continuous online monitoring of polymerization reactions
acoustic irradiation 155, 167
acrylonitrile butadiene styrene (ABS) polymers 217
activated complex 23
activation energy barrier 25
activation free energy 25
active pharmaceutical ingredients (APIs) 76, 142, 157, 176, 264
acylases 76
additive manufacturing 12
– 3D printing 220
– techniques, chemical reactors manufacturing 220
additive manufacturing (AM), polymerization 216
– melting-based techniques 218
– – electron beam melting (EBM) 218
– – fused deposition modeling (FDM) 219
– – laser sintering (LS) 219
– – material jetting (MJ) 219
– – selective laser melting (SLM) 218
– photopolymer jetting (PJ) 217
– physical binding 217
– stereolithography (SLA) 217
adsorbents 63

adsorption 28, 300, 320
– competitive 319
– properties 300
advanced-flow™ reactors 161
agglomeration 48
agitated cell reactor (ACR) 163
Ag nanocubes 54, 55
Ag nanoparticles 50
Ag nanowire 64
alcohol dehydrogenases 76
aldolases 76
aliphatic and aromatic hydrocarbons 259
alkoxyxanthones 22, 24
alkylation reaction 362
alkyl halides
– via phase transfer catalysis 140
alkyltransferases 76
American Chemical Society Green Chemistry Institute Pharmaceutical Roundtable 155
amine dehydrogenases 76
amines 74
amino transferases 84
ammonia lyases 76
amorphous Ni hydroxides 58
AM Technology 163
anthropogenic emissions of carbon dioxide 4
AP Miniplant 163
apparent transverse relaxation 296
applied mathematics 11
artificial intelligence (AI) 12
ART® reactor 161
ascorbic acid 63
asymmetric reduction of MLK-II to MLK-III 77
ATEX regulation 173

Handbook of Green Chemistry Volume 12: Green Chemical Engineering, First Edition. Edited by Alexei A. Lapkin.
© 2018 Wiley-VCH Verlag GmbH & Co. KGaA. Published 2018 by Wiley-VCH Verlag GmbH & Co. KGaA.

atom economy 38
atorvastatin 74
ATR-IR flow cell 231
attainable region (AR) 94
attenuated total reflectance (ATR) 265, 271
Au hybrid spheres 52
Au nanospheres 63
Au/TiO$_2$ catalysts 318
automated kinetic parameter estimation
– logic flow diagram 359
automated open-loop optimization systems 350
automated self-optimized reaction 365
automatic continuous online monitoring of polymerization reactions (ACOMP) 275
axial dispersion–exchange model (ADEM) 124

b

Baeyer-Villiger monooxygenases 76
barrier channels 136
base-catalyzed Knoevengel condensation reaction 353
batteries 47
Bayer Technology Services 158, 174, 178
Bayesian statistics-based design 11
benzaldehyde 345
benzyl bromide
– benzylation reaction of ethyl 2-oxocyclopentanecarboxylate 140
bespoke 3D printed device 223
big data approaches 335
bimetallic hollow PtAg nanoparticles 53
binder jetting (BJ) 218
bio-based chemical supply chain 11
biocatalysis 74, 76, 79, 80, 83, 85
biocatalytic
– reduction using a KRED 78
– retrosynthesis 75, 76
– transformations 75
biological catalyst 73
biorefining 12
– systems 9
bioseparations 12
biotechnology 12, 73
biotransformation 29
biowaste-derived limonene to isocarveol 7
biowaste-derived terpene feedstock 7
biphasic reactions 142
– esterification reaction 20
Bloch theory of NMR 295
Bloch vector model 291
Boltzmann constant 291

Boltzmann distribution 291
bonding electrons 293, 294
borohydride – TEA complex 272
boron reagent 77
branch and bound techniques 92
Bretherton's solution 130
3-bromothiophene-2-carboxylic acid 349
business environment 191
business models 12, 192
– in chemical industry 195
– innovation 196
2-butanone 316
tert-butyl chloride 27, 31
γ-butyrolactone 304, 305

c

CAD file 218, 221
CAD reactor configuration design 245
C$_5$ adsorption 301
calibration models 269
capillary microreactor 139
carbon dioxide 4
– rate of emissions of 2
carbon emissions 3, 4
carbon nanofiber (CNF) 120
carbon sequestration 10
carboxylic acid reductases 76, 80
Carr–Purcell–Meiboom–Gill (CPMG) sequence 297
cascade mixers 160
catalyst 47, 73
– behavior 63
– composition 47
– properties 73, 74
– surface 47
– wetting efficiency 300
caterpiller mixers 160
CCD experimental matrix 349
CCF design 350
CCF DoE model 351
^{13}C DEPT-45 NMR spectra of 1,4-butanediol 306
^{13}C DEPT NMR spectroscopy 301
^{13}C DEPT-45 spectra of coadsorption of C$_5$ hydrocarbon and hydrogen on Al$_2$O$_3$ 302
^{13}C-DEPT spectroscopy 320
^{13}C distortionless enhancement polarization transfer (DEPT)-45 NMR 304
central composite (CCD)– face-centered (FCD) design 346
cesium carbonate 22, 24
cetyltrimethylammonium bromide (CTAB) 63
cetyltrimethylammonium chloride (CTAC) 61

Index

CFD code 127
chemical analysis in manufacturing 259
chemical engineering-enabling technologies 12
chemical engineering science 9
– for sustainability 7
chemical industry 2, 194
– as energy storage solution 3
– sustainability challenge for 1
chemical leasing 12, 192–194
– model 193
chemical management practice 194
chemical manufacturing 5
chemical pollution 5
chemical processes 319
– conceptual design 89
chemical reactions 35, 38
– closed-loop optimization 350
chemical reactors 299, 300
chemical shift 293, 294
chemicals management 193
chemical wastes 259
chemistry of biofeedstocks 5
chemocatalysts 76
Chemtrix BV 170
chiral alcohols 74
chiral boron reagent 77
CHK-O cells 240
chloroform 267
chloromethane 293
chloroxanthone 24
chlorozanthone product 22
chronoamperometry 60
– measurements 55, 59
CINC Industries 164
circular economy 9
closed-loop automated reactor system 332
closed-loop discovery
– algorithm-based self-optimization 364
– design of experiments, application of 346
– high-throughput screening 342
– one-variable-at-a-time approach (OVAT) 344
– rate-based/physical organic approaches 350
– self-optimization 341
closed-loop experimental system 332
closed-loop optimization
– analytical techniques 332
– decision algorithms 334, 335, 337, 338
– system 331, 334
cloud-enabled automated continuous-flow reactors 368
^{13}C MAS NMR spectra 302

^{13}C NMR spectroscopy 294
CO_2-based technologies 4
CO_2-derived methanol 4
CO_2 emissions 4, 7, 8
cofactor recycle 84
Coflore® flow reactors 163
compact heat exchange reactors (Chart E&C) 161
complex multisubstrate kinetics 84
computational cost 26
– and accuracy 34
computational method 33
computer-aided blend design (CAMD) 35
– defined as 20
– formulation 37
– methodologies 21, 31, 34
– optimization problem 33
– problems 21, 33
– QM-CAMD 32
– techniques 20, 21
computer-aided mixture design (CAMxD) 34, 35
computer-aided molecular design (CAMD) approaches 20, 38, 97
computer-aided solvent design 31
computer-aided techniques 81
Co nanoparticles 49
conceptual process design 89
confidentiality 210
– and intellectual property rights 205
ConsiGma™ 165
continuous flow oscillatory baffled reactor (COBR) 162, 233
continuous-flow reactors 330, 334, 364
continuous-flow systems 332
continuous kinetic measurements 169
continuous lab-scale microwave reactor 167
continuous liquid interface production (CLIP) 249
continuous manufacturing
– asymmetric enone hydrogenation 176
– case studies 174
– classification 154
– continuous filter 178
– DFT/DFBA synthesis 178
– equipment for 158
– – downstream equipment 163
– – as enabling technology 166
– – microwave chemistry 166
– – photochemistry 168
– – process integration 165
– – sonochemistry 167

- equipment size/production capacity ratio 153
- F³ Factory case studies 180
- INVITE platform 179
- means of process intensification 153
- modular design approach 179
- modularization concept for versatile continuous production 175
- 2-nitro-ethanol, from formaldehyde/nitromethane 176
- process development and scale-up 169
- process development/implementation 168
- - changeable production systems, modularization 174
- - ecotrainer concept, for rapid process development 173
- - flexible implementation 172
- - heat of reaction 171
- - lab-scale development 168
- - mixing requirements 171
- - process analytical technology (PAT) 172
- - reaction kinetic 170
- - scale-up 170
- process intensification 153
- reactors with dynamic mixing 161
- - agitated cell reactor (ACR) 163
- - agitated tube reactor (ATR) 163
- - continuously oscillatory baffle reactor (COBR) 162
- - spinning disc reactor 162
- - tube/Taylor–Couette reactor 162
- reactors without active mixing 159
- - empty tube 159
- - intensified mixing 160
- - plate reactors 161
- - tube with static mixers 159
- small-scale continuous chemical production, modularization 173
- for small- to medium-scale chemical production 153
- thioether oxidation 175
- upstream equipment 159

continuous manufacturing and crystallization (CMAC) 83, 157, 158
continuous reactors 300, 320
continuous solid–liquid separation 164
continuous-stirred tank reactor (CSTR) 94, 352
continuum solvation models 24, 34
control hazardous reactions 330
conventional diffusion measurements 289
convex objective function 91
convex relaxation 91

- functions 91
Copiride 158
CO poisoning 49, 53, 64, 65
CO_2 recycling technology 4, 8
core–shell nanoparticles 50
core–shell particles 50
core–shell PtAg nanoparticles 50
corrosion resistant metal oxide 49
COSMO-based models 26
COSMO descriptors 31
COSMO-RS and COSMO-SAC methods 33
COSMO-RS/COSMO-SAC thermodynamics 34
COSMO-RS model 28
COSMO-RS/SAC formalism 33
COSMO solvent descriptors 31
cost-competitive technology 49
cost reduction 192, 196
costs of goods sold (COGS) 154
co-surfactant 63
CO tolerance 57, 60
coupling
- online analytical technique 364
- photochemistry 168
CPMG spin echo pulse sequence 298
cryo-NMR 309
crystallization 266
CSE-XR® mixer (Fluitec mixing + reaction solutions AG) 160
cubic nanocages (CNC) 61
Cu-catalyzed intramolecular amination 79
CuCl catalyst 344
cumene 301
cyanoborohydride ($NaBH_3CN$)
- in MeOH (1M) 231
3-cyanopyridine hydration 368
cyclic voltammogram (CVs) 49, 50, 52, 55
- Co/Graphitic Carbon (GC), Pt/GC, and PtCo/GC 58
- renormalization of 50
cyclization reaction 22, 24

d

database-screening approach 27
database solvents 29
3D computer-aided drawing program (3D CAD) 215, 227
dead zone (DZ) 250
decomposition approaches 32
deep reactive ion etching (DRIE) 116
dendritic hollow nanocrystals 61
density functional theory (DFT) 30, 319
design

- constraints 35
- engineers 93
- of experiment (DoE) 329, 338
- problems 91
detergents 3
deterministic methods 92
device setup 234
DFT + PCM calculations 30
diastereoselective catalytic hydrogenation 347
diazo coupling reaction 139
diazoketone 142
dichloromethane 29
2,4-dichloropyrimidine/4,4'-(2,4-pyrimidinediyl)bis-morpholine
- S_NAr reaction 358
dielectric constant 26, 28, 36
Diels–Alder cycloaddition 20, 27, 31, 34, 223
diethyl diallylmalonate 22
differential side stream reactor (DSR) 94
diffusion 307, 320
- attenuation curves of several hydrocarbons 309
- coefficients 306, 307, 309, 313
- competitive 301
- fluids confined in porous catalysts 306
- liquid hydrocarbons 309
- pathways 311
- reflectance Fourier transform spectroscopy (DRIFTS) 318
difluorobenzaldehyde (DFBA) 179
2,4-difluoronitrobenzene, with pyrrolidine in ethanol 355
2,3-difluorophenylboronic acid 179
2,3-difluorotoluene (DFT) synthesis 179
digital micromirror device (DMD) 248
1,3-dimesityl-4,5-dihydroimidazol-2-ylidene ruthenium complex 22
dioxygenases 82
Dipole moments 28, 265, 267
discrete variables 92
dispersion 124
disruptive - synthetic biology 12
distillation/extraction columns 166
2D NMR correlation spectroscopy (COSY) 300
DoE approach 339
D-optimal design 359
3D printed cell perfusion system 241
3D printed flat sheet membranes 237
3D printed flow reactors 228
3D printed fluid selector 242
3D printed IR sensor block 245
3D printed PDMS layer SEM image 238
3D printed PDMS photoresist 239
3D printed sealed sequential reactors 224
3D printer basic chemical reactor 222
3D printing
- applied to flow chemistry 226, 235, 239, 242
- combined reactor, and reaction design 222
- continuous flow millifluidic devices 231
- flow chemistry 220
- functional devices 226
- future trends 248
- integrated modular setup for multistep organic transformations 232
- keyword 216
- membranes 235
- of millifluidic devices 230
- in PDMS, membrane contactor 238
- processes, characteristics 216
- rapid prototyping of millifluidic devices 229
- shape memory polymer fibers 250
- smart materials through 4D printing 250
- techniques fabrication scheme 221
- ultrafast printing 249
DRIE silicon–glass microreactor 117
droplet device 249
drug manufacturing 157
drug product quality 177
drying processes 268
d-spacing 61
3D System Viper si2 SLA 3d printer 228
3DTouch™ 3D printer 229
dynamic mechanical analyzer (DMA) 250

e

Earths *vs.* biocapacity 2
EBM printing process 218
echo amplitudes 299, 316
ecological footprint 1, 2
economic gains 197
economic prosperity 1
economic viability 203
electrocatalysis 47–49, 55, 64
- devices 48
- reactions 47
- reduction 4
electrocatalysts 47, 48
- in fuel cells 47
- support corrosion 49
electrochemical
- conversion devices 47
- energy 49, 64
- measurements, of MOR on 60
- oxidation 63
- reaction 47
- surface area 55, 59

electrochemistry 47, 48
electrode potential 47
electrolyte composition 47
electrolyte ions 47
electrolytic cell 3D printed plates 228
electrolyzers 47, 226
electron beam melting (EBM) 218
electronegativity 293
electronic effects 48
electrons 47
electrospray ionization mass spectrometer (ESI-MS)
– high-resolution 232
emissions
– incurred 3
– savings 3
emulsification methods 48
emulsions 289
enantiomers 77
enantioselective amination 77
ene reductases 76
energy
– balances 91
– consumption 27, 91, 260
– demand 2
– efficient technologies 2, 3
– generation 3, 4
– resources 268
– savings 196
– storage 49
engineered
– biocatalysts 79
– enzymes 75
– microbial cells 73, 74
– transaminases 76
environmental, health, and safety (EHS) 28
– impact 38
environmental impact 27
enzyme ATA117, 234
enzyme discovery 75
enzyme limitations 81
equipment sizes 94
esterases 76
esterification 22, 29, 30
e_{surf} (NMR) against E_{max} (TPD) for water on different porous oxide surfaces 318
ethylene 4, 323
ethylene glycol 321
ethylene glycol/water–toluene flow 122
European Community competition rules 208
European framework for waste 207
experimental conditions 47

experimental design procedure, stages of 339
experimental systems
– self-optimization and requirements 330
extractive techniques 352

f

face centered cubic (fcc) 56
face-centered (FCD) design 346
falling film microreactor (FFMR) 115, 125, 133
FDA (US Food and Drug Administration) 156
FDM 3D printer 219, 234
feedback loop 330, 332
fermentation-based chemical production 12, 73, 74
F^3 Factory project 158, 179
filamentous structured material (FSM) 120
fingerprints, of molecular dynamics of molecules 289
Fischer indole synthesis 77, 79
flexibility 157
flow module, CAD drawing 243
flow rate, curves outlining 354
flow system, schematic diagram 353
fluid catalytic cracking (FCC) 311
fluidized beds 300
– reactors 166
fluorescence detectors 344
Fo number 126
Food and Drug Administration (FDA) 263
footprint 1
formic acid 53, 63, 64
– oxidation 63, 64
Fourier number 132
Fourier transform (FT) 292, 297
Fourier transform infrared (FTIR) spectra 345
free energy 23–25
free induction decay (FID) 291, 292
FT-IR analysis 334, 364, 366
fuel
– cells 47, 53
– oxidation 47
– technology 3
functional
– economy 191
– groups, combinations of 21
– modules, illustration and photographic images 247
– nanomaterials, design of 11
fused filament fabrication 219

g

galvanic replacement 64
– reaction 48, 51, 54, 60, 62, 64
gas chromatography (GC) 259, 319
gas chromatography-mass spectrometry (GC-MS) 259
gas-dominated flow 118
gas-expanded solvent (GXL) 34
gas flow passes 134
gas–liquid channeling 134
gas–liquid distributing principles
– two-phase flow uniformity 133
gas–liquid interface 127, 322
gas–liquid mass transfer coefficient 120
gas–liquid mass transfer, for Taylor flow 128, 131
gas–liquid–solid catalytic reaction 115
gas–solid reactions 170
Gaussian processes 340
GEA Westfalia Separator Group 164
generalized disjunctive program (GDP) 92
– formulations 104
– models 104
generalized reduced gradient (GRG) 91
generate-and-test approach 28
generic closed-loop automated system
– schematic diagram 331
generic pharmaceutical 77
genetic algorithms 91
genetic code 73
Genzyme 157
geothermal system 106
Gibbs free energy 98
GlaxoSmithKline (GSK) 158
global population 2
global warming potential 8
glycerol 28, 309
glycerol ethyl acetal 28
glycerolysis 20, 29
cis-glycol production 20
gold nanoparticles 52
gradient-based methods 91
gradient-based optimization algorithms 91
gradient-based quasi-Newton search algorithm 362
graphene 59
green and sustainable chemistry 5
green chemical engineering 8
green chemistry 5, 11, 47, 259, 260, 281
– principles 6
green engineering
– principles 9
– solutions 8

greener chemistry 21
green house gas (GHG) 4
– emissions 3, 4
greenhouse gas emissions 196
green solvents, in design approaches 35
Griseofulvin 274
group contribution methods 33
Grubbs-catalyzed metathesis 303
Grubbs II catalyst 22
gypsum 314
gyromagnetic ratio 291

h

Hahn echo sequence 297
Hahn spin echo pulse sequence 298
haloalkanes 22
halogenases 76
Hatta number 127
hazardous 5
– chemicals 197
– oxidants 82
– waste, management 193
HCOOH decomposition 63
heat exchange networks (HEN) 93, 101
– design 89, 101
– objective function 102
heat transfer coefficient 159
Henry's law 127
heterocycle 1-(4-methoxyphenyl)-1,2,3,12b-tetrahydroimidazo[1,2-f] phenanthridine 222
heterogeneous
– catalysts 47, 48, 289, 320
– catalytic microreactors 323
– catalytic reactions 47, 300, 320
– reactor 323
1-hexene 302, 303
Higbie penetration theory 132
highest-valued parameters
– pressure and temperature 348
high-fructose corn syrup (HFCS) 83
high-performance liquid chromatography (HPLC) 259
high-throughput screening (HTS) 225, 343
Hildebrand solubility parameter 26
HiTEC Zang 163
^{1}H NMR spectroscopy 323
^{1}H NMR spectrum of chloroethane 294
hollow Au nanospheres 64
hollow PdAg nanotubes 64
hollow PdAu nanoparticles 63
hollow PtAg heterodimers 54
hollow PtAg nanostructures 55

hollow PtAg popcorn-shaped particles 55
hollow PtAu spheres 53
hollow PtCo nanospheres 57
– synthesis via a novel galvanic replacement reaction 56
hollow PtCu$_3$ nanocages 63
hollow PtCu nanospheres 62, 63
hollow Pt metal alloys 51
hollow Pt nanoparticles 51
hollow Pt nanospheres 50
– peak current density of 50
– with positively charged ultrafine Ru nanoparticles 60
– stable during potential cycling 51
– synthesis via galvanic replacement using 50
hollow PtNi nanospheres
– ratio of forward to backward peak currents for 59
– supported on carbon 59
– toward MOR 58
hollow PtNi nanostructures, durability of 59
hollow PtRu nanospheres 60
hollow PtRu nanotubes 59
hollow Pt$_x$Ni$_{1-x}$ spheres 58
hollow spherical PtPd/C 62
homogeneous catalytic metathesis 302
homogeneous Grubbs'-catalyzed metathesis 302
human development index (HDI) 1
hybrid computer-aided molecular design technique 29
hybrid experimental/computer-aided approach 31
hybrid reaction-separation scheme 83
hybrid screening, based on predictive models 28
hybrid solvent selection methodology 29
hydration kinetics 314
hydraulic flow resistance 135
– ratio 135
hydrazine 54
– reduction 54
hydrocarbons 309
hydrodynamic
– dispersion 124
– regime 300
hydroformylation 20, 28
hydrogenation 170, 304
hydrogen bonding 268, 293, 309, 315
hydrogen evolution reaction (HER) 50
hydrogen mass transfer model 132
hydrogen peroxide 61
hydroprocessing catalysts 313
hydrothermal synthesis 224
hyperfine splitting 293

i
imine reductases 76, 80
imine reduction 233
implementation of a process change (ISPR) 83
Impulse 158
in situ product removal (ISPR) 83
indium tin oxide (ITO) 57
industrial trickle bed reactors 124
inhomogeneities, static magnetic field 296
in-line ATR FT-IR system 280
innovations 2, 191
– in unit operations and manufacturing in chemical industry 11
in silico model simulation 358
in silico synthesis 28
Integrated Computer-Aided System (ICAS) software 29
integrated process design, operation, and control 105
intelligent controller automated self-optimizing system 330
interdigital mixer 160
interfacial energy 63
interparticle distance 48
inversion recovery
– experiment 296
– pulse sequence 296
investment cost, redistribution of 197
iodo(nitro)methane 37, 38
ionic mobility 64
iron 47
iron(III) nitrate (Fe(NO$_3$)$_3$) solutions 55
irregular pore 311
isocarveol 8
Isomerization 301
isopropanol 77
ISO transportation 173

j
Jacobian matrix 90
Japanese auto industry 191
J-coupling 293, 294

k
Katapak-S packings 120
Keggin-type heteropolyanions 136
1,5-keto acids 80
β-keto esters, alkylation reactions 140
ketoreductase (KRED) 74, 77
ketoreductase (KRED) biocatalyst 77

kinetic
- analysis 84
- information 28
- models of biocatalytic conversions 84
- parameters 84
- process models 93
Kirkendall effect 48
Knoevenagel reaction 354
knowledge 210
- management systems 156
- sharing 202
Knudsen diffusion mechanism 308
Kolbe–Schmitt reactions 20, 30
Koutecky–Levich plots 57

l

laboratory-scale flow reactors 227
LabView 334
lactic acid 314
Larmor frequency 292, 293, 297
laser melting machines 218
laser sintering (LS) 219
layer-by-layer printing technique 219
LC-MS analysis 342
LCMS sampling 362
lean manufacturing 191
legal environment 191
legislations 5
lethal concentration 30
lethal dose (LD_{50}) 28
Levenburg-Marquardt algorithm 356
LeyLab communication array 368
life cycle assessment (LCA) 7, 8
limonene 8
linearity 309
linearizing nonlinear functions 92
linear programming 91
lipases 74, 76
liquid chromatography-mass spectrometry (LC-MS) 259
liquid film thickness 130
liquid–liquid bubbly flow coalescence 140
liquid–liquid extraction 176
liquid–liquid microchannel flow 122
liquid photosensitive inks
- polymerization 217
lithiation 20
- reaction 33
LMH model 128
Lockhart–Martinelli–Chisholm correlation 127
Lockhart–Martinelli models 130
Lockhart–Martinelli parameter 128
Lonza FlowPlate® reactor 161
low-cost electricity 3
low-temperature cleaning 3
lyases 76

m

machine learning 12
- algorithms 336, 340
- methods 336
macropores 48
magnetic field gradients 297, 315
magnetic moment 290, 291
magnetic quantum states 291
magnetic resonance imaging (MRI) 12, 320, 323, 324
magnetization 291, 292, 295, 296
manganese 47
manufacturing
- defined 191
- industry 191
- practices 191
- technical innovation in 211
mass and energy integration 6
mass specific activity 55
mass transfer 300
- coefficient 82, 131
- limitations 314
mass transfer-limited multiphase processes 161
mass transport 300, 320
- limitations 300
- properties 289
material erosion 167
material integration 100
mathematical programming 93
mathematical programming techniques 21, 89
MATLAB-based computer program 356
maximum oxygen transfer rate 82
maximum theoretical yield 73
Meerwein–Ponndorf–Verley reduction of propionaldehyde 301
melting plastic filaments 219
membrane reactors
- reaction and in situ separation 166
$MeNO_2$ 176
Menschutkin reaction 22–24, 31, 33, 35, 37
mesoscale baffled reactor (mCOBRs) 233
metal catalysts 304
metallic Pd 63
metal–organic framework catalysts 321
metal–organic frameworks (MOFs) 225
metals 48

metathesis 22, 302
methane steam reforming 166
methanol 4, 60, 63
methanol oxidation reaction (MOR) 48, 50, 57–59, 62
methyl acrylate 27
2-methyl-1,3-butadiene 27
2-methyl-3-butyn-2-ol (MBY) 121
methyl chloride 259
methyl iodide 31
methylnicotinate 350
3-methyl-1-pentyn-3-ol
– liquid-phase hydrogenation of 120
α-methylstyrene 301
– hydrogenation 116
Michaelis constant 83
Michaelis-Menten kinetics 84
microfluidic device 249
– assembled 248
microinnova 170
micropacked bed reactors 142
micropores 307, 314
microporous region 314
microreactor design 12
microstructural evolution 314
microstructured packed bed reactor (MSPB) 142
microstructures implementation 164
microwave
– heating 167
– irradiation 167
mid-infrared spectra 266
mid-IR spectrometers 265
mini-trickle bed operation 143
Minnesota solvation model 24
Miprowa® reactor 160
MIP support 194
miscibility 28
mixed integer dynamic optimization problems (MIDO) 21
mixed-integer linear programming (MILP) 92
– problems 21
mixed-integer nonlinear program (MINLP) 21, 92
– CAMD formulation 31
– mathematical problem 30
mixed integer optimization 92
mixed suspension mixed product removal (MSMPR) 176
ML mixers 160
model-based conceptual design 90
model-based methods 26
model-based optimization 168

model-based screening methods 27
model-based selection methods 34
modified rotational dynamics 315
molecular biology 74
molecular complexity constraints 35
molecular design 11
– concepts 26
molecular diffusion 298
– coefficient 131
molecular discovery flow synthesis 341
molecular modeling tools 25
molecular size 307
molecular structures 21
monometallic Pt 49
monooxygenases 82
Montelukast 76, 78
MPBR flow regimes studies 118
multibiocatalyst synthesis 80
multieffect distillation (MED) 105
multienzymatic/chemoenzymatic pathways 75
multienzyme organic synthesis 80
multijet modeling machines 219
multiobjective active learner (MOAL) 341
multiobjective optimization problems (MOO) 21
multiphase kinetics 84
multiphase microreactors 115
– barrier-based gas–liquid flow 135
– capillary microreactors 137
– – microstructured packed bed reactors 142
– – phase transfer catalysis 139
– – wall coated catalytic microreactors 137
– continuous phase multiphase microreactors 115
– – falling film microreactor (FFMR) 115
– – mesh contactor 116
– dispersed phase multiphase microreactors 116
– – fibrous internal structures 120
– – foam microreactors 120
– – microstructured packed beds 117
– – prestructured microreactors 118
– – segmented flow microreactors 116
– ethyl 2-oxocyclopentanecarboxylate with benzyl bromide 140
– falling film 136
– flow distributors 134
– flow regimes 121
– – capillary microreactors 121
– – structured packed beds 122
– gas–liquid flow regimes 123
– hydrodynamics 125
– – falling films microreactors 125

– interdigital mixer–redispersion capillary assembly 141
– liquid bridge nitrogen–water 119
– liquid film 126
– mass transfer 131
– – capillary microreactors 131
– – falling film microreactors 133
– micropillar device 119
– multichannel microstructured reactor 138
– octanal conversion in FFMR 137
– octanal oxidation with oxygen 136
– packed bed reactors 123
– – hydrodynamic dispersion 124
– – liquid holdup 123
– phase transfer catalyzed C–N cross-coupling reaction 143
– phase-transfer-catalyzed diazo coupling reaction 139
– picolinic acid, hydrogenation of 143
– pressure drop, in capillary microreactors 127
– – gas–liquid 127
– – liquid–liquid 130
– principles/features 115
– Taylor flow 129
– two-phase flow distribution 133
– two-phase flow regimes, in capillaries 121
multiphasic mixtures 84
multiphasic systems 84
multistage flash (MSF) 105
multistep reaction sequence, schematic diagram 223
multiwalled carbon nanotube (MWCNT) 64
mutagens 259
mutation 91

n
nanocages 63
nanocluster 49
nanoparticles 48, 49
nano-structured functional materials 11
National Institute for Occupational Safety and Health (NIOSH) 259
Navier–Stokes equation 243
near-infrared (NIR)
– CAD representation 247
– multivariate method 271
– radiation 268
– spectrometer 269
– spectroscopy 267, 270, 271
neural networks, training of 337
new business models, impediments in implementing 197

– quality assurance 198
– single supplier dependency 197
new energy technologies 4
new technology 8
Newton iteration 90
Newton method 90
– globalization of 90
– modifications of 90
Newton–Raphson method 90
NFPA health hazard classification 28
Ni hydroxides 59
Ni oxides 59
NIST MS spectra database 335
NiTech 157
NiTech® Solutions Ltd 162
nitrile hydrolases 76
nitroaromatic nitrile
– chemoselective catalytic reduction 346
– reduction of 346
2-nitroethanol 175
nitrogen–water two-phase flow 118
nitromethane 33, 37, 38
N-methylpyrollidone (NMP) 272
NMR toolkit 289
nondestructive analytical methods 270, 281
noninvasive Raman spectroscopy 353
noninvasive tool 319
nonlinear equation 90
– systems 90
nonlinearity 309
nonlinear problems 21
nonlinear programming (NLP) 90, 91
non-native biocatalytic reaction pathways 12
nonnatural products 83
nonnatural substrates 74, 75
non-Pt alloy nanostructures 63
nonreactive oxygenated species 59
nontoxic chemicals 194
nonuniformity factors 134
Normag 163
Novartis–MIT Center for Continuous Manufacturing 157, 177
n-propanol etherification reaction 366
nuclear magnetic resonance (NMR) 290, 319
– experiment 292
– methodologies 290
– pulsed field gradient 297
– quantitative ^{1}H NMR 294
– relaxation 295, 316
– – measurements 318
– – method 319
– – times 289, 315
– resonance 294

– – signals 293
– signal 294, 304
– – decay 310
– – enhancement 304
– spectroscopy 289, 293, 294
– – of liquids 293
– – resolutions 297
nuclear shielding 293
nuclear spins 291, 298
– alignment 304
– and bulk magnetization 290
nucleophilic substitution 20
numerical
– algorithms, challenges to 93
– optimization 93
Nylon 6 234
nylon-based 3D printed materials
– chemical modification 236

o

occupational health agencies 259
octahedral dendritic hollow nanocrystals (ODH) 61
octahedral dendritic nanocrystals (OD NCs) 61
octahedral nanocages (ONCs) 61
1-octane 320
octanol–water partition coefficient 28
1-octene 321
2-octene 320
oleylamine (OAm) 62
oligomerization 323
on-demand preparation
– concept diagram 362
one variable at a time (OVAT) approach 329, 339
onion model for process design 98
open-loop/closed-loop chemical discovery systems
– schematic diagrams 342
open-loop synthesis system 341
operating conditions 6, 94
operating costs 197, 198
operating variables 92
operational expenses (OPEX) 196
optimal kinetic parameter 361
optimal Pareto solutions 8
optimization
– based approaches 30, 33, 89
– based CAMD approaches 27, 31, 32
– complete flow sheets 11
– heat integration networks 11
– problem 91

– progresses, complex method of 367
ordinary diffusion 307
organic reactions 28
– catalytic discovery 343
organizational innovation 191
ortho-bromocinnamic acid 77
ortho-trisubstituted 2-hydroxybenzophenones 22
Ostwald ripening 48
outer approximation (OA) 92
overall flowsheets 97–100
overintensification 170
overlapping resonances 294
oxidases 76, 82
oxidation
– density 63
– 1,2-propanediol 314
– reactions 29
oxygen 82
– -containing species 60
– solubility 82
– transfer rate 82
oxygen reduction reaction (ORR) 47, 48, 51, 57
– activity 51, 59
– PtPd nanocubes 61
oxygen-sensitive reactions 362
oxynitrilases 76

p

palladium-catalyzed amination reactions 142
palladium chloride 63
palladium on carbon (Pd/C) 223
paper manufacturing 8
para-hydrogen-induced polarization (PHIP) 304
parallel pressure reactor (PPR) 346
pareto optimal solutions 7
pareto solutions 7
partial oxidation 49
particle size 48, 274
particle stability 48
PCA model 27
PdAg nanotubes 64
$PdAl_2O_3$ catalyst 116, 144, 301
PdAu catalyst 64
Pd catalysts 63
Pd lattices 64
Pd-M alloys 63
PDMS-based flexible membrane 241
Pd nanoparticles 63
Pd/SiO_2 catalysts 144
pentan-1-ol methylation reaction 365
2-pentene 302

Index

1-Pentene/hydrogen 301
cis-2-pentene/hydrogen 301, 302
perindopril 77, 79
pharmaceutical industry 84, 157
phase transfer catalysis (PTC) 139
phenacyl bromide 22–24, 27, 33, 35, 37
phenacylpyridinium bromide 23
phenylalanine ammonia lyase (PAL) 77, 79
photochemically generated singlet oxygen 168
photochemical reactions 168
photoinitiator (ethyl(2,4,6-trimethylbenzoyl) phenyl phosphinate) 237
photolithography 116
photopolymerization techniques 215
photopolymer jetting 217
– modeling machines 217
picolinic acid 143
piperidines 80
platinum (Pt) 47
– activity 48
– based catalysts 63, 64
– based nanoporous alloys 49
– as catalytically active phase 49
– diameter 48
– electrocatalyst 48
– electronic structure 58
– enriched surface layer 59
– hollow nanospheres 49, 50
– kinetics and corrosion stability 49
– mass activity 51
– nanoclusters 49
– nanosphere 49, 50
– nanostructures 55
– pristine 49
– surface layer 59
plug flow reactor (PFR) 94, 158
P450 monooxygenases 76
Podbielniak liquid–liquid extractor 163
Poiseuille flow 243
polarity 26, 28
polarizable continuum model (PCM) 26
polarization transfer techniques 295
policymakers 193
polyacrylates 229
polyamides (PA) 218
polycarbonate (PC) 218, 219
Polycat 158
polychromatic light sources 265
polyether ether ketone (PEEK) 219
polyethilenimine (PEI) 219
polylactic acid (PLA) 219
polymers 21
polymer stabilizer (PVP) 56, 57

polymethyl methacrylate (PMMA) 218
polyols 309
polypropylene (PP) 219, 224, 228
polysilane/SiO_2-supported Pd catalyst 142
polystyrene-co-methacrylic acid (PSA)-assisted templating method 58
polyvinylpyrrolidone (PVP) 54
popcorn-shaped nanoparticle 54
pore network connectivity 307
pore size distributions 48, 307
pore space 64, 289, 301, 309
porous 49
– catalysts 294, 301, 314
– catalytic materials, relaxation time analysis in 314–319
– materials 48, 306, 315
– matrix 312
– oxides 316
predicted maximum desired product yield 363
predicted rate constants 37
predictive capabilities 31
price setting 204
price volatility 191
principal component analysis (PCA) 27
printing polymeric active composites
– self-folding and opening box fabricated 251
process analytical chemistry (PAC) 260
– advantage at-line laboratory 261
– applications 270
– – for biodiesel production 276
– – of dairy industry 270
– – polymer industry 274
– – preparation of polymeric strip film unit dosage forms 273
– – synthesis of active pharmaceutical ingredients 271
– challenges to overcome 270
– concept and objectives 260–264
– definitions 262
– future trends in 279
– measurement system 261
– near-infrared spectroscopy (NIRS) 262
– nondestructive NIR spectroscopy 263
– noninvasive analyses 262
– off-line analysis 260
– off-line at-line laboratories 262
– productivity 261
– quality 261
– real-time optimization 263
– supervisory control and data acquisition system (SCADA) 263
process analytical technology (PAT) 172, 263
– challenges to overcome 268

– commercial manufacturing 172
– definition 262, 263
– future trends in 279
process and protein engineering methods, for improvement of biocatalysts 81
process-driven protein engineering 80
process/EHS model constraints 30
process equipment assembly (PEA) 174
process equipment container (PEC) 174
process flowsheet optimization 103
process intensification 11
process model constraints 21, 31
process optimization methodologies 330
process simulation 90
process synthesis 93
production-consumption pattern 192
product-service system 191, 192, 205, 206
– continuum 194
proportional-integral-derivative (PID) 331
propylene 304, 321
proteases 76
protein engineering 74, 75, 80–82, 84
– improvement 81
proton exchange membrane (PEM) 226
proton exchange with surface groups 315
PtAg dimers 55
PtAg nanoboxes 54
PtAg nanocubes 50, 51
Pt.Al$_2$O$_3$ catalysts 304
PtCo nanospheres 56, 57
PtCu$_3$ nanocages 63
PtCu nanoparticles 63
PTFE capillary microreactor 138
PtNi alloy 59
PtNi nanostructures 58
PtRu alloys, bifunctional mechanism 60
Pt:Ru ratios 60
Pt$_x$Ni$_{1-x}$-PSA core–shell structures 58
pulsed field gradient (PFG) 289, 298
– NMR plots 313
– NMR spectroscopic techniques 306
– PFG-NMR log attenuation plots for n-octane 313
pulsed gradient spin echo (PGSE) 298, 299
pulsed gradient stimulated echo bipolar (PGSTEBP) 298
pyridine 22–24, 33, 35

q

QM calculations 33
QM-predicted rate constants 33
quality assurance (QA) 197
quality standards 205

quantitative information 294
quantitative structure–property relationship (QSPR) 96
quantum mechanical calculations 25, 32, 34
quasihomogeneous behavior 309

r

rac-methylbenzylamine, with 3D printed bioreactors
– kinetic resolution 237
radio frequency pulse (RF) 292
Raman bands 265, 267
Raman immersion probe 266
Raman spectra 264–266, 269, 274, 302, 353
(R)- and (S)-amines 76
raspberry-like hierarchical PtAu hollow spheres (RHAHS) 52
– modified gold electrode 53
Raspberry Pi-based instrument monitoring and control 334
rate constant 24, 31, 33, 37, 354, 355
R&D (research and development) 156
REACH regulation 204, 207
reaction orders 355
reaction parameters
reaction rate 27, 47, 179
– constant 27, 28, 31, 38
reactionware 221
reactive–adsorptive process 28
reactive distillation (RD) 99
reactor design and manufacture 220
reactor networks 11, 93
– design 94
reactors internal volumes, and reaction mixture volumes 226
real adsorption systems 315
real time monitoring 268
real-time reaction 332
– monitoring 289
redox reactions 47
refocusing pulses 316
refrigerants 21
relaxation 91, 320
– constant 297
– rate 297, 315
– times 315
remote operator 367
renewable bio-based materials 73
renewable energy 9, 47
– installations 3
– technologies 3
Re$_2$O$_7$/Al$_2$O$_3$ catalysts 302
residence time distribution (RTD) 124

retrosynthesis 75
– analysis, application of 75
reverse osmosis (RO) 106
reversible process 296
Reynolds numbers 121, 160
RF pulse 295
robotic experiment 336
robotics 12
root mean square displacement (RMSD) 307
rotating ring-disk electrode (RRDE) 57
rotational and translational motions 315
Ru cluster 60

s

sacrificial template 63
scale-up strategy 170
scanning projection stereolithography (SPL) 248
Schwarz-P triple periodic structure 237
screening methods 27
(S)-DIP-Cl and biocatalytic route to 78
selective catalytic reduction (SCR) 318
selective laser melting (SLM)
– printers 219
selectivity 27
self-assembly concept 250
self-diffusion 313
self-formation 64
self-optimizing continuous-flow systems 364
sensors 47
– analytical techniques 333
sensory perception 336
separation systems 94–97
separation techniques, and relevant pure component properties 95
sequential ORR polarization curves 51
service industry 191
service-oriented models 192
shielding constant 293
signal-to-noise ratio (SNR) 294, 304
silicon–glass microreactors 118
silver nanoparticles, automated platform 235
simplex method 339
simultaneous parameter fitting
– concentration-time profiles 357
single resonance frequency 297
sitagliptin 74, 76
slit-plate mixer 160
small-scale continuous equipment 164
SMD solvation model 37
SMR™ (Sulzer Ltd.) 160
SMSIM (supermodified simplex) 366

SMX™ (Sulzer Ltd.) 159
sodium boron hydride ($NaBH_4$) 57
sodium chloride (NaCl) 55
soft (macro-) molecular and hard templating approaches 48
solar-thermal methanol synthesis 4
solid catalyst surface 315
solid foam packings
– gas-liquid mass transfer 120
solid–liquid separation 165
solid nanoclusters 49, 50
solid Pt nanoparticles 51
Solid State Pharmaceutical Center 165
solid waste 259
solubility 23, 24, 28, 33, 35, 37
solvatochromic 31
– equation 26, 27, 31–34, 36, 37
– parameters 26
– shifts 26
solvent 6, 27
– descriptors 31
– design 27
– – algorithm 32
– – reactions 38
– effects, on reactions 22
– generated from CAMD 29
– mixtures 35
– molar volume 33
– properties 31, 33, 34, 37
– and properties for 36
– recovery 27
– screening, alkylation reaction studied 362
– selection 27–29
– substitution 29
solvolysis 27, 31
spectroscopic process monitoring 12
spinning disc reactors 162
split and recombine (SAR) 160
Starlam mixers 160
state equipment network (SEN) 103
Stejskal–Tanner plot 299
stereolithography 217
– machines 217
structured packed bed microreactors
– drawback of 118
structure–property constraints 35
styrene 304
successive linear programming (SLP) 91
successive quadratic programming (SQP) 91
SuperFocus Interdigital Micromixer 160
superior CO tolerance 60
superlinear convergence rate 90

superstructure-based approaches 89, 101
superstructure-based design, of reactor networks 94
superstructure optimization 11, 94, 106
supply chain risk management 197
surface conditions, of PtRu catalysts/hollow PtRu assemblies 61
surface interactions construction materials 314
surface structure 47
surface-to-surface diffusion 311
surface-to-volume ratio 64, 315
surfactant/reductant ratio 52
surrogate model 37
sustainable 38
– chemistry 5
– – challenge of 11
– – OECD defined 6
– materials management 206
– process design 89
– production-consumption pattern 192
Synflow 158
systematic computer-based solvent design method 35

t
TaBaChem business model 194, 195, 199, 203, 207, 211
– advantages of implementing 196
– – continuous innovation 196
– – environmental impact and safety 197
– – long-term partnership 197
– – production and investment cost 196
– business confidentiality and protection of competition 208
– – intellectual property rights 208
– closing life cycle and preventing waste 205
– compatibility of service model with actual legislation 203
– focuses on 194
– general economic, technical, and management aspects 198
– – direct gains, indirect gains, and investments 198
– – pricing 199
– organizational and managerial aspects 202
– – knowledge sharing 202
– – logistics 203
– – quality assurance 202
– – sales 202
– – tendering and rewarding 202
– technical aspects 201
– – process optimization 201

– – reuse of chemicals 201
Taylor–Couette reactor 162
Taylor flow 132
– to bubbly 117, 132
– to churn 117
– regime 137
technical
– constraints 91
– innovations 192, 196
templating 64
terpenes 8
– cyclases 76
4-tert-butylphenylboronic acid 344
tetrabutylammonium hydrogen sulfate 140
tetramethyl silane (TMS) 293
tetra-n-butylammonium bromide (TBAB) catalyst 140
theory of sampling (TOS) 268
thermal degradation 319
thermal vapor compression (TVC) 105
thermodynamic 1, 31, 32
– equilibrium 166
– limitations 166
– motivated shortcut methods 93
– properties 28
thermoplastic 228
– filaments 219
thieno[2,3-c]isoquinolin-5(4H)-one-A (TIQ-A) 349
thioethers oxidation 175
tissue engineering 12
titania precursor spheres 52
toluene 259
tortuosity 307, 308
tosylfuranoside 272
toxic chemicals 5
toxicity 5
trade-off
– between cost and GHG 7
– between model accuracy and solution efficiency 34
trade-offs 29, 38
traditional model vs. TaBaChem situation 200
traditional sales business model vs. chemical leasing business model 193
transaminases 80
– enzymes 76
– technology 76
transamination 76
transesterification 30, 276, 278
transformations 75, 191
transition metals 47, 49
transition state theory (TST) 23, 32

trans-2-pentene/hydrogen 301, 302
transport kinetics 65
transverse magnetization 297
T_2 relaxation processes 314
trialkylamines 22
trichloroethylene 259
trickle bed reactor packings 124
triethylamine (TEA) 272
triethylbenzylammonium chloride (TEBA) 142
trifluoroacetic anhydride (TFAA) reacts 271
tripropylamine 31
cis,trans-1,3,5-tris(pyridine-2-ylmethylene) cyclohexane-1,3,5-triamine 233
T_1-T_2 correlation plots for water in 317
T_1-T_2 pulse sequence showing RF pulses 316
two-dimensional Laplace inversion 316
two-factor system, simplex progression 340
two-phase resistive network (2PRN) 134

u

UNIFAC-based models 32
UNIFAC group contribution method 33
United Nations Industrial Development Organization (UNIDO) 193
– sustainable resource management program 193
α,β-unsaturated aldehydes
– carbonyl bond, hydrogenation of 138
user-defined model 359
UV and visible spectrometry 319
UV curable polymer 246
UV light 216
– irradiation 250
UV protection 336
UV response 342

v

validation 34
value-added approach 193

valve design 240
vanadium 47
van der Waals interactions 315
vapourtec 170
variable progression
– local optimum parameter 345
verification stage 35
vibrational spectroscopic methods 264–268
– comparison 264
voltammograms 51, 52
volume selective spectroscopy (VOSY) 321
volume-translated Peng-Robinson equation of state 34
volumetric flow rate over time 356
volumetric mass transfer coefficient 118

w

Warnier model 130
waste
– legally defined 206
– legislation 197
– material 22
– treatment, permits 196
Waste Framework Directive 206
water and energy processes 105
Watershed resin 239
Watershed XC 11122 resin 239
We numbers 131
Wet stirred media milling (WSMM) 273
Wilkinson's catalyst 304, 324
– using THF 346
wind turbines 3
working capital expenditures 154

x

Xylene 259

z

ZnO as sacrificial template 59